ROUTLEDGE LIBRARY EDITIONS:
THE GERMAN ECONOMY

Volume 1

TECHNOLOGY AND WORK IN GERMAN INDUSTRY

TECHNOLOGY AND WORK IN GERMAN INDUSTRY

Edited by
NORBERT ALTMANN, CHRISTOPH KÖHLER
AND PAMELA MEIL

Routledge
Taylor & Francis Group
LONDON AND NEW YORK

First published in 1992 by Routledge

This edition first published in 2018
by Routledge
4 Park Square, Milton Park, Abingdon, Oxon OX14 4RN
605 Third Avenue, New York, NY 10017

Routledge is an imprint of the Taylor & Francis Group, an informa business

British Library Cataloguing in Publication Data
A catalogue record for this book is available from the British Library

ISBN: 978-1-138-29360-1 (Set)
ISBN: 978-1-315-18656-6 (Set) (ebk)
ISBN: 978-0-415-79121-2 (Volume 1) (hbk)
ISBN: 978-1-315-21248-7 (Volume 1) (ebk)

Publisher's Note
The publisher has gone to great lengths to ensure the quality of this reprint but points out that some imperfections in the original copies may be apparent.

Technology and Work in German Industry

The future of work organization and the role of workers in manufacturing has provoked a considerable amount of debate in light of the rapid technological developments of recent years. German industry in particular is a central focus for studying technical and organizational changes in industry due to its pivotal position in international markets, its technological sophistication and its well-established training systems.

This study was first published in 1992 and brings together contributions which contain both theoretical approaches and extensive empirical studies, on the manufacturing industry in Germany, including comparisons to other European countries. It looks at the developments of new technology, identifying trends in rationalization and the influences they have on organizational behaviour. It also anaylses the consequences of such changes on the work-force, asking whether they will become more or less skilled. As it discusses the relationships between technology and the work-force it includes discussions on flexible specialization, labour processes, union relations, small and large firms and training processes. The studies show that optimistic assessments on the future of industrial work should be viewed with some skepticism.

This book will be of particular interest to all those involved with new technology, industrial sociology and organizational behaviour.

Christoph Köhler is professor of sociology at the Friedrich Schiller University, Jena.

Pamela Meil is senior research fellow and member of the governing board at the Institute for Social Science Research, Munich.

ISFMÜNCHEN

Institut für Sozialwissenschaftliche Forschung (ISF München)

ISF München is the Institute for Social Science Research in Munich/Germany. It is an independent non-profit organization established in 1965 and ranks among the leading research institutions on issues of work and industrial sociology in Germany. The institute exemplifies the principles of self-organization, autonomy, and networking in its own organizational model. It has a board of governors and a management committee to oversee finances and daily operations.

ISF carries out research on new developments in society and within companies and the connections between them, especially such topics as rationalization and restructuring strategies (global value chains, technology design, work organization), personnel and training policies and their effects on the labor market, the educational and training system, sustainable employment and work forms, and industrial relations often from an internationally comparative perspective. ISF conducts critical research on work, carrying out empirical social research with a strong link to sociological theory. Research findings are prepared for scientists, the academic community, as well as practitioners from companies, associations, and public institutions. The institute maintains strong ties to local universities (Univ. of Munich (LMU); the University of Augsburg; the Technical University of Munich), cooperating on research projects and training students. It also engages in international cooperation with universities across Europe, the U.S. and China.

Currently more than 30 scientists and administrative personnel work at ISF as well as interns and doctoral students. Cooperation between project teams takes advantage of the interdisciplinary character of the staff and creates synergy effects in the research work. ISF receives some support by the Bavarian Ministry of Science, but most funding is project-based and comes from institutions such as the German Federal and State ministries, the European Union, a variety of foundations and public agencies, as well as companies in conjunction with publicly supported research projects.

Technology and Work in German Industry

Edited by
Norbert Altmann, Christoph Köhler, Pamela Meil

ROUTLEDGE
London and New York

First published 1992
by Routledge
11 New Fetter Lane, London EC4P 4EE

Simultaneously published in the USA and Canada
by Routledge
a division of Routledge, Chapman and Hall, Inc.
29 West 35th Street, New York, NY 10001

Typeset by LaserScript, Mitcham, Surrey
Printed and bound in Great Britain by
Biddles Ltd, Guildford and King's Lynn

British Library Cataloguing in Publication Data
A catalogue record for this book is available from the British Library

ISBN 0–415–07926–8

Library of Congress Cataloging-in-Publication
Technology and work in German industry/edited by Norbert Altmann,
Christoph Köhler, Pamela Meil.
p. cm.
Includes bibliographical references and index.
ISBN 0–415–07926–8 (hb)
1. Industrial sociology – Germany (West) 2. Technological innovations – Germany (West) 3. Employees – Germany (West) – Effect of technological innovations on. 4. Organizational change – Germany (West) 5. Industrial management – Germany (West) I. Altmann, Norbert. II. Köhler, Christoph, Diplom-Soziologe. III. Meil, Pamela
HD6954.G3T43 1992
306.3′6′0943 – dc20 91-41732
CIP

Contents

Figures

Foreword

Since the 1950s, industrial sociology has occupied a somewhat special position in the field of sociology in Germany: at the same time that it has concentrated on developing theory, it has also engaged, especially since the mid-1970s, in extensive empirically based research. Unfortunately, this work has just begun to be introduced and discussed in other countries. This is certainly due in great part to the fact that German industrial sociology has simply not been widely published in foreign languages. And, to be honest, one of the reasons for this is that German industrial sociology has only incorporated foreign research findings or theoretical perspectives to a very limited extent. We at ISF view this situation as unsatisfactory in every respect. Thus, one of the goals of this book is to increase the understanding, and we hope interest, of German issues and perspectives on industrial sociology in international circles and to introduce some of ISF's comparative research to an international audience.

Another reason we have undertaken the publication of this book is to celebrate. In 1990 our institute became a quarter of a century old. Since its establishment in 1965, the ISF has carried out a wealth of research in the areas of labour markets and employment, technical and organizational development and its effects in a number of different industrial sectors, occupational and professional training, and interest representation of workers. ISF, Munich is one of the few research institutes in Germany that conducts both theoretical and empirically-based applied research. Moreover, its theoretical contribution on these subjects has strongly influenced the general discussions on the future of work in Germany.

Most of the chapters in this volume contain results from empirical research projects; several present their results within a larger theoretical framework. Many are based on international comparisons. In reading this book you will notice that the general thrust of all of the research conducted at the ISF, Munich has similar intentions – understanding trends in the development of work in the context of larger societal developments and offering the bases for the realization of a more humane design of work. However, this volume also makes clear that scientists at the institute have different premises as well as different approaches to their subjects. There is no desire to lose these distinctions; they have always proved fruitful in generating better work and creating a stimulating atmosphere. A more detailed account of our thoughts, differences, and common interests is presented in the editor's introduction in the context of the larger German discussion.

A final important word. Since we have begun work on this book, a major historical event has taken place: German unification. Thus, all of the chapters here are based on the situation before unification. It is certain that new and challenging problems will arise in the areas that we study as a result. However, we are convinced that they will be strongly influenced by the structures and processes that characterized the 'old' Federal Republic. Therefore, our studies have not lost their significance. On the contrary, they can only add to an understanding of future developments.

Like ISF in general, the successful compilation of this book required a lot of team support. Therefore, we would like to thank all of our colleagues at ISF who helped with the word processing, graph preparation and corrections for this book, our translators,[1] and Christa Hahlweg for the preparation of the manuscript.

All of the contributions presented here were carried out in the framework of the 'Special Research Unit' (SFB 333 – Perspectives on the development of work) of the University of Munich. Publication and editing were done in cooperation with the projects B2 and B3 of SFB 333.

Munich, January 1991
Norbert Altmann
Christoph Köhler
Pamela Meil

1 Translations were carried out by the translation office Baker & Harrison (Munich), Peter Berg (Notre Dame, Ill.), Michael Schatzschneider (Frankfurt), Language Consultancy (Berlin), Anton Kirchhofer (Munich). Reworking, final corrections and editing were carried out by the editors.

Preface to the 2017 edition

This book was first published in 1992. Upon receiving the news that Routledge wanted to reissue this book, we were naturally pleased, but also aware of the challenge: So much has happened in the world of work and technology in Germany over the last 24 years – can we build a bridge between then and now? The book's focus then was explaining and understanding the transition from the old "fordistic" German model of production and employment, which was experiencing an era of change and entering a period of crisis, to a new more globalized, financialized and flexible one (chapters 1, 3). The analysis centered on the core sectors of German industry, particularly their organization of production and technology (chapters 5, 9, 10), organization of work (chapter 11) and qualification structure of the workforce (chapters 12, 16, 17).

Until well into the 1980's, German industry was characterized by a mix of mass production (chapter 14) and niche production of small series manufacturing (chapter 20, 21, 22); the companies were highly integrated and functionally structured. Given the large share of skilled production workers, trained in the dual apprenticeship system, and the institutions of co-determination, it was indeed a very specific German variant of fordism and taylorism (chapter 2). It was unclear which developments would predominate and which new structures would emerge: Would the new rationalization strategies set in motion an extensive de-skilling and standardization of production work or would a new industrial democracy founded on an empowered, qualified workforce be realized?

In going over the arguments and research documented in the book, we realized that, yes a lot has happened. However, we believe that a look back at the research of the 1980's and 1990's sheds light on a number of contemporary trends and positions on work and employment. Moreover, several of the concepts and controversies developed in this book are still relevant for understanding technology and work in German industry today. We briefly set out 3 examples to illustrate our point.

The chapters on systemic rationalization (chapter 4, 6, 15) portended the consequences for work organization, work content, skilling and interest representation through ongoing processes of distributing work across global value chains and the creation of an international division of labour. The "new" rationalization, facilitated th rough the use of new digital technologies, targeted subprocesses of company-wide and inter-company production flows in which less skilled, often labour intensive work tasks would be outsourced to lower cost locations with more precarious working conditions. The authors hypothesized a core-periphery gap, with high value added processes remaining in the core with a smaller, but better paid and often better protected workforce. These processes were seen as a dangerous threat to worker solidarity and collective bargaining. Today, many of the developments have indeed come to pass. However, dichotomies between core and periphery are not as clear-cut as predicted: pressures on core workforces as well as upgrading in peripheries, not to mention processes of crowdworking and digitally mediated platform work, have made the outcomes of "systemic rationalization" more complex than could be seen in 1992. Nonetheless, the general arguments as well as the conceptual underpinnings (concentration of technological and capitalistic value added and market-driven financialization strategies) are still highly relevant for understanding contemporary developments in restructuring work and employment across global value chains.

Another body of research dealt with the old question of to what extent technical-economical constraints determine the paths of organizational and work design. At the time the book was written, new computer aided and flexible manufacturing technologies (CIM – computer integrated manufacturing) were being introduced in German production industries. A public discussion on the completely automated factory and dramatic job loss among production workers blossomed. The similarities with current discussions on Industry 4.0 (a term that originated in Germany as a particular form of the Internet of Things, but which is rapidly spreading around the globe) are obvious. Industry 4.0 is characterized by highly automated production and logistic networks in which virtual and real processes merge on the basis of cyber-physical systems, aided by exponential increases in computer processing capacity (big data) and internet based communication technologies. Optimistic (new opportunities for the German economy, re-skilling and re-professionalization) and pessimistic (huge job losses, replacement of human labor by machines, extreme polarization of skill profiles) scenarios on the future of work are being debated. The analyses in this book showed that, not only is technology development itself socially shaped (chapters 5, 10), but moreover the institutions of education and training play a large role in the design of work systems (chapters 17, 18, 19). Additionally, authors in the book warn that there is risk in over-estimating the performance capacity of expert systems (chapter 7) and that experiential and tacit knowledge are important elements for a successful and optimal use of technological systems (chapter 8).

Finally, a third important area of analysis focused on new challenges for industrial relations in light of new rationalization strategies. The rather pessimistic projections in the book on the power of unions to exert a decisive influence on working conditions or reverse the erosion of standard work forms were unfortunately proven mainly true (chapter 13, 23, 24). Developing labor policy and strategies for improving work have been pushed in the background by pressing concerns of job security and employment protection. Nonetheless, the institutions of co-determination and the corporative governance structures characterizing German industrial relations still exist, albeit in a modified form. Some attempts to revitalize German unions with new orientations are visible and Germany's industrial relations system is still considered strong in international comparisons.

The Institute for Social Science Research (ISF), whose research fills these pages, was and is a leading research institute in the area of the sociology or work. The concepts and developments in labor and industrial sociology laid down here were shaped and advanced at the ISF. Our co-editor and co-author Norbert Altmann, together with Burkart Lutz, founded and built the institute, devoted the major part of their professional life to it and contributed decisively to its scientific profile and success. Burkart Lutz passed away in 2013 and Norbert Altmann in 2015. They are sorely missed.

Christoph Köhler
Pamela Meil
Munich, October 2016

Part I

Concepts

1 No End in Sight – Current Debates on the Future of Industrial Work

Norbert Altmann, Christoph Köhler, Pamela Meil

INTRODUCTION

This chapter is designed to give our non-German readers an overview of the development of industrial sociology in Germany as we on the 'inside' see it. We also touch on the current debate on the future of industrial production work in Germany, and place the positions represented at the ISF, Munich within this debate. This chapter lays out the shared framework as well as the different perspectives and concepts of ISF members reflected in the following chapters found in this volume.

WHAT IS '*INDUSTRIESOZIOLOGIE* ?' – THE DOUBLE CHALLENGE[1]

Industrial sociology has become an important part of the discipline of sociology in Germany in the post-war era. It has had both a considerable influence on sociology as a 'science' as well as some – although seemingly limited – effects on industrial practice and policy (rationalization concepts and strategies, health and safety, humanization of work, labour market analyses, occupational training, etc.). The range of empirically oriented research projects in the field is extremely broad in terms of subject (choice of technology and organization; intra- and inter-company division of labour; production and administration; conditions of work; qualification and training; worker consciousness and work attitudes, etc.) and research focus (from the auto industry to the food processing industry; from family-run craft industry to heavy industry; from dependent supplier to multinational enterprise).

The range of subject matter and levels of intervention underline the inter-disciplinary character of industrial sociology. One of its tasks is to integrate those scientific disciplines directly related to technology and work which are generally distinct from one another in Anglo-Saxon research traditions, such as industrial relations, (which does not exist as a separate field in Germany), labour economics, industrial engineering, labour psychology, labour law, etc. Industrial sociology also pushes other fields to adopt broader perspectives, such as the extension of ergonomics (mainly the areas of physiology and industrial medicine) to a 'labour science' (*Arbeitswissenschaft*) which is not limited to the physical sides of work organization, but addresses issues of qualification, self-development, long-term safety, etc. in a didactic manner. Furthermore, industrial sociology orients itself explicitly towards macro-sociology and economics as well as the political sciences which is reflected in the academic training of industrial sociologists. What this adds up to is that economists, engineers, psychologists and lawyers increasingly work together with sociologists on industrial sociological projects.

Despite the turn-around towards empirical research and the development of a complex method-mix that came about with the professionalization of the discipline, the current debate on the future of work can only be understood in the context of industrial sociology's high socio-theoretical and critical standards.[2] Since its beginnings, German industrial sociology has been, and continues to be, characterized by the double role of European sociology (Saint-Simon, Comte, Spencer, Marx, Weber): that is, it strives to be a theory of societal development, and at the same time, a means of change. This double role implies a connection between the results of theory and research on the one hand, and implementation and application in social practice on the other. However, this application normally does not take the form of either instrumental, manipulative interventions such as seeking better means of worker motivation (social technology), or compensatory social policy. In brief, with help from critically and empirically supported analyses, industrial sociology wishes to show the way to more humanistic social practices by elucidating the starting points and obstacles for this realization.

Industrial sociological theory and research is mainly directed towards the issue of the development of rationalization and dominance in the society, and the forces behind social change. This question is dealt with at two levels: firstly, through an explanation of the structure and development of industrial capitalism, and by making the possibilities for political intervention visible. Secondly, by focusing on the concrete changes in company production and work systems, especially the interests behind changes and consequences for the workers.

Due to rapid and far-reaching social changes, concrete political demands on industrial sociology's contribution to solving practical problems in companies and society have grown since the beginning of the 1970s on the part of the state, the firm and the unions (Bergmann 1982). Through this, the discipline found itself mediating between progressive interpretations of current developments for political actors and direct consultation and application. Along with its theoretical and analytical orientations, industrial sociology has always maintained its connections to practice. However, industrial sociology as a discipline has not engaged in either past or present internal company consulting nor direct intervention and does not have such an institutional infrastructure at its disposal. Consultation, or even influence over the technological and organizational design of work, generally takes place in the framework of (state-supported) research and action programmes which are mainly exemplary and demonstrative in character. The task, as understood by ISF members, is to inform, elucidate and clarify to the responsible actors, the conditions and side-effects of their activities, but not to directly intervene (Lutz, Schultz-Wild 1986).

In carrying out empirical research, the 'case study,' especially the company case study, is the 'model' which characterizes German industrial sociological research methods, in contrast to the survey or interview methods widely used in American sociology.[3] The case study has also played an important role in the work done at ISF. In the company case study, a number of different, mostly qualitative survey instruments, come together in a number of different ways.[4] Quantitative methods increasingly serve to support qualitative findings as well as to identify relevant issues, which can then only be meaningfully worked on in connection with qualitative methods. Through its particular application, the company case study becomes a research instrument that can stand on its own.

In accordance with the high complexity of object, field and research issue, the case study makes an open-ended procedure possible with regard to which research instruments should be used, which company areas, personnel and data should be included, and how long the research process should run. The case study is the basis upon which hypotheses and theories can be differentiated, revised, enlarged and further developed in the research process itself – thus making possible the decisive exchange between empirical work and theory building. This is a critical aspect of the case study because the problem of generalizing the results can only be solved through this exchange.

New tasks arising out of current industrial sociological debates, briefly described in the following section, are bringing new methodological demands with them, still only partly foreseeable today. These include: encompassing longer term developments, which is only possible with the help of accompanying longitudinal surveys; the inclusion of inter- and external company conditions (such as inter-company networks, labour markets, activities of state or intermediary bodies); the integration of findings of company surveys on macro-economic and macro-social structures and their changes. All of this leads ever more strongly toward the need for a greater internationalization of industrial sociological research. The rationalization processes to be observed and analyzed increasingly extend beyond the borders of a particular country. The significance and relevance of macro-structures in technological and organizational change at the company level, emerge more clearly in international comparison.

FUTURE OF WORK – THE CURRENT DEBATE IN INDUSTRIAL SOCIOLOGY

There is an overwhelming consensus in German industrial sociology today that the development of technology and work in industry has been long characterized through a consistent pattern of company rationalization, usually coined 'Taylorism' or 'Fordism'. The characteristics of these rationalization strategies at the micro-level of the company (chiefly for large corporations and mass producers) are: the separation of conception and execution in work; the extension of the division of labour according to functions, hierarchical levels and work tasks; the tendency to plan fully all production processes and to standardize the work process; careful control of performance, based on wage incentives; finally, and most strongly tied to the above-mentioned points, a widespread simplification of tasks and thus the deskilling and polarization of the labour force.

A consensus also exists that since the 1970s in several areas of industry some preconditions for the efficient functioning of these rationalization strategies have begun to erode, thus bringing about a new stage in the history of industrial rationalization which will doubtless have far-reaching consequences for work. In contrast to the overwhelming agreement on the past and present state of industrial rationalization, however, a large controversy exists on predicting exactly what the future holds.

Since the end of the 70s, empirical findings and analyses based on them have been produced for important areas of German industry. These have demonstrated that under certain technological, economic and social conditions, rationalization processes which are not directed toward the intensification of a division of labour and a deskilling of work, but on the contrary are based on professional competence in carrying out work, can be in the company's interest.[5] In other large industrial nations in which Tayloristic strategies exhibiting a highly advanced division of labour and polarization of skills appeared fixed, similar developments have been documented.

Some authors, such as the representatives of the post-Fordism thesis (Piore, Sabel 1984; Streeck 1991) consider the trend toward a 'professional' work pattern in which all workers have greater autonomy and responsibility in a more open work and company organization as an important component of a new, post-Tayloristic phase of industrial development. Several authors, the most well known being Kern and Schumann (Kern, Schumann 1984), judge these, at present still relatively rare, developments as a sign that a new pattern of industrial rationalization is forming. According to this view, the 'new production concepts' of an 'enlightened' management are striving for a comprehensive reprofessionalization of production work, which goes way beyond the work structures introduced in the 70s (such as job rotation, job enlargement, etc.). The vision is that of skilled production workers running flexibly automated technical systems and achieving high levels of productivity.

This more or less optimistic vision of the future, however, is contradicted by a number of empirical and theoretical findings. Several studies predict development trends leading to problematic effects for workers in general or at least for particular branches of industry and groups of workers.[6]

All in all the analyses of the structural relationships which affect the development of industrial work present a complex and contradictory picture. It is a fact that pressure coming from sales markets to produce more flexibly and the integration, interdependence and capital intensity tied to the use of CIM systems provide strong impulses to break with Tayloristic rationalization strategies. In this regard, political factors can play a major role (see chapter 15). At the same time, powerful forces remain which push toward the continuation of Tayloristic forms of work organization even under altered circumstances. Among these are the automation concepts developed by engineers and the interests of large manufacturers of CIM-components and systems. The automation and centralization of complex functions and tasks in CIM systems makes it possible in new – technically mediated and fixed – ways to further the Tayloristic strategies of separating conception and execution even under altered production conditions. The still far from exhausted technical potential of CIM systems is thus often set up to absorb a portion of the pressure for flexible production. Long-term trends in labour supply may also serve as an impediment for a generalization of 'new production concepts' through a shortage of skilled workers (see chapter 3). In any case, whether coming from general interests in power and control or the vested interests of particular management factions, strong impulses exist to find structurally conservative solutions to production problems.

POSITIONS IN ISF

The research done at ISF, as this volume reflects, tends to show that neo-Tayloristic strategies dominate even in the face of growing flexibility needs and the introduction of CIM systems, and in general display a marked vitality. This is the case for mass production industries with a predominantly semi-skilled work-force (see Part IV) as well as for those with specialized and small series production having mainly skilled workers (see Part III). We do not interpret experiments with group work in mass production industries at low levels of automation as a departure from Tayloristic strategies in work design (see chapter 13). And in work systems with predominantly skilled workers, there are indications of an erosion of this high skill profile (see chapter 10). The role of small and medium-sized firms in flexibilization is also viewed critically (see Part VI).

Although all of the future visions in ISF tend to be less optimistic than those of the 'new production concept' proponents, the chapters in this volume do differ in the estimation of what potential exists for the development of industrial work. Based on observed changes in today's industry, the position of the 'contradictions of post-tayloristic rationalization' project (see chapters 3, 10) sees possible alternatives to the development path of a modernized neo-Taylorism. According to this approach, it has to remain open whether one of the alternatives will win the upper hand in the long run and if this happens, which one it will be. The alternative paths are understood as varying reactions to divergent, in part even conflicting, impulses, constraints, and opportunities of company rationalization. These are determined by internal company structures and interests (strongly shaped by Tayloristic logic) as well as by the company's external environment (the educational system, labour and technology markets, industrial relations, etc.). Decisive for those with this view of the future is the thesis that a new, dominant rationalization paradigm (in contrast to the Fordistic accumulation regime) can only occur in the context of new and stable social and economic arrangements that display a high accordance between macro- and micro-structures. Such a new configuration is not in sight at the present time.

Another position in ISF (Sauer *et al.* in chapter 4) sees little alternative to the neo-Tayloristic development path at the empirical level. This position argues that a new type of 'systemic' rationalization is appearing with predominantly problematic consequences for workers. Rationalization strategies are oriented toward the 'elastic potential of technology' and not of labour, that is on fixed and not variable capital. More importantly, these rationalization strategies are no longer simply directed towards an individual company and the utilization of its capital, but on the relation between autonomy and control in the entire, including inter-company, production and distribution chain. It is in this way that companies achieve flexibility and economy of production.

The concrete character of this new type of rationalization and the forms of industrial work connected to it are not predetermined. With integration of intra- and inter-company subprocesses achieved through computer technologies, and with the decreasing weight of labour costs, work in production can be connected to differing strategies of labour use, influenced by internal and external company constraints. In the long term, however, the technological and organizational potential that exists in systemic rationalization tends to be made suitable for the flexible control of complex work tasks, and for organizational centralization or the central control of decentralized units. New demands on work which promise higher skills, greater autonomy, more decision-making power, and similar developments for the individual worker at the shop-floor level are less foreseeable, apart from very limited key positions, than polarization tendencies. Moreover, new problems are likely to arise for those workers who work in companies which are in the weakest or most dependent position in the hierarchy of the entire production chain.

The industrial and macro-economic structural changes reflected in this approach signal a new quality in developed capitalist societies. With 'systemic rationalization', a new accumulation regime, replacing the Fordistic–Tayloristic regime, is laid down in a nutshell.

Both positions agree that an important and, for the development of industrial work, central point of departure for company strategy lies in the design of the production process.[7] Production is understood in the general (Marxian) sense as a unit of the 'labour process' (material production) and 'producing surplus value' (the transformation of money into capital and back into (more) money). The technological and organizational structures of the production process and the forms of labour deployment decide to what extent it will be possible to secure productivity through the maximal use of the labour employed in an individual company.

In this connection, company strategies try to neutralize those external, environmental conditions which are not controllable (such as the situation on consumer and sales markets, economic fluctuations, legal and collective bargaining regulations, etc.) so that they do not directly affect the production process. This requires that individual capital succeeds, through the organization of its production process, in transforming external conditions into exigencies that are subject to internal company control. Technology, organization and labour are the elastic potential most easily controllable by a company; they can be coordinated 'in the company mould'. The technological and organizational integration of company-external production subprocesses (systemic rationalization) can be interpreted as a strategy to transform and neutralize changing sales market conditions. The basic objective of systemic rationalization is the expansion and control of relationships between internal and external company production and capital utilization.

Starting from these theses, the two basic currents developed in ISF discussed above, have differing accents in their theoretical assumptions, research areas, and in their political conclusions.

In terms of theoretical position, the differences centre on the weight that various factors have in determining the development of industrial work. In very simplified terms, one approach ('systemic rationalization') sees the most important determinants as being anchored in the economics of market and production of individual capital, taking into account the complex framework of considerations sketched below (see chapter 4). In selecting research topics and areas, this approach concentrates on the relationship between capital accumulation, technology, organization and work. Research focuses on the internal and inter-company (also international) division of labour in industrial branches with large series and more or less complex products in which the Taylorization of work is already at a relatively advanced stage (for example, the automobile and electronics industry). Also examined is the influence of the relationships between technology manufacturers and users on the technological and organizational rationalization of the production process.

The theoretical and empirical analyses of this approach have revealed very limited possibilities for political or union controls for the design of work. Thus, political measures and interventions for current, concrete situations tend to centre on improvements in labour standards rather than in the possibility of redirecting technological and organizational structures in the production process. In the longer term, one concern of this position tends to focus on newly defining union policy.

The second approach ('contradictions of post-Tayloristic rationalization) weights the factors determining the development of industrial work quite differently (see chapter 3). Company rationalization strategies and their consequences for work are tied together in a complex way at the macro-level mediated through 'meso-structures', such as technology markets and labour markets, and through political regulation. The development of industrial work derives from an interaction of the various levels. In this view, the production process does not a priori play the central role.

Therefore, in choosing the research topics and areas, the labour market, educational system and, most recently, macro-economic developments predominate next to analyses of technology and work. Empirical analyses concentrate on industries such as machine building or process industries such as the chemical, steel, or food industries in which skilled work (at least in Germany) plays a major role. The analyses of this research approach have produced evidence of strong structurally conservative economic and social forces, pushing toward the continuation of Taylorism even in changing contexts. However, alternative development paths of industrial work which make sense economically have also been seen in company rationalization processes. Opportunities for political intervention to induce changes in the way industrial work is designed exist in this historical period in which capitalism is undergoing a restructuring. Areas of intervention are, for example, technology policy, education policy, labour market policy and industrial policy (see chapter 12).

An important aspect in the analysis and evaluation of company rationalization measures is their effects on workers. The studies carried out at ISF are not limited to the 'quantitative' changes occurring from rationalization, but particularly to the 'qualitative' ones, such as the changes in skill requirements, physical and mental stress and opportunities for influence and control. Besides the positions in ISF, laid out above, there is a further focus of research. The point of departure is that new demands are arising at the workplace which cannot be adequately understood using conventional categories of analysis. In particular, these involve changes in tacit knowledge and demands on the experiential knowledge that workers possess. Another issue is the existence of contradictory work requirements and the mental stress resulting from this. The thesis is that these conditions are new and potentially growing problem areas in the way labour is used. An examination of these issues, as well as coming to terms with them on a practical level, demands a conceptual and methodological expansion in the research on the social consequences of company rationalization. Against this background it becomes clear that the current controversy over 'deskilling' and 'reskilling', and the displacement of work pressures, etc. do not take into consideration important aspects of the actual changes taking place at work. This necessarily leads to one-sided and incorrect estimations. In response to these deficiencies, ISF is developing a new position to better deal with the effects that company rationalization measures are having on workers (see chapters 7, 8).

NEW ISSUES THROUGH UNIFICATION

It is certainly necessary to make some initial remarks on the incorporation of the German Democratic Republic (East Germany) and what effect this will have on the development of industrial work. The perspectives that will emerge as a result of Germany's unification and the revolutionary changes in Eastern Europe could not be considered in this volume. Most of the contributions presented here were completed before the radical change that took place in East Germany and as of early 1991 – no one could predict what further developments will arise. On the topic of company rationalization processes and the future of industrial work in a unified Germany, only conjecture is possible.

For the West German economy, the wave of 'immigrants' represents a temporary expansion of labour supply and might support strategies of skilled production work. However, several questions exist concerning what significance this increased labour supply will have on the German economy. Firstly, workers coming from the East have exhibited certain deficiencies in skill, raising the question of whether these workers are in a position to fill weak spots in labour supply in the West. Secondly, it is not clear how long the stream of workers coming from the East, drawn to the West due to economic problems in the former GDR, wage differences, etc., will continue (see Köhler, Grüner 1990 on this issue).

In industry in the former GDR, potential for the development of alternative paths of development of industrial work most certainly exist. Firstly, control of labour policy within the company as well as political control in the society as a whole is at a highly critical point given the situation of crisis and radical change. Secondly, the technological and organizational structures of production must be completely modernized to meet the demands of international competition. These circumstances create the possibility for comprehensive organizational innovation, in contrast to step by step rationalization processes. Thirdly, the former GDR economy has a large supply of skilled workers at its disposal who, even given some deficiencies with regard to the mastering of new technologies, can represent an important resource in a societal modernization process.

However, when one observes the ongoing restructuring of the former GDR economy, the impression arises that existing opportunities for redirection will not be used due to the purely market economy-oriented incorporation strategy. There are several reasons why this approach is so strongly pursued: First and foremost, this is due to the strictly conservative-laissez faire (CDU and FDP) policy of the Federal government. Secondly, it has come about as a result of the self-imposed time pressure, and finally because of the opportunism of the new/old ruling elite of the former GDR. Also decisive is the takeovers of former GDR industries by West German firms and with them a relatively uninterrupted transfer of their rationalization models. It is still open to what extent GDR firms will become mere 'extended workbenches' or other types of dependent companies. The Tayloristic rationalization paradigm could be all the more easily transferable to these companies since it was a specific form of this paradigm that predominated in the planned economy of the former GDR.

Completely open is the question of what role faces those companies in the former GDR that survive, or put another way: how will these companies fit into the framework of systemic, that is inter-company oriented, rationalization processes of West German corporations. The repercussions that the GDR's incorporation will have on the industrial structure of Germany as a whole, including the supplier structures, internationalization of production, position in international competition, the relationships between companies and industrial branches, and effects on workers and industrial relations are still open: will there be an intensification of existing problematic trends of systemic rationalization, or an opportunity for a new openness in workplace designs?

The chapters in this volume should be read in the light of the thoughts on the current discussions on the future of manufacturing work in industrial sociology, on the relationship between theory and practice, and on the various theoretical approaches that we have briefly presented here. The chapters concentrate, in accordance with the research focus of ISF, on industrial production and only include developments in services and administration from this perspective, and only very selectively. Some are written at a theoretical, conceptual level; others have a more empirical, analytical character. Several chapters develop arguments on the relationship between technology and work based on ongoing research projects. Thus, the chapters of this volume concretize the positions of ISF in current debates described above.

The title of this chapter – 'No end in sight' – refers to the current debate on the future of industrial work. For the research approach of 'systemic rationalization', a change in form, but no end to Tayloristically designed work is in sight. For the 'contradictions of post-Tayloristic rationalization strategies' approach, several 'ends' to present ongoing processes of the restructuring of capitalism are possible. In the final analysis, no end to the controversy in German industrial sociology is in sight: a good basis for fruitful research in the future.

2 Stranger in Paradise – An American's Perspective on German Industrial Sociology[1]

Pamela Meil

This chapter looks at some aspects of German industrial sociology: the discipline itself, as well as the institutions and structures which it studies. The purpose is to provide an introduction to the concepts underlying German industrial sociology and thus make the following contributions in this book easier to understand for the foreign reader. Within this framework, some interpretations of the origins and significance of certain themes and concepts are offered, and comparisons with how these same concepts are perceived by American sociologists.

INTRODUCTION

Perhaps the best way to begin introducing the subject of this paper would be to explain the rather provocative title. The first part is easy enough: I am an American living in Germany, a sociologist trained and indoctrinated at an American university. This makes me a stranger both in my everyday existence and, to a certain extent, my academic orientations. At times it is difficult for me to understand the motivations behind behaviours, as well as the perceptions behind the sociology of my colleagues, and I am sure that this feeling is mutual.

Now, for paradise. Admittedly, the use of this word was meant to stir up some controversy as well as to pique interest, and it seems to have succeeded. In seeing the title, my German colleagues react with curiosity as well as a bit of nervousness. They assume, rightfully of course, that a certain amount of irony stands behind the use of the term.

I could take the easy way out and point out Germany's affluence, extensive welfare system, and the newly acquired unity as a motivation behind the use of 'paradise'. But I would not be doing my colleagues justice to imply that they perceive themselves in a paradise. Like most social scientists around the world, especially sociologists, they tend to view their society and its problems quite critically. Their job is to point out the blemishes existing in their respective paradise – in the end, making them strangers in paradise as well: a state perhaps all critical sociologists are doomed to find themselves in.

Nevertheless, I can pinpoint three factors that make this the 'paradise' of my German colleagues and me a stranger in it: one, their basic understanding of the structures and institutions that make up their society; two, linked to one, the natural practice of carrying out all research (including comparative research) and understanding phenomena, from the perspective of your own society and its particular characteristics; and three, the language (i.e. everyone here speaks and writes in German).

As far as industrial sociology in Germany goes, what this means is that, in the subjects, points of emphasis and methodology chosen for research, and in the understanding of certain basic institutions, we non-Germans cannot help but be strangers. Moreover, although considerable differences in opinion and approach certainly exist among German researchers, the outsider is quickly struck by an amazing solidarity in terms of research topic, methods, theoretical points of departure, and perhaps most surprisingly, an agreement about the legitimacy of fundamental social structures and institutions.

In sum, in carrying out their sociological research and analyses, German industrial sociologists bring with them a series of historical, cultural, and institutional legacies. My job in this chapter is to try and convey some of these legacies and point out the difficulties in doing so, difficulties that have cropped up repeatedly in translating and editing this book. Cleverly, I subtitled this paper, 'an American's perspective . . . '. This accomplishes very much the same thing as titles such as 'Impressions' or 'Some remarks on . . . '. It ensures that no one expects empirical research, or scientifically demonstrated theses. In fact it insures that I can present thoughts and information that are my – hopefully not uneducated or poorly observed – opinion.

Trying to translate not only words, but also meanings from one language to another, necessarily entails translating the context from which this language arises. Thus, not only German industrial sociology must be communicated, but also some basic aspects of German society. To begin communicating this paradise of shared understandings, I will briefly discuss some characteristics of German industrial sociology from an outsider's perspective. Then, I will focus on several German industrial structures and institutions with the goal of explaining them more fully to the uninitiated, thereby (I hope) making the chapters in this volume easier to understand for those not highly familiar with the German case.

INDUSTRIAL SOCIOLOGY

William Faunce, an American sociologist, wrote in 1967 that (paraphrasing Mark Twain) 'the reports of the death of the field (industrial sociology) have been greatly exaggerated' (Faunce 1967: 1). This was in the introduction to a selection of 38 readings on the subject. Unfortunately, Professor Faunce was wrong. In the United States, industrial sociology had always had a difficult time defining itself, and eventually was subsumed in a number of different disciplines such as the sociology of organizations and work, industrial relations, social psychology, etc. Originally, it comprised an enormously wide range of subjects – under the rubric of the study of industrialism – ranging from human relations, to union-management conflicts and relations, authority in industrial organizations, bureaucracy, work groups and structures, and the process of industrialization itself.

In Germany, on the other hand, industrial sociology is alive and well and a blossoming discipline. You can imagine how curious I was to find out exactly what industrial sociology is.

Two German sociologists with strong ties to ISF (one the retired, but still active, institute director, the other a former member) wrote the following on industrial sociology:

> What in a particular country at a particular time or from a particular author is understood as a characteristic subject of industrial sociology, . . . will be understood at another place, another time and by other authors to be special areas clearly delineated from industrial sociology.
>
> (Lutz, Schmidt 1977)

Yet, given the wide range of possibilities of what industrial sociology could encompass, what strikes an outsider when confronted with industrial sociology in Germany as it has been practiced up to the 1990s is the uniformity with which the discipline is defined. What do I mean by uniformity? Not that industrial sociology does not have a sufficiently broad interdisciplinary framework. In fact, German industrial sociology is highly cross-disciplinary in character in that it includes fields that are completely separate in American research traditions, such as industrial relations, labour economics, psychology, etc. Nevertheless, there is an amazing amount of research done all over the country specifically on the relationship between technology and work within industrial enterprises, especially in the production area itself. I am going to point out what I believe are three reasons for this state of affairs:

Firstly, as pointed out in chapter 1, German industrial sociologists have traditionally been oriented to the political sphere. This has led to the grappling of subjects which have also interested the unions, although from a more critical perspective. Therefore, it has been not uncommon to engage in research, the results of which can be discussed by unions in dealing with current problems. Clearly, in the 1970s and 1980s (the 1990s still being a question mark), the introduction of new forms of technology (computerization) and work organization presented an opportunity for German unions to reformulate their demands for codetermination of workplace design and qualification. This provided an impetus for German sociologists to concentrate on these subjects.

Secondly, in my opinion, some of the uniformity in research thrust derives from the way research is funded. There are not that many different funding sources for industrial sociological research, especially when one considers that most industrial sociologists shy away from private company funding to circumvent the possibility of becoming 'management consultants'. In turn, most companies have not been traditionally interested in academically oriented research. Industrial sociological research is organized into only a few institutes, some affiliated with universities – some not, and some state-funded – some not. Even in cases where salaries are secure through state-funding, researchers must apply for money to carry out their studies. A major source of funding comes from the German Research Fund (DFG), which designates special topics for defined time spans. Other sources are the Federal ministries which also distribute their funding on the basis of special projects. Most recently, a most important trend in this direction was the government research programme, 'humanization of working life', strongly advocated by the unions. This programme provided the material means for industrial sociologists to pursue long term research studies. As a result, obviously almost all industrial sociological institutes oriented their research toward this theme.

Thirdly, another reason to favour the industrial case study as instrument and the production worker within industrial enterprises as research subject is that they are better suited to the traditional Marxist and neo-Marxist theoretical orientations in industrial sociological research in Germany.

The traditional methodological tool – the industrial case study – leads researchers into the production area where they examine the effects of technological and organizational structures on workers in detail. German industrial sociologists have looked with disdain upon the positivistic orientations of their American colleagues, as they perceive most American social scientists to be. Positivistic methods are dismissed because of their superficial handling of a topic, proclaimed objectivity and neglect of historical developments.[2]

Of course, the more you are involved with something, the more detail and richness you see. Being involved with industrial sociological research in Germany, and going beyond first impressions, has made the differences in focus and orientation clearer to me. The nature of industrial sociology in Germany has made it possible to produce a book that is very coherent in terms of focus, but with differing approaches, opinions, and accents. However, in order to understand many of the issues, analyses of problems, and subtleties in this book, it is necessary to know something about the institutions and structures comprising industrial organizations in Germany.

INSTITUTIONS AND STRUCTURES IN THE SOCIETY

The majority of industrial sociological analyses that are written in Germany either deal explicitly, or at least touch on, issues of qualification at the workplace. Why is qualification such a central dimension of German industrial sociology? The basis for dealing with many issues revolving around the workplace, including worker autonomy, decision-making authority, disposition, power and job security, have been embodied in the subject of qualification in Germany. For the outsider, it is useful to have a brief overview of how the educational system is organized, especially *the dual system of training* for skilled workers, a fundamental aspect of the German deployment of labour.

In secondary school, a student decides between three alternative paths of education: a general study programme called *Hauptschule* (10 years of schooling), and further education in the form of the practically-oriented *Realschule* (11 years), or the scholastically oriented *Gymnasium* (13 years) (where students receive *Bildung*, that is, they get, so to say, 'cultivated'). The latter leads to a high school certificate based on a comprehensive examination called *Abitur* which is the prerequisite for entering a university. *Hauptschule*, followed by vocational training, was traditionally the path pursued by blue-collar workers. In the 1970s, about 75 per cent of those in training programs (*Ausbildung*) had graduated from the *Hauptschule*. In 1990, this figure was only about 35.5 per cent. Many who continue to take this road are the children of 'foreign' workers. An increasing number of secondary school students are choosing *Realschule*, formerly the means to some kind of middle level white collar work, as preparation for vocational training or *Berufsausbildung* in a blue-collar or craft occupation. A *Berufsausbildung* generally lasts about three years.[3]

All in all, vocational training in Germany proceeds in a nationally standardized system of occupations. There are presently 378 state-recognized occupations in this system from violin maker to salesperson; in the metal and electrical industries the number has actually decreased from about 400 to about 17 in the last decade by combining some very specific occupations into more general categories, in part to achieve greater flexibility on the job (see chapter 12). The basic learning content for every occupation is both predetermined and government-controlled. At the end of the training period, adequate ability in a given occupation is tested by the appropriate trade association.

For blue-collar workers, who are of main interest to us here since the chapters in this volume deal exclusively with industrial work, a training program and successful completion of an examination leads to certification and the title *Facharbeiter.* (In this volume we have translated this term as skilled worker or skilled tradesman and sometimes added the German term in parenthesis when extra clarity was necessary.) Generally speaking, a *Facharbeiter* is trained in the dual system of vocational training, a term also used in several of the papers of this book and perhaps one which could use a bit of clarification.

Most prospective *Facharbeiter* enter apprentice programmes run by a firm whose training programme has been approved and whose instructors meet government requirements. Their training, carried out by the company in conjunction with guidelines set by the state, is composed of both theoretical and practical aspects of the occupation they will be certified in – thus the 'dual' in dual system. This mix of theory and practice is generally divided up so that one or two days are spent in the classroom and the other days are spent on-the-job. Students attend a *Berufsschule* (vocational school) for their theoretical training. All students in apprenticeship from a particular industrial branch meet at the vocational school for their classroom training sessions (in large firms the *Berufsschule* may be internal). The first year of training is supposed to be somewhat broad-based, allowing an overview of an entire field such as electronics, construction, metalworking, etc. In the following two years, the training gets increasingly specialized in one particular area (see chapter 16 for a detailed comparison of vocational training systems).

The idea behind the dual system of vocational training is to make the training general enough to go beyond a single company or even industrial sector and at the same time make the training specific enough to permit immediate implementation of what was learned on the job. When workers with a particular qualification are recruited, companies know exactly what training their recruits are bringing with them. Part of what *Facharbeiter* are expected to have acquired is 'social qualifications', i.e. discipline, dependability, the ability to express themselves clearly and to define and lay out problems. The whole system of occupations is characterized by the comprehensive, regulated and structured nature of training as well as the coordination between different types of actors in the determination of training: unions, companies, government agencies. The other side of the coin is that those working in industrial production who are not *Facharbeiter* (the majority) are either unskilled or semi-skilled (*Un- and Angelernte*, literally not trained and trained on-the-job). The *Un- and Angelernte* receive some learning-by-doing training for work on specific machines or procedures.

From an American perspective, the German system of occupations contains a great deal of rigidity. The benefits of well-rounded, extensive occupational or professional training that is nationally uniform are obvious, but the disadvantages in lack of flexibility for an individual's career are more subtle. At about 16, or perhaps younger, the German going the vocational route selects an occupation and enters an apprenticeship. It is difficult to reverse or change this decision later in life because if you enter a different type of job in industry, you do so as a semi-skilled worker.

The argument often made by industrial sociologists in Germany is how effective the *Facharbeiter* system is in attaining flexibility in the production system (see Parts III and V in this volume for more complete analyses and critiques of this topic). In this book, there are really only two aberrations from the generally positive evaluation of this system. The critiques observe that unskilled and semi-skilled workers, especially the women and foreigners among them, have few opportunities to enter special training programs to improve their positions or adapt to new production needs. Thus, although companies' *Facharbeiter* are well-trained and in a good position to receive further training, firms with large amounts of semi-skilled labour cannot easily adjust their work-forces to new requirements, creating potential flexibility problems (see chapters 14, 19).

Another position well-represented in this book sees the *Facharbeiter* as having a key role in an optimistic scenario of the future of work (see chapters in Part III). Given the *Facharbeiter's* broad training and potential for improving flexibility in production, the vision is to have more and more workers in production being 'Facharbeiter', doing more and more challenging jobs. The proponents of this position argue that this would be good for new production requirements, thus for productivity, and at the same time good for workers from the viewpoint of job enrichment and self-fulfilment. Obviously, this situation is not presented as the status quo, but as an alternative path of development in industrial work. Observed realities, in fact, show a number of negative trends for *Facharbeiter* in current work arrangements.

In contrast to this view, the systemic rationalization approach in ISF (explained more fully in chapter 1) predicts a much less rosy future for the *Facharbeiter*. Advances in technology and automation and the way these are being implemented in production bring with them the potential for a segmentation of qualifications which can open the way for a polarization of skills among the work-force (see chapter 15). In addition, the concern for the work experience of the *Facharbeiter* is extended to the loss of 'subjectified action' or the ability for a skilled worker to use tacit knowledge on the job due to the introduction of centralized, automated production processes (see chapter 8).

It is difficult for someone not brought up in the German training system and rehearsed in the research traditions of German industrial sociology to really comprehend the role and significance of the *Facharbeiter*. What is striking to the outsider, whether German industrial sociologists mean to communicate this or not, is the acceptance of, even pride in, the industrial training system. Of course, the *Facharbeiter's* relationship to other workers and the way their labour is used are points of critique, but not the institutional form itself. Where does the motivation for the encouragement and defence (in the case of deskilling) of the *Facharbeiter* come from? Certainly it derives in part from the way existing structures leave their imprint on how a scientific discipline defines itself and its goals. German industrial sociology has traditionally had the 'two-sided challenge' to be both a social theoretical (in the sense of German *Gesellschaftstheorie*) as well as a pragmatic science (see chapter 1). On the pragmatic side, one of the central political orientations of German sociology has been to use its theoretical perspectives to help provide the potential bases for promoting government reform policies and labour and union policy, concretized in examples such as the humanization of work. In a country with such a strong tradition of the artisan or craft worker, it is hardly surprising that the concept of humanism in the world of production work would get embodied in the concept of the *Facharbeiter*.

Another dual system repeatedly mentioned in this volume and potentially requiring explanation for those of us not familiar with German institutions, is the dual system of interest representation. This refers to worker representation which exists at the national/industry branch level in the form of the union (*Gewerkschaft*) and that which takes place in the firm in the form of the works council (*Betriebsrat*). In any firm that has more than five employees, a works council has to be established, elected by and representing only the employees in that particular firm. The works council's rights and constraints are regulated by the Works Constitution Act of 1972 (see Weiss 1987).[4]

The industrial union is responsible for negotiating with the appropriate employers' association on the terms and conditions of employment for all employees, both hourly or salaried, covered by a collective agreement. The works council is responsible for operationalizing the collective agreement at the plant level and is both institutionally and legally delineated from union activities above the plant level. The works council can even enter into its own agreements with the company as long as they don't contradict the provisions of the collective agreement. Thus, what makes the union–works council configuration an especially 'dual' system in Germany is the works council's independent status.

From the union's perspective, it is difficult to make meaningful policy for the special problems that arise within a particular plant. Therefore, the works council was meant to take on an active role of codetermination in the firm. This was to be done in the spirit of cooperation (in any case, works councils have a no-strike clause in their constitution) even though, by definition, it was often conflicts that were being worked out. However, up until quite recently, the works council's role has mainly comprised routine problems. They have mainly dealt with issues connected with the collective agreement or labour laws, such as wage classifications and bonuses, general working conditions (noise, heat, etc.), work safety, dismissal, or personnel 'matters,' such as holiday plans, transfer, etc. At present, in the framework of designing new workplaces in conjunction with new computer technologies and flexible manufacturing systems, the unions are in a position to redirect their negotiation strategies to influence some fundamental conditions of work: job enrichment, job rotation, qualification, retraining, etc. The tendency is for the unions to formulate their programmes or demands in general terms, leaving the details to be worked out by the works councils. This puts greater requirements on the works councils, potentially leading to a difficult position in terms of capability to negotiate on these issues and in determining exactly what their role is (see chapters 23, 24 for a more detailed account of these problems).

The dual system of interest representation makes it possible to have, on the one hand, a national union negotiating at the national level with political power to back up the implementation of demands, and on the other hand, company level representatives that are familiar with company level problems. However, the position of the works council is not without difficulty in this arrangement. Works council members are company employees, elected by the work-force, and therefore represent the work-force (or at least certain groups in the work-force). Although the works council carries out union policy in the company, in the final analysis the works council's loyalty lies with the company in terms of protecting the interests of their constituents – the work-force.

Another problem is that although inter-union competition is alleviated through single union representation in one plant, inter-worker competition is not necessarily eliminated. Depending on the industry, plant, region and other factors, a union and its inner company counterpart, the works council, could perceive their base to rest quite identifiably with a particular worker category or skill group. This essentially leaves the other categories or skill groups de facto with limited representation at the collective level. Indirectly, this brings us back to the *Facharbeiter*. Unlike in the United States, where semi-skilled workers are the backbone of the industrial unions, in Germany it is the *Facharbeiter* who have hegemony. Not being as easily replaceable as unskilled workers, they have greater leverage. White-collar workers, although they benefit from union agreements, take a lesser part in union activities; fewer join the union, fewer participate in union meetings etc. Likewise, unions have traditionally not actively represented the interests of white-collar workers.

The example of the dual system of interest representation once again demonstrates how strongly the national context in which research is conducted and the structures and institutions making up that context influence what are defined as industrial sociological problems, and how they can be resolved.

INSTITUTIONS AND STRUCTURES INSIDE INDUSTRY

Moving away from the societal level to the company level, the first subject is the *evaluation or rating of the workplace* in German industry. It is necessary to know something about the concept of workplace evaluation in order to grasp the way issues of skill and deskilling are analyzed in this volume.

Often, German industrial sociologists refer to the American system of job classification as individually based, in contrast to the workplace-related German system. I find the distinctions between the two very subtle, seeing as how both define jobs, search for people with particular qualifications to carry them out, and provide compensation for negative working conditions. But the differences are there and can have important effects on how labour relations proceed and what the future of work looks like.

In both the United States and Germany, wages are, in principle, determined by qualification and the level of strain (for example, physical burdens) at a particular workplace. However, in the United States the role of classifications for job assignment differs from labour deployment in Germany. In the United States, a certain task gets classified as the responsibility of a specific type of worker at the plant level. Then, these tasks belong, by definition, to a certain type of worker and cannot be carried out by another type. For example, removing the protective closures on machines might be assigned to the fitters based on local union and plant management agreement. This means that all over the plant, whenever a closure has to be removed for repair, in principle, only the fitters can carry out the task even if it is in preparation for a mechanic, electrician, or regular production worker to engage on some work on the machine. Obviously, it would not be difficult to train other workers to perform this task in addition to their regular duties. However, under American practices of job demarcation, this is not possible. In practice, the rules of job demarcation are not so strictly followed; usually workers only begin adhering rigidly to job classifications when there is a disagreement ensuing with company management. However, job classification does build rigidities into the system by making it extremely difficult to rearrange tasks and responsibilities (see Jürgens *et al.* 1989 for a detailed account of this topic).

In contrast, in German plants, technical planners have more room to determine which tasks are required at a particular workplace. A workplace (often translated as job although the more concrete direct translation from the German actually describes the concept more accurately),[5] is then evaluated in terms of task requirements and negative working conditions and someone is sought whose qualifications match the demands of the workplace. Thus, if a workplace, in the light of its various tasks, requires an electrician or welder, such a skilled worker, who has a state-recognized certificate of training, will be chosen. This workplace will have a particular wage connected to it, as agreed to by the union and the appropriate employer's association, and as adapted by the work council and company management.

In the United States, if you redefine which tasks a certain worker category, such as welders, should perform, or if you change the tasks previously carried out by this group, the new demarcation rules have to be agreed upon at the plant level. The chances are good that the workers with a particular classification are not going to easily accept either having tasks being taken away from them, which could endanger their wage and potentially their job security, or being assigned new, additional tasks.

In Germany, if a workplace's requirements change due to, for instance, technological innovation, then the definition of which type of worker needs to be fulfiling the tasks at the workplace changes. The works council can respond with attempts to protect the wage of the worker in the case of downgrading, regulated for a particular time period through both collective and company agreements, but the integrity of the workplace is gone. For this reason, many of the chapters in this volume refer to the dangers for skilled workers in the face of changing organizational and technological structures.

Due to the German system of labour relations and the way jobs or workplaces are delineated, the discussion about reskilling and deskilling of skilled workers takes on a special character. It does not mean more or less conflict or negotiation between union and employer association or work council and management than in the United States – only that the arena for these discussions are different. In Germany, the issue for skilled workers focuses on the rating of workplaces and what skills are required at them, whereas under other systems, such as in the United States, the issue centres on the defence of particular job classifications and the demarcation of tasks defined by them.

A related issue to that of workplace evaluation is how *performance is regulated or measured.* For mass production, this refers mainly to the Tayloristic system of time and motion studies. As far as I know, the significance of this issue for German industrial sociologists has no counterpart in Anglo-American sociological traditions.

In the United States, incentive wage systems are rare. Performance is regulated through definitions of output and manning levels which are determined by a number of different methods, including time and motion studies. The union generally has little say in the process and can only respond through grievances, and in cases of extreme conflict, by striking (see Jürgens *et al.* 1989). Under this system, the use of time and motion studies represents a transparent form of company control. In Germany, however, the whole issue of performance measurement in mass production has a political and theoretical relevance quite different from that of the United States.

Politically, the German unions and works councils have always had more involvement in the implementation of time and motion studies than in the United States. As a means of having a so-called objective (i.e. 'scientific') basis by which to measure performance, the unions pushed for the use of time and motion studies. The works councils, as part of their codetermination rights, were responsible for overseeing the setting of standards and methods of measurement within the companies.

The works councils' equal voice with company management not only involves the means by which standards are carried out, but the whole definition of 'work performance'. In Germany, incredible efforts are made to try and define what 'normal performance' (100 per cent) is as the basis for determining wages (anything above normal performance involves additional incentive wages). The trade-off for greater codetermination is the prohibition of strikes on issues of the setting of performance standards. At the same time that this system ensures control by worker representatives in the process, it also legitimates time and motion studies and adds credence to the idea that they are scientific and non-partisan (see Jürgens *et al.* 1989: chapter 6, for more on this issue).

The unions and works councils adhere to these systems, not necessarily because they believe so strongly in performance levels, but because the Works Constitution Act only gives them codetermination rights on these issues. Obviously, it is extremely difficult to 'objectively' determine what normal performance is at any given time. Therefore, in the final analysis, behind all of the 'scientifically' based methods for defining normal performance, looms the issues of power and control. In Germany, these issues are played out in the arena of negotiating 'scientifically' determined performance levels.

The combination of the union role in promoting time and motion studies and the long term discussion over how they should be regulated, and, at the theoretical level, Marxist influence on industrial sociological orientations in Germany, has produced a whole discussion on the 'economy of time' or 'time regime' in German industrial sociology which does not exist in Anglo-American labour process debates (see chapter 14).

The 'economy of time' is basically the determination of how much labour will get transformed into labour power, or how much labour power it takes to produce a commodity, and the company's strategy for controlling this transformation. Time and motion studies are a technical means for controlling this transformation process. The implementation of time and motion studies under 'time regimes' or 'time management' makes the uses and intensification of labour transparent. From the management side, the name of the game, under Tayloristic production principles, is to intensify labour as much as possible to produce as many standardized commodities as cheaply as possible. From the (German) union perspective, under Tayloristic production principles, the objective is to have as much co-determination right as possible in setting the performance standard.

Several problems develop with time management given the nature of changing markets and the possibility for new technological and organizational structures in the production area. For management, in any industrialized country, a contradiction arises in that their 'economy of time' is pursuing the ever increasing productivity of standard production, while the market is demanding more diversity and higher quality products. For unions, German ones in particular, the presetting of standard times or doing away with Tayloristic methods of time management altogether could potentially contribute to their pronounced goals of greater worker autonomy and more humane and enriching forms of work. However, it simultaneously removes a traditional power base in the form of codetermining standards, and thus in influencing wage setting, *vis-à-vis* company management (see Sohn-Rethel 1978 and Benz-Overhage *et al.* 1982). Thus, when authors in this volume refer to the 'economy of time' or even Taylorism as a form of regulation, all of these considerations represent the starting point of their analyses. 'Economy of time' can only be understood in the German discussion from the background of the term's development in both Marxist theoretical orientations and union politics.

CONCLUDING REMARKS

Hopefully this paper has served to explain some of the intricacies of German society to better understand German industrial sociology, specifically the papers in this book. But a problem still exists, recognized by my German colleagues, by my American colleagues perhaps less so, and for the other countries I unfortunately cannot say. In sociology, the nature of the field is to pinpoint problems within given structures and institutions, ones that are so specific to national frameworks, such as training, wage systems, union agreements, labour market trends, etc. that the discussion gets limited to analyses or solutions that are completely inappropriate outside of the national context. Moreover, it is difficult enough to understand each other in the face of the obstacles different languages pose. But, within our countries, we make it even worse by communicating in a sort of 'new speak' which can only be understood by those well-versed in a particular system and in the discussions that have preceded the current one. Furthermore, often, when we carry out comparative research, the entire orientation is either that we have little to learn from other systems, or what is interesting about other systems is only what it can tell us about our own. This leads to unfruitful discussions in international groups.

Given the internationalization of markets, growing numbers of joint ventures, and the integration of Europe, to name a few, this seems like a highly irrational orientation. Perhaps a change in attitude will occur, leading to the realization that we really can learn from the problems and solutions in other nations, and that other traditions, both theoretical and methodological, which form the basis of sociological research in other nations, might indeed benefit us. If this happens we may not achieve more of a paradise, but we will be able to communicate less as strangers.

3 The Contradictions of Post-Tayloristic Rationalization and the Uncertain Future of Industrial Work[1]

Burkart Lutz

INTRODUCTION

This chapter argues that realistic assertions about the future of industrial work can only be made if social and economic meso- and macro-structures and their historical development are taken into account. European Taylorism is seen as the result of specific conditions and constellations existing in post-war Europe. The present erosion of the 'Tayloristic syndrome' has not yet been replaced by any new dominant pattern of enterprise rationalization and social reproduction. In the ongoing process of the restructuring of capitalism, different paths for the future of work have become apparent leading to what is described here as computer-aided Taylor, computer-aided Drucker,[2] and computer-aided Huxley.

Current research on the development of industrial technology and work is characterized by two indisputable findings:

1 Extensive and rapid changes in technology and work have taken place under the influence of various factors, such as the international division of labour, changing consumer behaviour, and the availability of new technological components and systems.
2 It is clear, especially when one looks beyond the borders of a single country, that the observed processes of adjustment lead in very different directions. While some far-reaching technological and organizational innovations are, without a doubt, an expression of 'new production concepts', which orient themselves to the principle of the 'reprofessionalization' of production work, hundreds, in fact thousands, of highly qualified engineers and computer scientists work on models of an 'unmanned' factory and expert systems, whose use is only justified by the definitive elimination of skilled production work.

The controversy about the future of industrial work, as described in chapter 1, consists of diverging and openly conflicting positions that are not based on academic speculation; rather these positions can at any one time refer to tangible – albeit selectively interpreted – empirical evidence.

This highly diverse empirical picture is not merely the result of too little observation. A much more intensive description of current adjustment processes in technology and work and particular enterprise policies would not bring greater clarity. The diversity of the observations and expected trends must be understood as a fundamental characteristic of the historical situation in which the advanced industrial nations presently find themselves. In complete agreement with authors such as Piore and Sabel (1984) or the French 'regulation school' (see Boyer 1989), I maintain that realistic assertions about the future (or the futures) of industrial work can only be the starting point of research which combines the macro-conditions of technology and work, and the historical processes that have been generated and changed by these macro-conditions.

This chapter will outline this formulation in broad strokes. Therefore, it begins with a look back at the time following the Second World War in order to understand the historical context of the use of work and technology in the enterprise and of the decisions made to transform it. Obviously, the use of technology and work and the decisions surrounding it are critical for the present and foreseeable future.

ECONOMIC EXPANSION AND THE SUCCESSFUL ADVANCE OF TAYLORISTIC RATIONALIZATION IN POST-WAR EUROPE

The Welfare State and the Economic Miracle

After the end of the Second World War, most European industrial nations found themselves in an absolutely desolate economic situation, despite the fact that the worst destruction from the war had been repaired and the most flagrant crises overcome.

The export of industrial goods to non-industrialized or partially industrialized areas, which for centuries had been the most important force behind European development, was depressed and not likely to be restimulated in the short term.

Relief through growth based on expanding domestic demand also seemed to be hindered. There was an immense demand for necessities, such as clothing, furniture, housing and infrastructure. The standard of living was clearly below the level of the pre-war period, and was near subsistence for many segments of the population. What was destroyed or worn out during the war had to be rebuilt or refurbished. Transforming these immediate needs into buying-power, and thus putting into motion a self-regulating dynamic of growing income, was made impossible by the cost structure of industrial production and the huge supply of labour, acute or latent, which was willing to work at almost any price.

In this situation (which included the imminent threat of communist expansion), a totally new social strategy arose in all European industrial nations. This strategy can be expressed as 'welfare capitalism'. A new 'accumulation regime' (in the sense of the regulations school), combining social-political, fiscal, institutional and political-economic innovations, was established after the war and served as the foundation for an economic upturn unparalleled in history.

This accumulation regime received its, at times extraordinary, dynamism from the interaction of two components:

1 The first component was found within a historically new type of 'mode of regulation'. At the crux of this mode was the extensive political neutralization of the 'law of wages' (or, one could say, the price-quantity mechanism of the labour market). The functions of the law of wages were, as the classicists said, 'brazen', but they could hardly be dismissed from capitalists' economic development up to this point.
2 The second component of the constellation of prosperity that characterized post-war Europe was found in the opportunities opened up to the industrial-capitalist sectors of the economy through the new mode of regulation. This mode broke up and extensively absorbed economic sectors that had been very important and stable, i.e. the traditional, farm-craft economy and life-style. In this context, two powerful mechanisms interacted, both of which were based on continuously large increases in wage income: the first mechanism was the emergence of a 'new milieu', which served to satisfy the need for fundamental material, and a large part of necessary non-material, goods and services. This new life-style could only function, be produced, or furnished through industrial techniques and organization. The second mechanism was the mobilization of large quantities of workers previously employed in the traditional farm-craft economy (such as proprietors and family members) into wage labour for the modern industrial-tertiary sectors of the national economy.

Strong macro–micro congruence

The above-outlined European prosperity in the decade following the Second World War could develop and stabilize only if a large number of firms in the economy followed their fundamental interests in a way that was congruent with the newly established macro-structures, the emerging new mode of regulation, the newly generated economic cycles and the flow of money and goods.

The principles of Tayloristic or Fordistic rationalization and the forms of enterprise structure, production, and work organization clearly corresponded with the functional requirements of the accumulation regime and its model for growth, which were progressively established in post-war Europe. These principles include: the use of scale effects through mass production as the most efficient compensation for massive wage increases (which is necessary to generate new massive buying power); the extension of the division of labour in technical, hierarchical and functional dimensions; and the intense standardization of work, creating the preconditions for large quantities of non-skilled labour to be quickly and easily integrated into the production process and thereby allowing the firm to react quickly to the growing demand for goods and services.

However, after the Second World War, European industry found itself in a situation that made a quick expansion of Tayloristic–Fordistic rationalization policies and production methods unlikely. The industrial structure of the economy was characterized by a strong oversupply of raw material and investment goods industries. The industrial manufacturing of consumer goods played altogether only a small role in the economy since a large part of private consumption was satisfied by the traditional farm-craft sector (with a high share from household production). In 1950, large firms in West Germany with products geared toward mass production (the automobile and electronics industries) represented less than 8 per cent of the employed in industrial firms with more than 100 employees. This figure was much lower than the coal-mining or the textile industries. Also, in the 'modern' industrial sectors, such as the metalworking and electronic industries, the dominant forms of enterprise structure, production and work organization as well as personnel structures were not compatible with the principles of Taylorism and Fordism. Rather the forms were predominately characterized by a high level of in-house production, a large variety of products, an organizational structure on the shop-floor where the foreman (*Meister*) had a strong and independent role, and a clear oversupply of skilled workers in many sectors.

Nevertheless, the new rationalization norm expanded very quickly throughout Europe during the 1950s and into the 1960s. Increasingly, it shaped the structure of industrial production and industrial work. This can be explained for the most part by a series of extremely favourable conditions which combined to sharply reduce the costs and risks of shifting to the new norm of rationalization and organization.

Consumer demand was both structurally stable and predictable. The goods appealing to consumers, such as furniture, household appliances, radios, stereos and autos (which were gaining increasing economic importance), were not difficult to produce from a technical standpoint. The level of technology and the international price and cost level generally corresponded to a production method that, at least from today's perspective, was highly labour-intensive and required a small amount of capital. Workers, who up to this point were bound to traditional economic structures and traditional work relationships, responded to financial incentives being offered by industry and provided, on average, a highly motivated work-force. These workers showed a strong tolerance for physical labour and brought to industrial work a wealth of abilities; at the same time, they accepted relatively low wages.

The Tayloristic concept or the forms, principles and processes of a high division of labour represented by mass production, reached a high degree of popularity and maturity in the time following the Second World War in German and European industry. In part, this can be attributed to pursuing the American example and Germany's own experience in weapons production.

Since the 1950s, these very favourable conditions caused a rapidly growing number of companies to orient their organization, production and work process increasingly toward the principles of Tayloristic rationalization in order to capitalize on the opportunities for expansion and profit.

The Tayloristic syndrome

Two aspects of the new welfare-state mode of regulation crystallized in a short time into a very powerful historical constellation in Europe in the 1950s. The domestic demand dynamic, which made possible the 'appropriation' of traditional parts of the economy and society, formed a growth and accumulation constellation that merged with the dominant enterprise rationalization strategy based on Tayloristic and Fordistic principles. The resulting historical constellation is referred to here as the 'Tayloristic syndrome' (a synonym for the term would be 'Fordism', which is widely used in the regulation theory literature).

The power of this syndrome is derived from strong interactive effects with distinct interrelationships on two levels: the macro-level consisting of the domestic economy, social relations and institutions, as well as political processes, and the micro-level consisting of individual firms with their production and work processes and their technical-organizational structures.

Enormous buying power, as a result of rising wages, opens new opportunities through the use of scale effects to reduce costs and to secure additional room for increasing wages and/or reducing prices. At the same time, this increases the attractiveness of wage labour as compared to employment in family enterprises in the traditional sector which were still very numerous and seemingly stable by the end of the war. Rising wages also exposed traditional products and services to intensive competition from mass-produced substitutes. Higher productivity arises from work intensification under Tayloristic work organization, but it also leads to the dissipation of workers' strength and abilities. The system of social security compensates for the strain on workers through payments which are generated from rising wages and benefits. Moreover, through the creation of independent 'social-political' markets and economic activity (one thinks here of the enormous growth of the 'health economy') it is guaranteed that a large part of this compensation is recycled back to the firm in the form of demand.

Driven by such mechanisms of positive feedback, the 'Tayloristic syndrome' develops a dynamic which is without precedent in Europe and can be compared with certain phases of US development that were shaped by similar constellations. The dynamic of economic growth and well-being, the dynamic of industrial expansion and the dynamic of rationalization and mechanization of firm production and work processes all mutually support and strengthen one another.

THE POWER OF THE 'TAYLORISTIC SYNDROME' TO SHAPE STRUCTURES

A historical constellation of such power and dynamism cannot be established and survive over a long period of time without causing far-reaching structural changes in institutions, authorities and areas of society. To better understand the current situation, three structural effects of the Tayloristic configuration are examined more closely: changes in the production and work process, changes in the organization and power structures of the firm, and changes in the environment surrounding the firm and the firm's relationship to this environment.

Production and work process

In the first instance, rationalization measures in accordance with Tayloristic–Fordistic principles can be understood as simple tools or options offered to a firm in certain situations which it undertakes, uses and, under changed external and internal conditions, can give up (whereby the firm loses at most a certain amount of investment in physical or human capital in the process). Over time, the progressive accumulation of steps in rationalization and mechanization form structures of production and work in a growing number of firms. These structures are deeply shaped by the fundamental goals and principles of Tayloristic rationalization, and (without great difficulty) they can no longer be reoriented to other goals and principles.

The fundamental points of restructuring and processes of change have been thoroughly described in the literature; however, let me refresh your memory: the gradual reorganization of industrial structure and factory-design through the systematic transition to mass production; a technology oriented to this reorganization centred on products which make maximal use of large batches and production facilities with extreme specialization in products; technical and organizational innovations that rely on a continual increase in productivity from direct work and result in increasingly selective and rigid automation, supplemented by a few qualified 'stop-gap' jobs; an organization of work shaped by a deep horizontal, vertical, and functional division of labour and the apparent unending decline of skilled work; a performance and wage policy whose primary goal is short-term maximization; etc.

There are two types of structural consequences of Tayloristic rationalization that have received much less attention in the scientific debate. These structural consequences take place in both the internal firm structure and the relationship between the firm and its social-institutional environment as well as this environment itself.

Firm and company organization

A firm could only take full advantage of the opportunities for expansion and profit in Europe's post-war prosperity constellation, when it not only secured the efficiency created by the new Tayloristic rationalization structures in production technology, work organization and quality control, but also secured their continual increase over time.

If firms want immediately to exploit every chance of increased sales and transform external pressure to raise wages into increased productivity, so that this is consistent with constant or even falling per-unit labour costs, a more or less regular increase in labour productivity is inevitable. Achieving this benefit, however, requires continual innovations in production and work processes and performance policy. In this regard, there must be adequate problem-solving and organizational skills continually available in these areas of the enterprise. This should not be regarded as a sideline of general management responsibilities, as it was in the traditional *Meister-firm.*

With the takeover of Tayloristic principles of rationalization, firms put a process of the functional division and differentiation of labour in motion which basically appeared as the simple application of well-known organizational principles or as an obvious continuation of long trusted trends. Yet these principles and trends gradually arose into something extensively new, something like a 'meta-structure' of enterprise organization. The actions and tasks of the administration and service areas (such as process planning and operations scheduling, controlling and quality control, data management, as well as important parts of the personnel department) of this structure are no longer geared directly to the production of goods. Rather they ensure that in the long term the conditions which enable particular parts of the firm to make an optimal contribution to these purposes are present.

The institutional areas of the meta-structure increasingly take over technological and organizational responsibilities and technical knowledge that have, up to then, been a more or less given, although inexplicit component of production work skills. This meta-structure, with more and more refined rules of procedure, embodies a large part of what used to make up the dynamic of a company's personality. One could call this the 'strategic capacity' of the firm: the ability to recognize new external forces and problems that arise or opportunities and resources that offer themselves, so that the firm can react correctly and hopefully, prior to its competitors; and the ability to use the technical and organizational knowledge necessary for this reaction as well as to possess enough power inside the enterprise to carry out necessary changes.

It is clear that these meta-bodies in the enterprise, which promote the rationalization process and increasingly monopolize the necessary qualifications for this process, could not arise overnight. The provision of technical and personnel resources needed to ensure the effectiveness of these institutional areas can only come about in the course of a long term development (see chapter 18). Because of the nature of its duties, the work-force requires a special skill that is not readily available on the external labour market, but which must be built up slowly. The applicable, especially technical-organizational, knowledge comes about first through a long process in which partners outside the firm (business consultants, business associations, researchers, etc.) take part. In the short or long term, these partners have to find a professional milieu which can serve as a consulting body for internal firm discussions. This increases the risk, however, that strategic options of the firm will be maintained long after their premises cease to apply.

Environment and the environmental relationships of the firm

An integral component of the welfare-state regulation mode which arose after the Second World War in the European industrial nations is a network of institutions. From these, policies are articulated, given shape, and implemented, norms are established and executed, and public services are provided. The functional relationships and circulation of the new accumulation regime is founded on this network. These institutions reflect the particular, and often strongly, nationally specific characteristics that result from historical tradition and represent the starting points of the individual countries after the war. Nevertheless, these institutions have similar, and in certain constellations, essential meaning for enterprise action within the Tayloristic syndrome. This is the case in at least three ways:

1 First, they furnish services which the firm, in the context of Tayloristic rationalization, directly or indirectly benefits from. They take over costs if need be, such as easing obligations or opening additional room in the policies of the firm. A large number of examples for this exist. The services of the educational system, which can be used by the firms to employ trained workers; the research results of universities or publicly financed research and development arrangements that are available to firms as they are or in the form of newly recruited labour; and the above-mentioned social security services which the firms at least indirectly benefit from.
2 Additional institutions (or other activities within the same institutions) affect, above all, the mobilization of resources available for the firms use. The most obvious examples of this are found in labour market policy. The public policies geared toward labour mobilization, however, are not limited to the activities of the state labour authorities such as the German Federal Labour Office. A large number of programmes and projects concerning regional and structural development partially serve the same purpose. These projects help ease the resettlement costs of industrial firms with large numbers of laid off workers.
3 Finally, institutions of this type guarantee the uniform behaviour of all potential competitors through the establishment and use of regulations and minimum standards, even when these interrelationships require firms to act in a way that does not directly comply with their interests. The most prominent example of this is collective bargaining and its effect on raising wages; the major collective bargaining associations (employers' associations and labour unions) are also a fundamental component of the Tayloristic syndrome's institutional network.

Thus, in the first phase of rapid Tayloristic rationalization expansion these institutions and their effects stand outside of the individual firm as entities which cannot be influenced, at least not enough to speak of. Furthermore, enterprise practices have adjusted around them in such a way that it is in the best interests of the firm to use them. This changes to the extent that a growing number of firms learn, as a result of the process of the institutionalization of specialized, strategic capacities as sketched above, to build stable relationships with institutions that are especially important for them and whose services increasingly further their own interests. Then a more or less continual interaction (directly or through associations or similar institutional bodies) between firm policies, actions and programs, and services of public or semi-public institutions takes place. This results in an alignment of these external organization's maxims, practices and instruments with the specific objectives, procedures and problems of Tayloristic rationalization.

For example, general legal terms, such as 'efficiency', 'equity' or 'level of technology' on which procedures of such institutions have oriented themselves, are largely compatible with Tayloristic rationalization. This, however, leaves other rationalization strategies, such as 'flexible specialization', at a strong disadvantage. In the same way, the requirements for social security benefits are being increasingly defined in terms of relationships that are typically generated from Tayloristic rationalization.

The orientation of public intervention toward the model of Tayloristic rationalization contributes to the supply structure of factor markets being determined within a Tayloristic logic. In this way, it strengthens the forces pressing for the expansion of Tayloristic rationalization and for stabilization of already created structures. Thus, the markets for production technology are strongly and increasingly dominated by machines, equipment and systems which can only be used efficiently and profitably (with intensified use of computer technology) in the context of a Tayloristic organization of work and organization of the enterprise. Markets for professionally skilled labour and their corresponding training courses increasingly dry up in the light of the decline of traditional skilled occupations and the high vertical and functional division of labour which dominates in the economy. This makes it more advantageous for younger generations to make use of the growing opportunities for formally acquiring a higher qualification in schools and universities. In this way, the pressure grows for those firms that have, for whatever reason, gone down a different path to adopt the Tayloristic model of a high division of labour.

THE EROSION OF THE TAYLORISTIC SYNDROME: NO NEW DOMINANT PATTERN OF RATIONALIZATION IN SIGHT

The mid-1960s brought the first signs of an incipient slackening of the dynamics of the prosperity constellation based on the welfare mode of regulation and the 'appropriation' of the traditional sector of the economy; this trend then became increasingly evident in the 1970s.

This eliminated the central preconditions for the efficient functioning of a rationalization strategy based on Tayloristic–Fordistic principles. The conflict – largely neutralized at the peak of the Tayloristic syndrome – between 'production economic' interests directed at cost control and cost reduction, and 'market economic' interests aimed at maximizing sales opportunities, began to break out on an ever broader front. The more and more turbulent markets demanded a higher degree of flexibility and innovation; measures to meet these demands contradicted the maxims and practices of increased productivity and cost reduction through utilization of scale economies and progressive standardization of products and production processes.

The 1980s doubtlessly marked a break in the history of European industry, as decisive as that which heralded the development of the 'Tayloristic syndrome' in 1950. To this extent, the concept of an historic turning point, an industrial divide (as formulated by Piore and Sabel 1984), is by no means exaggerated.

What does this now mean for the development of rationalization in the enterprise and the future of industrial technology and industrial work?

At first glance, it seems quite obvious that enterprises would from now on gradually adapt their rationalization strategies to the new historical conditions. Accordingly, new forms of work organization and production technology would have to arise. This would entail the creation of relatively small operational units with considerable professional and functional autonomy, high levels of flexibility, fast and intelligent reactions to unforeseen problems and challenges, and a clearly improved ability to innovate, as well as a necessary dismantling of the types of division of labour which were typical of Tayloristic–Fordistic rationalization.

Following in the tradition of the tried and true social science analyses on the development of technology and work, many researchers are convinced that by examining typical cases of enterprise conversions in technology and production organization, they will be able to identify at an early stage the step by step implementation of post-Tayloristic rationalization logic and the changes in industrial work associated with it.

With their thesis formulated in 1983–4 on the change in the rationalization paradigm and the coming of the 'new production concepts', Kern and Schumann are characteristic representatives of this conviction. However, their position has not gained in credibility in recent years. The picture so courageously sketched by Kern and Schumann of a new generalizable post-Tayloristic rationalization pattern has not taken on more precise contours or stability in the light of developments in the last years and intensive research on the subject. On the contrary: the non-uniformity of empirical findings is increasing, especially if one looks beyond the borders of an individual country. The impression among quite a few researchers is that at the moment a large (increasing?) number of enterprises are confronted with diverging constraints and opportunities to which they can and do respond in very different ways.

The assumption that the most important industrial nations are currently in the middle of the transition from an old Tayloristic–Fordistic phase of development to a new one with similarly stable interactions between macro- and micro-levels is becoming more and more risky. To answer the question of what the future of industrial work will be, it would seem more useful and promising to assume that the non-uniform picture provided by empirical studies does not simply reflect mere transitional phenomena, but is rather an indication of essential features of a new, but lasting, historical constellation. In this constellation many, perhaps an ever-increasing number of enterprises, are forced to enter a strategy of trial and error over a long period, whose results and outcome are not predictable without being purely arbitrary. With this perspective, it is possible to make statements on the future of industrial work and technology to the extent that the logic of these experiments and identification processes, the most important internal and external forces influencing them, and their characteristic patterns can be pin pointed and analyzed.

A first attempt in this direction is sketched below in very rough terms. There are two premises underlying the core of the argumentation:

1 the fact outlined on pp. 31–3 above that Tayloristic rationalization at its peak had a far-reaching effect on internal company structures as well as many external societal structures important to the company; and
2 the thesis that precisely because of this fact, the erosion of the Tayloristic syndrome caused a massive reduction in the ability and opportunity of many companies to redefine their rationalization strategy in accordance with their current interests.

The logic of 'identification and experimentation' can be derived from the efforts of companies to break away from the structures created under the dominance of Tayloristic principles and from the difficulties, bottlenecks and constraints which they attempt to overcome in the pursuit of new paths.

CROSSROADS AND PATHS OF DEVELOPMENT OF POST-TAYLORISTIC RATIONALIZATION

Three paths of development of work and technology

Existing research findings allow the identification, without excessive stylization, of three ideal-typical development trends of rationalization in industry (and very similar ones in many organizations in the service industry, although this will not be dealt with here); each of them corresponds to a very characteristic path of development of work and technology:

1 A first path is characterized by the continuation or further consolidation and entrenchment of a strong division of labour of a horizontal, vertical and functional type. This path differs from the classic Tayloristic–Fordistic rationalization pattern in particular through the large-scale use of computer-aided information and communication technologies to increase companies' flexibility margin, and in this way, compensate as far as possible for the increasing inefficiency of the structures within the companies and society created by the Tayloristic syndrome. This path of industrial work and technology may be labeled 'computer-aided Taylor'.

2 A second and sharply contrasting path is characterized by an extensive and lasting elimination of the Tayloristic division of labour. In this model, at least partially autonomous groups of very qualified 'reprofessionalized' production workers without hierarchical differentiation form the basic unit of enterprise organization. Modern technology also plays an important role here: technical resources eliminate most of the physical and stress-related work burdens which had been inseparably linked to production work; the cohesion between the basic autonomous unit and the integration of the entire company is guaranteed by highly interactive information, communication and control systems.

   Taking up older, long-gone traditions of organization, this development path (which corresponds in many respects to the 'new production concepts' of Kern and Schumann) can be termed 'computer-aided Drucker' (see note 2).

3 A third path is characterized mainly by the fact that in connection with a partial abandonment of horizontal and functional divisions of labour by means of strong decentralization and the creation of organizational units with considerable autonomy to execute tasks, a new type of industrial manpower is created, that of the formally high-qualified 'production engineer', who tends to belong to the social and professional milieu of academically qualified experts. In the field of direct production (in which he usually spends only one phase of his working life in the course of his career) this production engineer monopolizes what may be labeled 'production-intelligence'. He draws into his own domain – as the control centre, the system leader, the supervisor or acting in some similar role – all the tasks which presuppose a certain degree of competency and responsibility; the complement he needs is therefore an unskilled work-force and one which has even fewer upgrading

opportunities than in Taylorism, workers who have to assume all the remaining heavy and unattractive tasks and carry them out according to strict instructions.

With reference to Huxley's famous novel, *Brave New World*, and his strict segregation into 'alphas' and 'epsilons', this development path may be labeled 'computer-aided Huxley' (although few examples of this path exist in Germany, there are many cases in other countries).

Comparison of these three development paths – which can also be viewed as scenarios – prompts the questions research must take up if the assumption of a long-lasting identification and experimentation stage of company rationalization is to be realistic: What are the factors determining which of these paths a company or a group of companies takes? What factors have the greatest influence? What consequences would a far-reaching implementation of one or other rationalization paths have on the meso- and macro-levels? And how would the structural changes triggered on these levels effect the enterprises and their rationalization strategies?

To illustrate how an answer must be sought to questions of this type, a very simplified schema with two alternative crossroads of company rationalization can be developed.

The first crossroad: structural conservatism or reprofessionalization of production work?

The more powerful a historic constellation has been, the more likely it is that the structures it gave rise to will continue to exist and stay in effect far beyond the point in time in which the constellation itself begins to disintegrate. For this reason, a large number of enterprises are still subject to strong influences compelling the continuation or even the intensification of a rationalization in line with Tayloristic principles which persistently hinders a fundamental reorientation.

The reasons for this lie, first of all, in the strong influence Tayloristic rationalization has had on internal enterprise structures. A key role is played here by the 'meta' structures of enterprise organization in which specialized strategic competencies have crystallized into a quite specific orientation: on optimal use of accumulation and profit-making opportunities offered by the Tayloristic syndrome. Even with the erosion of this syndrome, authorities, departments, planning and decision centres, which make up these meta structures, are eager to draw the newly arising complex of problems into their own areas of responsibility, i.e. to define them so that they appear to be genuine cases of application of their respective areas of competence. As a result, enterprises are so often pinned down to the typical pattern of action defined by Tayloristic rationalization that it is difficult to explore new paths, much less pursue them.

The conservative effects created through the inner company relations which stem from Tayloristic rationalization can be reinforced by the characteristic dynamics of the decline of the Tayloristic syndrome. The erosion of this historic constellation, and the slackening of the positive feedback mechanisms it is supported by, creep in over a long period of time. For the company, these changes manifest themselves in the form of an at first barely perceptible and then slowly more massive deterioration of the conditions characterizing the predominant rationalization strategy. For this reason, dramatic changes in company strategy cannot be induced; at best, pragmatic reactions along the line of practices and policies tried and proven in the past which may be conceived and implemented without any great modifications in the existing organizational structures seem to be the most appropriate plan of action.

Other conservative effects repeatedly pinning down or retarding company action to the basic Tayloristic pattern stem from circumstances in the external environment. Thus, the tendency of most enterprises to avoid any far-reaching changes in organizational structure (which would be inevitable if they really reoriented their rationalization strategies) is very often reinforced by the aggressive marketing strategy of the manufacturers of information technology systems. The increased utilization of their information technology in an otherwise unchanged organizational context is presented as an adequate response to the enterprise's problems. In fact, the problems could only be solved through a rejection of the basic principles of the rationalization they have practiced to date. Moreover, the limitations of such purely technical solutions usually only become visible after highly consolidated, integrated system structures have arisen, only permitting enterprise reactions which correspond to their inner logic – in as much as the enterprise will not risk neutralizing central functional areas for a considerable length of time. Far-reaching strategic reorientation is also quite frequently rendered difficult by normative rules and public intervention. In recent decades their premises and maxims have increasingly been defining the specific conditions created by Tayloristic rationalization patterns as normal. In this way, they have fixed company policy (in particular, but not only, personnel policies and labour) to certain lines of action which highly conform to company interests under the Tayloristic syndrome, but which may well be totally incompatible with any other strategy.

For both internal as well as external conservative factors, the availability of skilled labour for production work (and similar types of activities) and the forms and systems of education and training conditioning them take on central significance. It was, of course, one of the main objectives of

Tayloristic rationalization to create as large a percentage of jobs as possible where the work might be performed by any one of many people without specific training or competency. Any serious departure from Tayloristic principles opens the door to a 'skill gap' in production departments and requires that the company radically and rapidly closes this gap (perhaps even with proactive measures). This aspect will be treated in more detail in the discussion of the second crossroad.

The forces, trends, structures and relations discussed above might well restrict the scope of many enterprises to a redefinition of their rationalization strategy which at best only partially modifies and at certain points corrects the basic pattern that arose in the post-war period. As a result, these enterprises are tied down to the development path of 'computer-aided Taylor'. They must therefore invest considerable effort to improve, through maximum use of information technology systems, the totally inadequate flexibility and innovation potential of their structures which derives from the persistently marked division of labour and strict segmentation of unskilled-execution and skilled-conceptual work. This path is usually burdened with considerable losses in efficiency due to the remaining massive barriers which impede flexibility and innovation, the possibly even rising costs of a division of labour, and the growing dependency on the often underestimated costs and the often overestimated capabilities of information technology. This may well jeopardize the economic survival of the companies to the extent that competitors are not also subject to the same constraints and restrictions. If the latter is the case, adherence to a rationalization pattern corresponding to quite different historical conditions restricts not so much the profitability of the enterprise as the overall economic productivity and welfare of the country.

The second crossroad: homogeneity or dichotomy of (reskilled) production work

A series of impressive evidence indicates that enterprises do manage, often as a result of an external shock such as the threat of bankruptcy or a takeover by new owners, to break free from the web of structural conservative forces and open the way for truly strategic innovations. The path of technology and work which these enterprises (may) embark upon obviously depends essentially on the way they can make or buy, produce, train or hire the competencies needed for their new pattern of rationalization.

In the sense of the 'new production concepts' of Kern and Schumann, basically the most obvious answer to the continuing loss of efficiency of Tayloristic rationalization and division of labour should be a return to forms of skilled labour in industrial production which were common in many economic sectors prior to the hegemony of Tayloristic–Fordistic principles. In this respect, the development path labeled above as 'computer-aided Drucker' simply takes up a long tradition which has survived in many areas even under the dominance of Tayloristic rationalization (especially the manufacture of high-quality products in small series).

But a simple return to this tradition is no longer possible for most companies in most industrial nations. The radical breakdown of horizontal, vertical and functional divisions of labour opens the door to very high flexibility and innovation under usually very favourable economic conditions. However, it necessarily presupposes that the enterprises can recruit and employ a particular type of worker fairly easily, one for which the German industrial skilled tradesman (*Facharbeiter*) can serve as the prototype, and can keep him long-term. Workers of this type must combine considerable intelligence with the willingness to perform physical, at least at times, monotonous and taxing (and for this reason not particularly high status) work; he must be willing and able to gather experience in practical work over a long period of time and to supplement this experience with ever new technical knowledge.

In most industrial nations before the historic constellation of the 'Tayloristic syndrome', a work-force of this type in more or less large numbers was available virtually as a matter of course. Since their origins, heavy industrial enterprises found a stable and reliable reservoir for procuring skilled labour or workers who could be easily trained to perform production work in the social and economic structures of the traditional farm-craft sector. Thus, the industrial skilled tradesman – including the traditionally and still very significant portion of technical personnel in the German-speaking countries who have been promoted from the skilled worker ranks – is embedded very deeply in the social structures and cultural traditions of the pre-industrial 'folk'. It is here one finds workers willing to perform physical, often very taxing and not particularly well-paid work, and at the same time accept responsibility for people, machines and material being dealt with, and who are able to learn many techniques, put them into practice and develop them. This phenomenon is only imaginable against the background of conditions in which hard physical labour was the inevitable fate for children in this social sector and skilled production work in industry was regarded as a real opportunity (see chapter 17).

The coming of the Tayloristic–Fordistic pattern of rationalization and economic growth brought the total erosion of the conditions which ensured industry's supply of intelligent and competent production workers. Two processes interacted in this development.

For one, the highly efficient mechanisms with which the Tayloristic syndrome mobilized manpower for industrial wage labour, slowly but surely destroyed the social and economic structures in which the manpower had grown up and been socialized in traditional rural-craft (or, in the case of women, domestic) living conditions and production modes. In industrial nations today (even in the German-speaking countries, which were able to keep the tradition of industrial skilled work alive longer than the others), virtually no children who are normally gifted would consider industrial work as an acceptable prospect under the conditions largely applicable in industry today. And, massive immigration from less developed countries has been receiving more and more resistance for a long time.

Second, through the generalization of behaviour oriented to utility maximization prevalent in modern industrial nations, the human capital calculus has become the dominant factor in selected educational and occupational paths. At the same time, Tayloristic rationalization has destroyed the previously dominant position workers trained in practical skills have enjoyed in industry. The massive rush for high-level education (towards university or quasi-university education) unleashed by these two trends, has made the schools, forced to engage in ever more scrutinous selection procedures, the main means of distributing professional and social chances in just a few decades; in the short or long term, graduation from institutions of higher education is going to be the necessary prerequisite for access to all the attractive and above-average paid jobs. Furthermore, in the course of a generation, the traditional forms of social reproduction of the skilled or highly skilled industrial work-force will lose all meaning through practical training and practical experience.

To the extent that these two processes converge and are not counteracted by political measures, which would require both great courage as well as exceptionally favourable circumstances, enterprises will more and more frequently find intelligent employees with utilizable industrial skills and knowledge exclusively among the graduates of higher semi-academic or academic institutions.

Under these conditions, if enterprises, nevertheless, wish to overcome the inefficiencies of a structure characterized by a marked division of labour and avail themselves of the advantages of an organization based on autonomous units of production, capable of reacting flexibly and innovatively, then they must create the necessary conditions (in work organization, career prospects, payment, etc.) to attract graduates from institutes of higher education. These potential employees come mainly from the middle-class or at least have assumed middle-class orientations and expectations for jobs in industrial production. In designing a system of work organization to fit these conditions, enterprises are inevitably led to the path of 'computer-aided Huxley'.

Even if it is possible at the beginning to employ qualified experienced workers side-by-side with young graduates, the massive utilization of engineers (or technical personnel with comparable academic qualifications) in production will not be possible in the long term without giving way to the century-old societal distinction between manual and mental work (and the status and pay differentials linked to it). Only within the context of a strict dichotomy between qualified (engineer's) work and unskilled (worker's) work can the differing earnings and career prospects be legitimized. And without these, highly qualified personnel cannot be drawn into the production process anything other than intermittently.

For this reason, the chances are quite high that at the second crossroad, the majority of enterprises in most of the large industrial nations (specific national characteristics will almost certainly play a role here) will opt willy-nilly for the path of 'computer-aided Huxley'. This will rarely occur through an explicit strategic decision; it is far more likely that a multitude of isolated decisions and incidents, each on its own seemingly insignificant, will push, over the course of time, internal enterprise rationalization and personnel policies in a new direction which will then be difficult to modify.

FINAL REMARKS: SOCIETAL EFFECTS AND THE QUESTION OF LONG-TERM STABILITY

The thoughts posed so far on the realization chances of the three ideal-typical development paths of work and technology place the societal macro-structures as developed in the 'Tayloristic syndrome', as external conditions and factors influencing company rationalization. However, these very structures will change to the extent that one or the other of these paths gain a certain degree of stability and generality. The decisive question is then whether the interaction between company rationalization and external societal structures will make it possible to attain long-term stability or will cause the eruption of instabilities time and again.

No one is in a position to give a credible answer to this question today. However, much evidence speaks in favour of:

1 This answer being very different depending on the prevailing development path;
2 the only one of the three development paths which could mean real long-term stability in a society with our values being the path of 'computer-aided Drucker';
3 without radical changes in enterprise and societal frameworks, this development path (and the radical departure form a strong division of labour which it represents) has clearly fewer chances of success than the

two other paths although in many respects, it appears to be superior not only socially, but also economically and technically.

The development path of 'computer-aided Huxley' would, were it to achieve greater generality, unleash in all probability a process on the societal level which would – since it would very quickly have an intensive effect on company rationalization – inevitably result in a 'society of experts'. A societal structure in which a minority of about a quarter or a third of the population monopolizes all knowledge, competency, power and influence, reproduces itself and keeps the rest of the population in a more or less comfortable state of immaturity, is hardly compatible with the basic principles of a free and modern society.

Finally, the generalization of company rationalization patterns corresponding to the category of 'computer-aided Taylor,' would most probably emphatically reinforce the 'meritocratic' structures existing in virtually all developed industrial societies. The hierarchy of functions and positions in the persistently bureaucratic structures of the enterprise would be coupled more and more closely with the complementary hierarchy of educational graduates, which would help to protect the occupational and social status associated with the most attractive educational paths from the growing rush to higher education. If the need for stricter and more differentiated selection created by this development is not to fundamentally jeopardize the schooling and training function of the educational system, an extensive (re)segmentation of the school population will have to come about in the short or the long term.

This, in turn, would mean that in the long run and as a result of new micro–macro interactions, the originally quite autonomous development path of 'computer-aided Taylor' would also end up as 'computer-aided Huxley'.

4 Systemic Rationalization and Inter-Company Divisions of Labour[1]

Dieter Sauer, Manfred Deiß, Volker Döhl, Daniel Bieber, Norbert Altmann

For some years, increasing market saturation, stronger national and international competition, shorter product life cycles and the pressure exerted by rising internal costs have been forcing companies to develop *new strategies to make their production, administration and distribution processes more flexible and economical.* Of crucial importance in these strategies is the availability of computer-aided organization and control technologies in technology markets and the ways they get implemented. This chapter is concerned with the technological and organizational changes that companies have initiated since the mid 1980s in response to these new conditions. Based on the findings of our research, the assumption is made that these changes point to a new thrust in technological and organizational rationalization measures. What is decisive for these rationalization strategies is that they are no longer directed toward individual production sections or individual companies, but toward the production and distribution process as a whole. These developments are analytically and theoretically interpreted as a 'new type of systemic rationalization'.

INTRODUCTION

Given the reality of rising competition and the push to produce more efficiently, companies no longer confine their strategies for increasing productivity to the manufacturing area. Assisted by computer-aided organization and control technologies, system-oriented integration measures representing a new dimension in rationalization are being introduced.

This 'new type of rationalization' entails the integration of company-wide and inter-company production processes by means of information technology networking and the strategic use of technology as a way to secure flexibility. Present approaches in industrial sociology are inadequate for assessing this new kind of rationalization strategy. The changes likely to result from the 'new type of rationalization' and their effects on the structures of personnel, skill and decision-making within companies are presented in this paper, using typical cases. The theoretical and empirical analysis of these rationalization processes and their implications give rise to new questions and point to new areas of research. An important focus of our research is the new division of labour and the reorganization of intra-company and inter-company networks arising through systemic rationalization processes.

CHARACTERISTICS OF SYSTEMIC RATIONALIZATION

New rationalization strategies geared toward the integration of company processes

A first characteristic of the 'new rationalization type' is that rationalization increasingly occurs at a level of the reorganization of *overall* company processes. It is directed toward gathering company subprocesses in data-based systems, rearranging them organizationally, and integrating them into an information technology based network. Of great significance in this process is the development of computer technology, both as a precondition for and consequence of systemic rationalization.

Until now, rationalization has been directed mainly toward particular work processes or subprocesses; rationalization effects were primarily sought in the area of processing costs, especially with respect to (direct) personnel costs. Accordingly, rationalization measures usually focused only upon single processing stages, on increases in performance of individual or subprocesses, often even only on individual tasks, machines or equipment (e.g. on increases in processing speed). In cases where rationalization was directed toward increasing the technological availability or capacity of processing equipment, breakdowns and downtimes (induced by delays, faulty components, search processes, etc.) emanating from downstream or upstream processes (e.g. storage, transport, organizing orders) were only made objects of rationalization to a limited extent. Indeed, even processes of assembly-compatible product design played a negligible role.

Systemic rationalization measures, on the other hand, aim at a computer-based *linkage and integration of each individual subprocess*. If the company's rationalization measures are still largely applied one after another, in an isolated and piecemeal fashion, their intended effects are increasingly designed to encompass the overall company production process. This orientation is of central importance, even given that real integration potential is presently only latent and will only take effect in the course of time.

Systemic rationalization is directed primarily toward the control of sequential dependencies of various subprocesses, toward the organization in each individual subprocess of concrete technical, material and information technology (IT)-based interfaces, and their data-based interdependence to other subprocesses in production and administration (e.g. planning, design, inventory, shipping). The new rationalization type thus aims at increasing the performance of the overall production system by way of integrating and *optimizing coordination* of each individual subprocess. In concrete terms, this entails, for example, reduction of total throughput time, increase of capacity for several operating stations within the framework of overall company production flows, and reduction of overall company inventory through an integration and flexible organization of inter-departmental processes. The most important rationalization instruments for purposes of organizational and computer-based networking are the new *information, organization* and *control technologies.*[2]

Of course, company organization has hitherto mostly been oriented to process-specific and product-specific forms of production organization. These traditional forms of rationalization were predominantly based on material requirements (related to processing, means of transportation, product type, etc.). A consideration of the determining factors of integration themselves (pattern of time management, data processing, etc.) was considered a subordinate task *vis-à-vis* the organization of concrete manufacturing operations in individual subprocesses.

New rationalization strategies incorporate inter-company linkages

The 'new rationalization type's' systemic character does not stop 'at the factory gates'. Rather, rationalization measures of this type incorporate company-*external* delivery, processing and distribution activities. The data-based integration of company-external processes (relating to, for example, suppliers' manufacturing processes or purchasing departments of trading partners) imply changes in the inter-company division of labour and inter-company relations hitherto shaped primarily by market mechanisms. In this way, the societal consequences arising from the use of computer technologies grow: The specific organization of inter-company relations do not only have indirect effects for employees (e.g. of other companies), but touch upon the very basis of free-market societies (e.g. competitive mechanisms).

Also, measures aimed at optimizing overall company processes by integrating company subprocesses, lead either to the removal of non-integratable company subprocesses or to the inclusion of processes previously organized externally. However, decisions of exclusion and inclusion are no longer primarily based on the company's potential for performing certain tasks (the capacity to process certain materials or to apply certain technological processes), but on whether, considering process technical or cost criteria, certain administrative and manufacturing processes should be integrated at an intra- or inter-company level. The issue for the centres of capital accumulation is not merely one of productivity increase of the entire system. Rather, it must be ensured that control is not endangered by complications generated by greater recourse to external resources. In this way, know-how and exclusive access to the heart of company-specific technology for reasons of securing innovation and (thus also) market positions is crucial.

The new quality of inter-company relations comes into effect when these become an object of *strategic purchasing decisions* ('make or buy') and are carried out within the framework of a computer-based supply, production and distribution system. The increased inclusion of supplier firms allows buyer companies to shift market-induced flexibility requirements to them.[3] Thus, the relationship between companies gradually extends beyond the determination of characteristics of goods to be supplied (material, quality, supply schedules, etc.). The technological and organizational structure of suppliers' processes (machines, methods, means of transportation, etc.) also gets included into the rationalization measures. With the trend toward computerization growing, this can lead to *direct data-based linkage of individual subprocesses between companies*. This means that other dependencies between companies, between manufacturing and retailing companies etc., (for example, through rules concerning the nature and application of quality control methods, determination of material sources, direct control of company facilities, etc.) ones that could not be constituted within the framework of conventional market relations, are coming into effect.

New rationalization strategies concentrate on the elastic potential of technology

The 'new type of rationalization' pursues contradictory goals: the increase of flexibility in company administration and manufacturing processes in order better to fulfil constantly changing market requirements with respect to quality and quantity, and the achievement of a more cost-effective production system under conditions of fiercer competition. Coping with market requirements through qualitative and quantitative product variability on the one hand and through standardized and cost-effective mass production on the other has been largely irreconcilable until now given conventional technology and organization. Companies had to choose whether to follow a strategy of mass production or flexible craft production.

New information and communication technologies offer a rationalization instrument which makes it possible to resolve conflicting company goals. The strategic orientation of rationalization measures is no longer primarily concerned with the elastic potential of human labour (as was still valid for the 'New Forms of Work Organization' in the 1970s – see chapter 13 – and as is still often asserted as the dominant principle behind 'new management concepts'). Rather, these strategies are directed toward the utilization of technology's flexible potential, especially computer-based control and networking of company and inter-company processes.[4]

The increasing application of computer-based technologies as the new central instrument of flexibilization has also brought about new strategies of *economization*: the increase of the share of fixed capital induced by the increased application of technology has further pushed companies toward measures aimed at a better utilization of its fixed capital investments. This is also the case for rationalization measures that are predominantly directed at reducing inventories and increasing rates of turnover.

Within the framework of systemic rationalization, all strategies summarized under the term 'economization' are directed toward company cost factors, all of which belong to *constant* capital and therefore do not directly aim at reducing elements of variable capital. This is also valid – at least in terms of company goals – for flexibilization strategies in which the application of computer technologies is central. Even if productivity increases are effected through savings in labour, this is normally not the decisive strategic goal but more of a secondary effect. Of course achieving this effect is part of the vision of every rationalization measure; there is even a tendency for indirect labour to enter the picture. Also, reductions in labour, when they do occur, can be used as a legitimizing factor for costly rationalization investments.

ON THEORETICAL POSITIONS

Orientation on structures, not concepts

With the form of systemic rationalization sketched above, we have described a trend we presume will increasingly assert itself. The 'new type of rationalization' thus does not depict the present transitional situation in companies; rather, based on present trends, the term attempts to sketch the outlines of future developments. The term means to suggest neither a conscious, directed, systematic plan nor an inevitable realization (see Döhl *et al.* 1989; Deiß *et al.* 1989; Altmann, Sauer 1989; Bieber, Sauer 1991).

It is decisive that such rationalization is, of course, not sweepingly and systematically planned, but, that it ultimately does act upon all subprocesses of company-wide and inter-company production flows. We thus distance ourselves from an approach contending that processes of systemic rationalization are thoroughly planned and then instituted in one sweep (Baethge, Oberbeck 1986). In industrial production, the case seems to be that the effects of systemic rationalization enforce themselves, as it were, behind the backs of the actors. They derive from a process in which intended and non-intended effects of individual rationalization measures are gradually linked together to form a consistent pattern of systemic rationalization. Saying that a tendency exists to implement this new type of rationalization does not exclude the possibility that very different forms of rationalization can arise concurrently in different industrial sectors or companies, and that these different rationalization forms can be increasingly linked with one another in terms of the implementation of comprehensive production systems including both upstream and downstream production stages (suppliers, distribution outlets). This thesis, which stresses that the common moment of differing forms of the new rationalization type is their systemic integration in a comprehensive production system is what differentiates our position to those which see the future of rationalization strategies as theoretically open. These positions largely disregard the trend toward intercompany rationalization strategies (see chapter 3).

Of major significance in the approach presented here is that we do not deal with management concepts of personnel and labour policy. On the contrary, our task is to analyse the integration potential embedded in new organization and control technologies as well as their often hidden structural effects. Also of central importance is the analysis of power relationships and influence within companies – such as the role of engineers, their technology-oriented organization concepts and their importance as a key group in rationalization (see chapter 24).

New forms of dependency and domination

For us, the most crucial aspects of new rationalization strategies point to a vision of the future of company rationalization and its effects that is distinctly different from those currently being propagated: in contrast to ideas connected with the end of Taylorism or division of labour (Kern, Schumann 1984), and the crisis of mass production (Piore, Sabel 1985), and the hopes of positive societal changes connected to them, our perspective produces a more differentiated picture of future developments.

In our view, Piore and Sabel underestimate the opportunities that mass production companies have to establish a system of flexible mass production by using new organization and control technologies. On the one hand, they misjudge the fact that the restructuring of traditional organization and production structures as well as the transformation of supply and distribution structures make it possible for large-scale mass production companies to respond to a flexibilization of demand. On the other hand, they overestimate the autonomy and stability of smaller companies, whom they view as the primary agents of flexible production. What is misunderstood in this context is that, in many sectors, the observable trends toward decentralization of company organization, autonomy for certain company units, and reduction of in-house manufacturing, also means that smaller companies are used by large-scale enterprises through the incorporation and direct control of their economic, technological and organizational structures (see chapter 21, 22; see also Semlinger 1989; Mendius, Wendeling-Schröder 1991, and Brandt 1990: 315).

We can also not support positions that attribute the extensive transformation of company rationalization strategies to technology developments and new management philosophies which no longer view labour as an adversary, but rather perceive workers as a source of previously unused skills (Kern, Schumann 1984; for a critique see Malsch, Seltz 1987). For one, it is not clear what is new about the 'new production concepts'; forms of using broad skill profiles have always existed in certain areas of industry. Also, just in the 1970s a company performance policy based on a wide utilization of worker potential – even for those with few skills – (without however, corresponding compensation) became popular (see chapter 24; also Altmann *et al.* 1982, 1982a; Altmann 1984).

The question also arises as to how far the generalization made by Kern and Schumann, who, after all, forecast the 'end of the division of labour', is actually backed up by their own findings. Our own empirical research, in any case, concludes that one cannot speak of a paradigm change in rationalization strategy, at least not in the sense of Kern and Schumann's visions of 'reprofessionalization' or even 'requalification' of industrial work (see chapter 15). The question also arises as to whether the enlightened managers' influence can be assessed to be as high as Kern and Schumann do, or if the previous structures of industrial production do not in fact strongly influence the strategies of labour utilization.

There is no dispute that skill demands of selected key positions in the company will increase. This is especially so where adjustments need to be made in quasi permanent implementation processes, in order to utilize the structurally available potential of new technology. Such key positions are to be found particularly at the intra-company and inter-company interfaces of various production processes. According to our empirical findings, however, both the efforts of producers of manufacturing technologies and the demands of companies intending to use them concentrate on replacing workers with technology, or on reducing the comprehensiveness of workers' skill profiles with the aid of various computer aids (not to mention an overall savings in personnel).

The new form of systemic control of production processes, which we are seeing as a result of the strategies of reorganizing company structures and processes that have developed in companies signifies, in the final analysis, a reorganization of capitalist industrial structures. It is characteristic of systemic rationalization – as explained above – that it is not oriented to the direct production process alone, but incorporates other functions within and outside of a company (QA, R&D, controlling, logistics) into the rationalization process (see Bieber, Sauer 1991). In this context we can observe a tendency toward the declining importance of the direct production process for company decision-making centres, while other processes, namely those directed at more abstract value calculations (organization and finance departments, marketing, controlling), gain in significance.

Data-based integration of company subprocesses developing in the wake of new rationalization strategies increase the transparency of company processes and enlarge possibilities of centralized control. However, both new forms of decentralized organization and company-internal processes mediated by the market develop in accordance with the need to react more flexibly to market demands. We see both factors as belonging together. Centralized versus decentralized forms of company organization cannot be regarded as merely alternative lines of development: only the growth of opportunities for central control (in company-external spheres as well) creates – within a company perspective – the necessary prerequisites for the establishment of decentralized modes of organization.

The application of new computer technologies has strengthened the trend away from the material qualities of company production toward greater abstraction: its unspecificity with regard to product, process, company, and economic sector, and, therefore, its vagueness with regard to specific application in concrete production processes, has the effect of standardizing production processes among companies and sectors with respect to both their structure and to forms of process control. This computer-mediated abstraction from company and material qualities of production also explains the plurality of forms in which the new rationalization type manifests itself. What these forms have in common, though, is their strong a priori orientation toward company and company-external process integration.

No general upgrading of workers

In contrast to Piore/Sabel and Kern/Schumann, we see a trend toward the reduced importance of workers within company rationalization strategies.

Especially in new, predominantly technology-based flexibilization strategies, the work-force is no longer management's primary reference point.

This thesis should not be misunderstood to mean that problems of labour are becoming irrelevant or that the 'unmanned factory' and the age of complete control over company processes is dawning. On the contrary, the contradictions within systemic rationalization make the work-force less relevant, but also generate problems that require skilled workers in key positions. Therefore, the implementation of new production technologies and the more acute interface problems in the wake of systemic rationalization further – and in part increasingly – necessitate the application of skilled workers in certain stages of the production process. However, labour does not directly serve as the means of securing flexibility within the process of systemic rationalization. Rather, it predominantly serves to implement and utilize the flexible technology. For this reason, we regard hypotheses which assert the existence of a general trend toward higher qualification and a reduced division of labour as rash as those which anticipate a future predominance of small firm forms of work organization.

Complementary to the growing importance of skilled production work in some key areas, changes are occurring for the majority of workers in direct production. These can, for instance, lead to the integration of functions which were traditionally excluded from direct production processes (quality control, logistics). The resulting effects of the organizational and data-based integration of different production (and distribution) stages which cause an increasing abstraction of labour, make their mark on the structure of working conditions and the skill profile. And, because of the growing proportion of product and process-neutral manufacturing and organizational structures arising in the wake of the establishment of systemic rationalization, a worker's product and process-related skills are also becoming less important. This, in turn, has effects upon the exchangeability and mobility of remaining workers. We are not postulating a comprehensive deskilling of those workers not in key positions or not installed at particular interface areas in the production process; they too are forced to become familiar with the new organization and control technologies. However, we stick to the term polarization, commonly applied in German industrial sociology, to describe the drifting apart of skill requirements. Also, we do not see the opportunities of lesser skilled workers improving on the labour market because there is a trend toward the removal of company-specific and sector-specific skills (see chapter 15).

Market and integration

We have described the tendency toward data-based integration of company-internal and company-external processes as a special feature of the new rationalization strategies. We posit that this also infers changes in the inter-company division of labour and in the structure of inter-company relationships traditionally mediated by the market.

In our conceptualization, systemic rationalization is characterized strongly by its double effects: within a company, the various functions as well as the individual manufacturing and service processes are reorganized according to comprehensive company objectives; between companies, the structures of their traditional division of labour change, and, therefore, the forms of their mutual reference points. From the viewpoint of the individual companies, both aspects of systemic rationalization form a whole. They are the reflection of a new and changed orientation of company strategies: the focus turns to the coherence of entire production lines or chains in terms of material characteristics. Under conditions of stronger competition, it is no longer sufficient to merely increase profitability through direct productivity increases within an individual company. Rather, it is necessary to influence and utilize upstream and downstream manufacturing, service, or sales processes in such a manner, that they, indirectly, generate increases in profitability for an individual company. By exploiting their market position, large companies then attempt to control increasingly large sections of the manufacturing chain, without directly undermining or destroying the autonomy of the supplier firms.

The novelty of these company structures is, however, not the goal of controlling upstream and downstream stages of production; rather, the novelty lies in the manner and the extent in which control takes place. At the same time, computer integration between companies (for instance, between production companies and suppliers on the one hand and marketing companies on the other, but also between goods producers, or equipment or machine manufacturers, etc.) makes the functional areas and work processes of the involved companies transparent, opening spheres that were previously out of bounds for reasons of competition.

The reduction of in-house manufacturing in large companies goes hand in hand with a broader view regarding the control of processes not taking place inside their company any longer. For such purposes, the new information and communication technologies are important tools: control and domination are no longer only mediated by the market and by economic dependency, but also through various forms of direct and permanent technological and organizational control. Various production processes formed at the company level seem to melt together on a technological and organizational plane without actually touching upon the formal economic autonomy of respective companies on the market. Thus, market and competition appear as societal modes of organization which seem to be inadequate in dealing with integrated production processes. A contradiction thereby comes to light: due to the new rationalization strategies, the accumulation of capital expands and focuses on several stages of value creation or on the whole logistic chain, while, in organizational form, it remains bound to the individual company and to mechanisms mediated by the market.

CHANGED FORMS OF INTER-COMPANY DIVISION OF LABOUR AS A TOPIC OF RESEARCH IN INDUSTRIAL SOCIOLOGY

The supply industry as a symptomatic case

Today, changes in the inter-company division of labour, particularly the transformation of relations between manufacturing and supplier firms, play a central role worldwide. It is precisely these developments which cannot be grasped using conventional analytical and theoretical research approaches oriented to rationalization concepts of individual companies.

Based on the characteristics of systemic rationalization laid out above, the following can be said: in many cases, companies react to aggravated marketing problems and the resulting demands for flexibility and economizing in their production structures by attempting to *transfer* those heightened demands to suppliers, thereby freeing their own production process of such burdens. In principle, the strategy of transferring demands to upstream production stages is not new; its massive occurrence in the past several years, however, has led to considerable restructuring in the supplier industry. Such transfers are not merely strategies for unloading insecurities, but are an important adjustment of supply to the changed internal company manufacturing structures of purchasing companies.

The data-based integration of both purchasing and supplier companies' functional areas is not only a prerequisite – it is also a consequence of this process. Data-based integration, mediated by processes of coordination and standardization, grow into large-scale networks (as can be already evidenced in the automobile industry). They fundamentally alter the *structure of previously market-based relations* between companies. They affect not only dependencies between individual companies, but touch upon further-reaching structures of competition and the market. However, it is not only the case that market relations are to a large extent cancelled out by (mutual) dependencies, or that the autonomy of independent (supplier) companies is eroded by company-external control. Parts of companies are re-exposed to the market and 'released' to (international) competition through, for example, reductions in in-house manufacturing or the reorganization of company-internal supply commitments.

Complexity of factors as a research problem

The analysis of this process of integration requires the inclusion of a chain of influences. In doing this, the complexity and range of these factors point to new problems with which research in this area is confronted (see chapter 1). Here, we would like to point out several interrelated factors, which must be considered in order to make general statements about rationalization processes.

1 In accordance with their specific standing in (sales) markets, manufacturing companies follow widely differing *rationalization strategies.* Demands upon suppliers and respective consequences for workers can thus only be adequately understood and evaluated when the specific situation of manufacturing companies is analysed with regard to their industrial sector, their outlet markets and their ensuing rationalization policies (see Döhl 1989a; Deiß 1989).
2 The increasingly systemic design of rationalization strategies in manufacturing and supplier companies has induced the development of *'encompassing solutions' of machine and equipment manufacturers.* These concepts are especially demanded by those companies, which possess only negligible engineering and systems engineering know-how and appropriately qualified personnel. In this regard, systems engineering has experienced a significant growth of influence. The application of integrated systems and the technological and organizational models incorporated within them have important consequences for the form of the division of labour between marketing and supplier companies, for the form of integration and networking, and, of course, for respectively affected groups of workers (see Döhl in Chapter 6).
3 *Foreign organizational models or production concepts* and the influence of foreign companies play a significant role in the trend toward new forms of supplier integration. Due to international competition, Japanese organizational models as well as organizational and integration concepts (Kanban, Just-in-Time, etc.) have attained special significance. In part, these concepts are rejected outright, and in part they influence company activities in a haphazard and unreflected way. Thus, despite widespread knowledge of these models' social preconditions and (side-)effects (upon skill, wage structure, labour deployment, representation of

interest, etc.), the latter are afforded only negligible consideration – whether in a positive or negative sense – by those carrying out rationalization in the company.

4 At present there is a growing trend toward the *globalization of supply*, particularly in the automobile industry. The use of location advantages, leads to changes in the international division of labour which have significant effects on national structures of inter-company divisions of labour. Also, the enhanced potential for international companies' headquarters to influence decentralized branches or subsidiaries has significant effects upon national supply industries and their employees (quality standards, cost competition, etc.).

5 Along with the trend of the globalization of supply, there are discernible tendencies toward a *growing utilization of regional resources* and a regional concentration of supply companies. In international competitive struggles, the consolidation of regional industrial structures, i.e. networks of regional supply companies grouped around a purchasing company – seems to be gaining significance. Research must also take into account the fact that regional and structural policies by the state are a vital factor in this process.

6 A large number of suppliers that are to be integrated into a comprehensively rationalized logistics network are *small* and *middle-sized companies*. Their personnel structure, their know-how, their investment potential, etc. differ widely. Thus, widely differing preconditions as to their ability to fulfil purchasing companies' demands exist (regarding supply quantities, time schedules, quality, computerization, etc.). Rationalization strategies and concomitant tendencies toward computerized networking as well as toward the reorganization of supply (single sourcing, internationalization, location change) will therefore radically transform the traditionally mid-sized structure of national supply industries (concentration, selection, hierarchical processes, etc.). In order to grasp the nature and consequences of their integration into logistics networks, it is necessary to analyse the structure and adjustment potential of suppliers (see chapter 22).

7 All of the above listed factors refer to crucial aspects of change in the inter-company division of labour. At the same time, they are crucial variables of research seeking to analyse the diverse indirect, and thus also hidden, effects of purchasing companies' rationalization measures upon working conditions in supply companies (see chapter 15). *Employment, employment status, the structure of working hours* and *skill* seem to be the central dimensions in which a transformation of working conditions is to be expected.

8 In addition, there is the separate problem of the changed conditions of *interest representation in supply companies*. Their different and often reduced opportunities to influence working conditions bring to the fore the question of new forms of interest representation within industrial structures which are organized into networks (see chapter 23).

To understand new rationalization strategies, it is necessary to grasp the central importance that the restructuring of inter-company divisions of labour has for fundamental social and economic transformations in capitalist industrial societies. Social science research seeking to adequately address the new developments and forms of rationalization needs to orient itself to rationalization's increasingly systemic character.

Part II

Development and Application of New Technologies

5 On the History of NC-Technology – Different Paths of Development[1]

Hartmut Hirsch-Kreinsen

This chapter compares the development of NC-technologies for machine tools in the USA and Germany since their inception at the end of the 1940s to the beginning of the 1950s. Shortly after their introduction, clear differences in development paths arose between US and German NC-machines especially with regard to potential for shop-floor programming. It is argued that the US controls' centralistic orientation compared to the increasingly greater significance of workshop programming in the German case can be traced to social factors rather than technological conditions. The factors pointed out here are the very different conditions in industrial structure, labour market and systems of occupational training and science characterizing the two countries.

INTRODUCTION

The idea that technological development is not determined by technological forces alone, but to a great extent determined by social processes, is relatively well-established. Quite clearly a great deal of choice exists in technological development between different alternatives. These choices open up a number of possibilities for the influence of economic, political and social structural conditions and factors on the design of technological systems. These connections were repeatedly exhibited in the past, especially for large technical systems in the areas of transport or communication (see for example Mayntz, Hughes 1988). D.F. Noble (1979; 1984) provides one of the few instructive examples of development processes in the area of production technology in standard systems, in contrast to large technical systems, with his investigation of the development of the NC-controlled machine tool industry in the United States. He shows that the development of this basic industrial technology did not proceed in a single direction. Rather, NC-technology displayed quite diverse paths especially at the beginning of its development, around the end of the 1940s and in the 1950s.

According to Noble, differences between paths of development are not only found in the performance, complexity or technological flexibility of various control systems, but also in their programming potential. The possibilities range from programming confined exclusively to the office, to control systems that can be programmed on the shop-floor. As Noble goes on to show, the developmental path of NC-technology that quickly became dominant was that of office programming. For Noble this meant that the fundamental interest of engineers and scientists to control and dominate technological development was the driving force behind the choice of a particular path. It is the case that office level programming meant in principle that through the introduction of NC-technology a large share of planning work would have to be removed from the workshop and placed in the office area. This had the further consequence that the elements in relatively self-contained tasks that allowed for some autonomy were taken away from machine workers. Through this, operators' functions went one step further under the control of engineers and managers than before.

However, in contrast to Noble's analyses and interpretations of American NC-development, it is impossible to ignore that NC-development in Europe, especially in Germany, has proceeded quite differently. A range of empirical evidence, mostly information from research engineers and researchers of company practices, but also a few passing observations from Noble, shows that NC-development in the Federal Republic was not so clearly directed toward office programming. Rather, the development path that made workshop programming possible played a relatively large role.

These differences in development and their social origins will be examined more closely in the following chapter. The first step will be to compare and trace the development of NC-technology in the United States and Germany. The second step concerns an analysis of the effects of social factors and conditions on technological development. These sections will argue that it is not so much interests for domination and control that explain NC-technology development as nationally specific constellations of institutionalized development interests, which are embedded in specific industrial and social structures.

FOCUSING ON AUTOMATION (USA)

From a historical point of view there can be no doubt that the most important path of NC-development aims for the extensive and complete automation of processing sequences carried out by machine tools. This approach is found mainly in the United States and right up into the 1980s has been almost exclusively oriented to the concept of enhancing the 'inherent capacity' of machine tools by making maximum use of available control technology. Such a technological concept reduces human work at the machine tool to a factor of merely marginal significance.

In 1952 the first prototype of a NC-machine, a vertical milling machine, was presented to the public at MIT near Boston. This unit typified the on-going development of NC-technology in the United States which is dominated by relatively complex and extensive control concepts. In the early days, NC-technology consisted mainly of continuous path control for the milling of parts with a particularly complicated geometrical design. The first NC-machines were employed in the aeronautics and defence industries, where manufacturing problems such as the machining of spatially curved surfaces of high performance aircraft could not be adequately dealt with by conventional control technology.

The programming of the sequences of the machine operations and the machine functions for these processes was extremely time consuming and complex. What this amounted to was a new type of work process set off from the machines on the shop-floor in terms of both time and space. This paved the way for forms of work organization based on a division of labour and strongly determined by technology. For the machine operators few possibilities remained for intervention into the pre-programmed machining process.

In order firmly to establish and increase the efficiency of automation and centralization of programming, the computer-based programming system APT (Automatically Programmed Tools) was developed from the mid-1950s on. The utilization of this system was dependent on the use of a mainframe and was strongly oriented to the machining problems of the aeronautics industry. Moreover, during the first developmental stages the system could only be used by programmers with mathematical skills (Noble 1984: 142).

Initially NC-technology was rather slow to spread in the United States because of the great complexity and expense of the systems. Yet from the 1960s on, increased efforts were made to expand the chosen course of NC-development. As a result some of the obstacles of technology and work organization were overcome – an example being the introduction of inexpensive point-to-point numerical control position systems to the market (Hough 1975: 352). Another factor consisted of introducing selected control types that had limited intervention potential for machine operators so that they could use NC-technology for manufacturing jobs that were not completely programmable in advance. As Noble demonstrated by using the case of a Bendix-control system (Noble 1984: 215), the control systems had overrides for varying some of the technological parameters of the programmed machining process such as tool forward feed and cutting speeds.

In no way, however, did this represent a major departure from the course of automation-oriented NC-development in the United States This is particularly evident when one regards the later development of CNC-control systems. Microprocessors and LSI-technology offered greater potential for the design and layout of controls than the NC-technology of the 1950s based on pipe and relay technology or the transistor and semiconductor technology of the 1960s. Nevertheless, the potential offered by microelectronics for direct programming of control systems was slow to find use. It was only at the end of the 1970s and at the beginning of the 1980s – under the pressure of Japanese and European competitors on the US market – that many, yet by no means all US manufacturers, followed suit and offered control systems allowing the relatively easy direct input of program data via operator elements. Exemplary in this context are elements such as key pads for manual data input relating to subprogram technologies and determined cycles involving repetitive machining operations, or alphanumeric data displays.

Even in the light of these developments, the majority of US-manufacturers adhered to their complex control concepts and continued, in many instances, to build them on APT computer-based central programming. As a result, possibilities for shop-floor programming held, and continue to hold, a merely subordinate and complementary function to central programming operations.

FLEXIBLE AND COMPREHENSIVE UTILIZATION OF NC-TECHNOLOGY IN GERMANY

In Europe a course of NC-development emerged which deviated considerably from the development in the United States. This can be seen primarily in the German and Swiss machine tool industry, but also in Scandinavia and partially in France and Italy. This path is more strongly oriented to rationalizing machining processes under widely differing manufacturing technological conditions. As in the United States the attempt to automate machining processes is of paramount importance. However, greater priority is given to control systems which can be utilized with a higher degree of flexibility, both in terms of technology as well as organization. Apart from the appropriate design of technology, direct human interaction with the program sequence is absolutely indispensable. In this system labour is not reduced to a subordinate and secondary factor, but instead is regarded as being a factor fulfilling an important, complementary function to automation. Direct intervention into the program sequence is necessary to cope with machining situations which cannot be calculated or programmed in advance.

This form of technology crystallized as the main path of NC-development in central Europe relatively quickly despite an initially strong orientation to the concepts pursued in the United States. Since the beginning of the 1960s a number of characteristic differences to the American path of development have emerged.

More technology than geometry

The first characteristic was the early development of simple and comparatively easy to use point-to-point controls and line motion control systems. On the one hand this can be attributed to the limited amount of technical know-how that existed in Germany at that time, a necessary prerequisite for the development of complex control systems such as continuous path control. On the other hand there was the undisputed interest in designing easily employable control systems for various areas of application in metal-working, i.e. not just for drilling and milling as in the United States, but also for turning, for example.

In comparison with the United States, the interest in such applications resulted in a stronger orientation to different cutting operations. Thus the focus remained less on geometrical problems than on technological problems of machining, such as the determination of tools or the specification of material properties and the resulting machining data for machine forward feed or running speeds.

Open systems

The above-mentioned orientation was associated with another developmental characteristic, namely the creation of programming languages and programming systems which were closely related to manufacturing conditions and which could be utilized on a broad basis. As opposed to the United States, very different programming languages with differing processing and technology orientations were developed at the beginning of the 1960s in the Federal Republic of Germany.

The programming language EXAPT (Extended Subset of APT) which is still in widespread use today, is a typical example (Grupe 1974): using the formal and syntactic basis of the American APT system, EXAPT, created in 1965–6, represented a technology oriented programming system which, after a number of developmental problems and initial running-in difficulties, finally resulted in three processing oriented system variations.

Alternative data input

Lastly, these particular characteristics point toward a third and central distinction in the comparison of NC-development in the United States and Europe. In the Federal Republic attempts were made at a relatively early stage of development to allow direct data input at the control systems level in order to increase organizational flexibility. Although early standard control systems were basically designed for central programming and data input via punched tape, and the automatic program sequence, especially for the first systems in the 1960s could not be influenced, it did not take long before a number of control types were available that had a limited scope for operating the NC-machine tools on the shop-floor.

Clearly such concepts went a lot further than similar steps in the United States – the position control systems of various large NC-developers such as AEG or Philips, for example, had already been fitted with manual data input via switches and keys as alternatives to punched tape operating at the beginning of the 1960s (Werkstatt und Betrieb 1961: 874). While the above-mentioned systems were designed for relatively simple machining processes, subsequent line motion control systems, and even later complex continuous path control systems, were generally provided with possibilities for modifying and augmenting the programs directly at the control system on the shop-floor.

This main path of Federal German, and in part West European, NC-development which emerged in the 1960s was pursued throughout the first half of the 1970s, particularly in the continuous optimization of 'hardwired' (*fest verdrahtet*) NC-control systems. The gradual transition to freely programmable CNC-control systems did little to change this basic direction of development. In the continuing process of development it is clear that this main path is evolving towards a third path of NC-development.

Work process-oriented path of NC-development

The third path of development of NC-technology can be called work process-oriented (*arbeitsprozeßorientierter Pfad*) since it offers a lot of possibilities for shop-floor programming. Central programming is either not contemplated at all, or fulfils only a complementary function. Although the objective remains the rationalization of machining processes, the skills, empirical knowledge and know-how of human labour form the foundation of this path of development. The basic concept itself opens up a greater degree of technological-organizational flexibility and seeks to avoid a multitude of obstacles and typical problems of NC-utilization.

Initially work process-oriented NC-technology emerged as a minor side line in the course of NC-development: in the United States sporadic, yet repeated attempts were made with different types of directly programmable control systems. The best known of these systems was the so-called playback method with which the first experiments were made as early as the mid-1940s. Although they proved flexible in use and were apparently also employable for relatively complex machining processes – as long as these could be coped with manually – the playback method and other methods of shop-floor programming remained, in comparison with the main path of NC-technology virtually without significance for a long period of time in the United States (Noble 1984: 143).

In the Federal Republic of Germany work process-oriented NC-technology was pursued in a far more systematic manner than in the United States. In Germany the first experiments with playback methods and mechanical control took place back in the 1950s. These initial models were continuously developed during the 1960s. Notable in this context are not only the further development of the playback method in the form of playback control as a comparatively cheap 'semi-numerical' control (Stromberger 1966), but also the increase of manual input control systems. Mechanical storage components such as decade switches, plug boards etc. allowed direct data input, whereby in some cases punched tape operation was offered as a complementary function. In the Federal Republic these concepts were apparently pursued by a number of manufacturers of control systems, particularly in the machine tool building sector. In some instances there were also control system developments, such as those opted for by the Swiss company Dixi for high-precision jig-boring machines, which made possible data input as well as playback methods, and tied to that, the generation of punched tape directly at the control system (Simon 1971).

Manual data input as option

At the latest, from the mid-1970s on, the work process-oriented path of NC-development in the form of manual data input indisputably became the basis of a highly successful development surge of NC-technology in Germany. Manufacturers of machine tools and control systems recognized the new possibilities offered by microelectronics and put them to use for an extensive ongoing development of existing shop-floor programming concepts. With the introduction of work process-oriented CNC-control systems the deviation from American NC-technology was immense: with this new course of development manual data input became – as interviewed experts put it – a European, if not a 'German invention'.

By the beginning of the 1980s those programming aids which are still in use today had clearly emerged, such as dialog-oriented programming languages, methods of operator and programming guidance, extensive subprogram methods and later the possibility of simulating the machining processes at visual display units integrated into the control systems. The NEF-turning machine concept launched by the turning machine manufacturer Gildemeister in 1975–6 played a pioneering role in this leap in development, which in the opinion of many experts was instrumental in securing acceptance of the work process-oriented path of NC-development, which had previously held a measurably subordinate position.

These developmental tendencies are continuing today. There is no question that, as before, extreme differences exist between various types of control systems, differences which can be categorized by the well-known differentiation between control concepts which are computer specialist oriented and those geared to skilled workers (Corbett 1985; Weber 1988). These differences lie behind the various developmental paths of NC-technology described above and they remain of great importance for the question of work design and the potential and operational scope of skilled machine work.

It is true, however, that there has been a certain amount of convergence of NC-technology paths across countries in the last few years. Whereas for West German NC-manufacturers the development of control concepts has been strongly directed toward work process-oriented solutions, it is only recently that American manufacturers of control systems, partially with Japanese support, have been trying to catch up on this development, on the European, as well as their own domestic market.

ACTORS TAKING PART IN DEVELOPMENT: THE ROLE OF THE MACHINE-BUILDING INDUSTRY

The question of which conditions and factors have been responsible for the different developmental trajectories of NC-technology leads us initially to consider the institutions, enterprises and individuals involved in the development of technology.

The predominance of military influence in the USA

In the United States NC-development proceeded during the Second World War through the growing network of interests between government and military, and large scientific-technological research institutions and certain sectors of industry, particularly electrical engineering. The machine tool building industry was represented by a few large enterprises cooperating closely with government and research institutions. Military-influenced control and computer technologies based on specialized and standardized machine tool concepts for series and mass series manufacturing formed the foundation of American NC-technology.

The application interests of American NC-developers were initially directed exclusively toward the military aeronautics industry, where it was possible to work with the new technology with a substantial degree of independence from considerations of economic efficiency and the risks of the private market. Therefore the interests and orientations of the NC-developers were more strongly geared to scientific and technological criteria for the most comprehensive and perfected utilization of computer technologies possible, while economic development requirements played a subordinate role. As Noble (1984) has extensively documented, these technology concepts were supported by computer specialists and engineers who had specific interests in maintaining and enhancing their position and competency over the shop-floor.

This constellation in the Unites States was absolutely characteristic of NC-development right up into the 1960s. Although a greater degree of market orientation did set in around the mid-1960s, the original concept has been basically adhered to up to the present. For many years NC-machine tools were almost exclusively developed for use in large-scale enterprises; initially in the defence industry, yet later also for applications in the automotive and electronics industry as well as in machine building: all of which had comparatively standardized manufacturing conditions (Schultz-Wild, Weltz 1973). It was only very much later that the broad field of general metal processing, consisting also in the Unites States of a large number of small and medium-sized companies, began to use NC-machines, a development which was mainly initiated by the offer of more flexible and universally applicable control systems, mostly of German or Japanese origin.

Europe: oriented to the market

In the Federal Republic of Germany and in other European countries, particularly in Switzerland, the machine tool building sector in general played a far more meaningful role in NC-development than in the United States. This was already apparent in the 1960s and in the first half of the 1970s. From the mid-1970s on, however, the machine tool industry became a major innovative and decisive factor in NC-development. Due to the strong position of this branch as a purchaser of control systems for their manufactured machine tools, the machine tool builders also exerted a strong influence on the electrotechnical companies as well as on computer manufacturers.

Therefore NC-development in Europe was strongly oriented to market requirements right from the very start. It was determined by pragmatic, manufacturing-technological criteria which were directed toward achieving the broadest and most problem-free utilization possible – especially through a high level of organizational flexibility in the use of NC-technology as a means of rationalization. The sales market aimed for by the Federal German, and by the Swiss NC-developers in particular, consisted of the domestic and foreign metal processing industry characterized by its generally diverse processing requirements. NC-technology was based on machine tooling concepts which were only partly scientifically derived and which most likely to this day are based on a high degree of empirical and practical experience. The concept of the universal machine tool formed the general point of departure for NC-development in German-speaking countries – a good example in this context is the universally utilized centre lathe, designed in a German-speaking country, in contrast to the specialized turret lathe used in the United States.

SOCIAL AND INDUSTRIAL STRUCTURES: HIGH-TECH VERSUS MECHANICAL ENGINEERING

The different courses of NC-development have also been influenced by social and industrial structures. In the United States industrial structures have been particularly affected by the economic weight of the aeronautics and defence industries, which are part of the high-tech sector that in turn plays a central role in US industry as a whole (Junne 1985). Supported by traditionally high government support, particularly in the form of defence and aerospace programmes, these sectors absorb tremendous developmental resources – which as a result are lacking in other areas (DiFilippo 1986). Moreover, they present the image of perfection and complexity in manufacturing technology development.

It is in these branches that we find a high and expensive level of automation compared with the metal processing branch. The fact that these industrial sectors are to a large degree shielded from market influences and can also operate free of the competitive conditions of the world market, has played an important role in this respect. This meant that there was little pressure to bring about comprehensive and cost efficient rationalization of manufacturing procedures. Right from the very beginning, American NC-development has been integrated into these basic structures.

No pressure toward rationalization

Up until the 1980s large areas of the metal-processing industry, and particularly the mechanical engineering sector in the United States, showed little inclination for innovative change. The majority of the companies had a relatively low level of automation, while moderate technological demands were made on the production process. The reason for this can be found in the large American internal market which remained very stable for a long period of time. At the same time there was a very low degree of orientation to the world market. The resulting manufacturing structures consisted of series or mass production systems which could easily be controlled by various forms of the so-called 'Detroit-automation' (Fordism), so that up until recent times there was little need for a reorientation of rationalization strategies. Currently, however, the American metal processing branch has experienced an increasing need to rationalize because of the poor economic growth of various markets since about the middle of the 1970s and the emergence of new competition from Asia (Piore, Sabel 1984). This can be regarded as one of the reasons for the considerable sales success of Japanese, Federal German and Swiss machine tools on the American market (Rauner, Ruth 1988).

The lack of skilled workers

In accordance with the manufacturing structures found in US industry there has been a predominance of mainly large-scale manufacturing processes in the United States along with forms of organization characterized by a high degree of division of labour and manufacturing personnel with a low skills average. This applies to large areas of the metalworking branch as well as to the aeronautics industry. The highly standardized manufacturing procedures distracted attention from the new processing problems arising in small batch manufacturing and custom production. Instead the emphasis continued to remain on the inherent potential of technical facililities and on the marginalization of labour. The typical American NC-path was indeed well adapted to existing conditions.

These conditions in US companies were supported by specific labour market structures in the United States. The labour market provided a very limited supply of qualified skilled, or technically competent workers. Training skilled workers takes place to a highly restricted extent, and then is mainly tied to in-house company measures. General occupational training in the US is based on on-the-job training within rigidly hierarchical mobility structures. It would be difficult to imagine the acceptance of relatively flexible job structures with few hierarchical grades an organizational precondition for shop-floor programming.

The predominance of computer science

In the end, these conditions refer back to the influence of the scientific system in the United States, where there is no such thing as the 'culture of mechanical engineering' (*Maschinenbaukultur*) such as exists in the German speaking countries (see chapter 17). By comparison, there is a much sharper dividing line in the United States between the practical application of technology and the development of technology at the theoretical level. There is a tendency to immediately regard problems of manufacturing technology from the standpoint of computer science and subject them to corresponding instruments and methods. Computer science plays a dominant role in scientific systems and there are close links between the personnel in the computer and defence industries (Kreibich 1986).

Machine building as a key German industry

The industrial and social conditions surrounding NC-development in the Federal Republic of Germany and Switzerland are of a completely different nature. With its comparatively diversified industrial structures, the capital goods industry has traditionally held a great deal of significance for the economy, in which machine building has played a key role (see chapter 10). Because of this industrial structure, the machine building industry is not only a manufacturer, but also the largest user of manufacturing equipment. Therefore NC-machines must be designed to cope with manufacturing conditions which are difficult to calculate in advance, and must display a high degree of technological-organizational flexibility. In order to achieve this kind of flexibility it is necessary to be able to engage in programming of the control systems themselves at the shop-floor level.

Meeting the requirements of the world market

Traditionally a large share of Federal German and also the Swiss machine tool production has been exported throughout the world. This has meant that the demands and requirements of the worldwide NC-market have had a very direct effect on NC-development. It is only recently that the aeronautics and defence industries – with their specific manufacturing conditions which deviate from those in general machine building – have begun gradually to play a more important role in Germany and Switzerland. Up to now these industries' requirements have had little effect on the development of manufacturing technology. It is therefore no coincidence that the European aeronautics and defence industries use manufacturing-technological equipment and systems either designed according to American specifications or that originate from there.

All in all the German machine building sector not only represents an extremely heterogenous area of application for manufacturing technologies and methods, but also exerts considerable pressure to innovate in the development of flexible forms of technology due to its orientation to small batch and custom production and its wealth of diversified interconnections with the world market. The central problem of this branch is its inherent 'rationalization dilemma', namely bridging the gap between cutting production costs while maintaining or increasing flexibilization of its production processes.

A broad foundation of qualified, skilled workers

Linked to this rationalization dilemma is the fact that in machine building, medium and small size firms which have a relatively low division of labour predominate. In most cases the work on the shop-floor is in the hands of well trained and qualified skilled workers. The process engineering departments often have had very restricted means for planning at their disposal in the manufacturing sector. Therefore, in the past, rationalization in the German machine building sector did not lead to the disintegration of qualified production work. On the contrary, given actual operating conditions, the control systems in work process-oriented paths proved to be a highly suitable means of rationalization.

Practical nature of training and studies

Interacting with these circumstances are the unique features of the German system of vocational training whose main characteristic is the close relationship between classroom studies and practical occupational training in the company. This system brings about a situation in which workers with a relatively high level of qualification get trained and an ample supply is available on the labour market (see chapter 16). The high availability of qualified production workers in Germany is one of the main requirements for the continued existence of labour processes that are relatively 'open' with a low division of labour. These processes, in turn, are a central reference point for the development of work process-oriented NC-systems.

Moreover, this system of vocational training is closely linked to the specific German system of manufacturing. The system of manufacturing sciences can be regarded as a continuation on a higher level of the institutionalized interaction of theory and practice which characterizes the system of vocational training. The relatively close relationship to actual manufacturing practice is reflected by the special characteristics of the study of engineering science in the Federal Republic and Switzerland. In the Federal Republic there is a particularly close exchange of personnel between engineering sciences and the manufacturing companies. Career patterns of engineering scientists, which in most cases proceed exclusively via the companies, as well as the company determined cooperation between science and enterprises, play an essential role in this context.

A SOCIALLY DETERMINED PROCESS

Looking back on NC-development in an international comparison, we may say in conclusion that a number of very different social and economic conditions and factors contribute to technological development. If we followed Noble's reasoning, we would have to consider the 'ideology of control' of managers and engineers as the overriding driving force behind technological development. However, with this approach it cannot be explained why NC-development in the Federal Republic, despite the same extreme technocratic orientations and controlling interests of technical personnel and engineers (see Bergmann *et al.* 1986), followed a different course. Even given the assumption that domination and control play a greater role in industrial relations in the United States than in Germany, Noble's explanation fails to see that the centralization of programming in American NC-technology – and inversely workshop programming in the Federal Republic – is part of a technically, socially and economically determined process.

Due to the complexity of the outlined inter-relationships, the development of production technology seems like an entangled, almost inadvertant, process running its own course. A better awareness of the determinants of technological development makes it possible to use their influence in a more desirable economic and social direction.

6 The Role of Manufacturing Technology Markets in Systemic Rationalization Processes[1]

Volker Döhl

INTRODUCTION

The application and design of technology are often dealt with from the perspective of performance policy or the humanization of work in German industrial sociology. In these contexts, the processes underlying the development of technology and technological concepts have also been studied. However, the way in which they get diffused in user companies is rarely touched on. This process is precisely the subject of this chapter. Using machine manufacturers of companies in the consumer goods industry, the very complex relationships of dominance and dependence on the technology market are analysed and the consequences for manufacturer and user companies outlined.

THREE THESES ON THE RELATIONSHIP BETWEEN THE MANUFACTURERS AND USERS OF TECHNOLOGY

The *first thesis* posits that if particular manufacturing technologies are to be successful (whatever their form or purpose), their development, production and supply have to be oriented to enterprise applications. These applications are largely determined, in turn, by the measures pursued by the user companies for the design of their own products and processes. This means that altered market and sales strategies of the user companies and the resulting product and process innovations generate new technology requirements. Indeed, in the final analysis, it is these new requirements which determine the direction that manufacturers take in the development and production of technology, and thus the technologies that are available on the market. According to this thesis, user application processes dominate the processes involved in the development and production of technology. Thus the user/manufacturer relationship is characterized by a dominance of the user over the manufacturer. This means that the user obtains the techniques and technologies needed to pursue his goals and fulfil his purposes, in the bounds, of course, of what is technologically feasible. However, this does not preclude the possibility that difficulties and problems could arise in the course of implementation.

The *second thesis* states that the users of techniques and technologies are largely dependent on what the manufacturers place on the market. The possible applications and uses of technologies are largely predetermined by the concrete design of the technology marketed by the manufacturer. This means that the general direction of product and process innovation for the user is closely linked to the technology available, which can strongly restrict the user's potential for strategic action. According to this thesis, the technology manufacturing process dominates the application process; thus, the user/manufacturer relationship is characterized by the domination of the manufacturer over the user. In concrete terms, this means that the user has to take what he gets and what he is able to afford in terms of technology and is forced to adjust his measures for product and process innovation accordingly.

Although it is evident that these theses are fundamentally contradictory, the validity of both has been confirmed by the empirical findings that have been carried out to date. This fact leads, in turn, to a *third thesis*. The fact that companies have to react to changes in market conditions leads them to develop rationalization strategies. These strategies are taking on an increasingly systemic character, i.e. they are moving more and more towards both intra-company and inter-company integration and networking. The manufacturers of technology adapt their products to these various systemic rationalization strategies by basing their own product and sales strategies on the intentions of the user companies, which leads to a growing (and market-dominating) supply of technologies which incorporate an increasingly systemic character. These technologies have a profound effect on the production structures of the users and influence the further course of product and process innovation in these companies. What follows from this systemic character of technology is that certain groups of users end up in a position of dominance over other users, due to the sales strategies pursued by the technology manufacturers. In other words, the dominance of certain large and/or market-controlling user groups over the manufacturers of technological systems is extended to other usually small, not very market-dominating user groups, who experience it, however, as manufacturer dominance. According to this thesis, both forms of dominance exist simultaneously.

We will now examine some of the aspects of this complex relationship in closer detail.

THE PRODUCT AND PRODUCTION STRATEGIES OF USER COMPANIES

In discussions by scientists and practitioners on issues concerning industry, one question has always been of central importance: How can companies cope with intensified price competition and expand their product range and service functions at the same time? In other words, how can companies deal with the drastic increase in flexibility demands and the increasing complexity and difficulty of production and administration while simultaneously producing in a more economical manner? What role do new technologies play in this process and which role could they play (i.e. in the form of flexible automation). The assumption usually is that diverse market and consumer demands affect all companies and their production and administration structures alike, and so all companies, without exception, are forced to find ways to deal with these demands.

Our investigations showed, in fact, that companies try to develop targeted sales strategies for their products in accordance with their respective structures and conditions. In this way, they seek to come to terms with the contradictory market demands which expect product differentiation on the one hand and price and cost reductions on the other. The possible response to these demands varies, but falls basically into two categories which have been typified here for analytical reasons: companies react either by concentrating production on low-priced mass products or by offering individualized, expensive and exclusive products entailing a number of variants. In this way, companies try to divide the sales market into two extremes in order to tackle the demands of cost reduction and increased flexibility independently of each other and thus ward off demands that cannot be mastered. This division of the sales market, which probably goes beyond national boundaries, is to be found in most sectors of industrial production, especially in the consumer goods industry. In some circumstances, both strategies are pursued within a single firm or conglomerate.[2]

Which direction a particular company takes depends not only on its sales expectations, but also on its basic situation, for example, its technical capacity, personnel, skill structure, the existence of a well-established brand name, and whether it is firmly positioned on regional and national sales markets, etc.[3]

It should be pointed out that many companies do not manage to secure such sales strategies for themselves and are thus unable to limit their activities to one particular option. As a result, their position on the sales and purchasing markets is extremely unstable and insecure.

Implementation of either of the two extreme options necessitates the introduction of respective innovations in company production and administration processes which determine the direction and form of rationalization, as well as demands made on the techniques and technologies to be used. The question that arises is how do technology manufacturers react to such specific user needs and how does this bring about the dominance of a particular form of technology on the market? Moreover, what consequences does this have for companies who are not in a position to pursue one of the two sales options? Before we can answer these questions, it is necessary to briefly outline the general thrust of the rationalization strategies followed by the two different (ideal-typical) kinds of user companies.

The producer of low-priced, mass produced and standardized products primarily pursues one goal in his rationalization policy: the drastic reduction of costs. In doing so, he places emphasis on two main points.

The first consists of increasing the level of utilization and productivity of the existing or planned production techniques (including manpower of course). These measures are generally directed toward specific technological systems or areas of production and work operations in which underutilized capacity and performance can be tapped. The measures do not deviate greatly from classical forms of rationalization, and their effects on cost reduction are not impressive because the unused potential they seek to develop has usually been exhausted by previous rationalization measures.

It is for this reason that the second point of emphasis obtains greater significance. In this case, the manufacturer departs from his preoccupation with isolated areas of production and the technological systems (including manpower) in them and seeks to rationalize the production process as a whole instead. Toward this end, he bases his strategy on the expansion of existing data processing systems in order to optimize his administrative and organizational structures. In doing so, his general goal is to increase the degree of integration of production and administration in order to reduce unproductive time and costs while intensifying the production process of the entire company. At the level of production this means the integration of processing, handling and transport functions by means of machine and technological control systems both within individual production areas, and more importantly, across different areas, thus achieving a reduction of throughput times (horizontal integration). There is the simultaneous attempt to implement an information and control system to accelerate the transfer of data and information between the individual areas of production and enable centralized access to production and administration processes (vertical integration). Administration processes, such as resource management, sales and purchase order processing, accounting, etc., are also increasingly subject to rationalization measures and are more and more integrated into the reorganization of all processes through data processing systems.

In this way, a process of systemic rationalization (see chapter 4) which seeks to achieve control of planning, managing, and implementation processes is set in motion early on, for the purpose of cost reduction. This system is oriented to centralized control of an entirely integrated production process. Generally speaking, these systems are usually suggested and pushed forward by central administration or data processing departments which possess the know-how and also the power to implement new measures of this kind. The fact that central administration departments are the driving force behind these measures has an effect on the design and control of the technological systems in production. The main effect is the limitation of possibilities for intervening into the system at the level of production as well as the fostering of a strict division of programming and operating tasks with a correspondingly high functional division of labour, etc.

The producer of individualized and exclusive products, on the other hand, pursues a rationalization policy which aims to improve the flexibility of his production and administration processes in order to cope with the increased demands being made in his segment of the market. The activities associated with this process are primarily concerned with the production process, especially assembly. In approach, the strategies are typified by the attempt to achieve the closest possible link between reorganization of the production process and the intensified application of flexible, NC-production-technology.

Organizationally, this approach seeks to cluster those production tasks which are particularly subject to flexibility demands into separate areas of production (in other words, to disconnect them organizationally from areas of mass production). An example is the concentration of all pending special order-related work tasks into or close to the final assembly area. While this reorganization and integration of flexibility-relevant functions into designated production areas allows the formation of concentrated areas for flexibility measures, it also creates production bottlenecks, making flexible automation technology in these areas even more of a necessity. From a financial standpoint, (flexible) mechanization or automation in new, highly consolidated production areas is now getting feasible.

Thus the companies try to concentrate on a few areas of production in which flexibility demands can be brought under control with new flexible processing and assembly technologies. This growing utilization of flexible technologies and the increasing outfitting of machinery and facilities with computer-aided control technology in other areas of production (such as in series production) make new demands on companies. Defined interfaces have to be set up, gaps in mechanization and automation have to be bridged over or eliminated, etc. This means that at the production level the companies have to achieve an extensive network of processing machines, handling systems and transport facilities as well as a comprehensive integration of different production areas (horizontal integration). At the same time, the new potential for flexibility can only be adequately utilized when sales order handling processes and production planning and control processes are also made more flexible. This, in turn, generates additional pressure to implement integrated technologies that control the production process in a comprehensive manner (vertical integration).

There is also growing pressure for all enterprises to invest in systemic oriented technologies in order to deal with increasing flexibility demands. Nevertheless, our findings have shown that the producers of exclusive products take a different approach in their efforts to implement comprehensive planning and control system processes from the one pursued by mass production companies. In the mass production of simple and inexpensive standard products, the significance of (especially skilled) work in the areas of manufacturing for adjustment and quality control tends to decline, replaced by a (centrally controlled) computer-aided manufacturing and control technology system which oversees the entire process. The manufacturers of exclusive products, however, at least given the present state of planning and control technologies, tend to rely much more on the direct intervention of (skilled) workers in the manufacturing and assembly process in order to secure the necessary flexibility. In particular, the know-how and power position of supervisors involved with the production process and job planning and scheduling departments still play a significant role. They also have a decisive influence in the implementation process of the planning and control systems. Unlike mass producers, they aim to implement decentralized systems. However, our investigations have revealed that the 'decentralized' systems are not situated at the direct manufacturing or workshop level, but in areas that precede the production process.

These differing strategies of horizontal and vertical integration which are pursued in concert with systemic rationalization policies in both kinds of user companies have an effect on the requirements, collaboration, and design options of the respective manufacturer of technological systems.

HOW TECHNOLOGY MANUFACTURERS RESPOND TO USER DEMANDS

The West German machine building industry sees its success as depending on its ability and willingness to produce innovative and high-quality products and, whenever possible, to provide customized solutions to the production technology problems of its user companies. Thus, its global sales strategy is more oriented to products than prices. Intensified competition among machine building companies has led to an even more concentrated pursuit of this production strategy, at least on the domestic market. The cornerstones of this product-based strategy have grown in significance with the following results: first, the rapid developments taking place in (potential) areas of application in the user companies demand a rapid response in innovation; second, the competitive situation enables market leaders among the user companies to have technological innovations designed specifically for their own particular production problems.

This sales strategy, which is characterized by the close relationship between 'innovation' and 'adaptation', means that the manufacturer's product range and service-related functions are oriented to the special demands of both mass and flexible producer companies. As for the user companies, they themselves develop and implement aggressive rationalization strategies and corresponding innovations in technology and organization. They see themselves as being confronted with clear-cut problems and tasks that have to be dealt with in order to pursue their production and sales strategies. Therefore, the demands they make of the manufacturers are defined in a fairly explicit form. Users at both 'extremes', namely the mass producer and the manufacturer of exclusive products, play a leading role in shaping technology precisely because they take the initiative for tackling the especially troublesome problems for their industrial branch. In this way, they define paths of future application and utilization and thus open up new possibilities for ways to link techniques and technologies for which the manufacturers develop products in response to present and future needs. For these reasons, it makes sense for the machine manufacturers to orient their innovations and sales policies towards the problems and the development of just these kinds of user companies in order to secure long-term sales opportunities.

However, this leads to a change in the range of products and system-related functions that the manufacturer provides. In the past, the 'custom tailoring' of technical solutions, their 'adjustment', was traditionally concerned with aspects of particular machines and technical procedures that corresponded to functions and performance needs in concrete areas of application. At present, however, new demands are being made on the manufacturers due to the increasingly systemic character of rationalization strategies being implemented by user companies. This prompts the manufacturers to offer a range of product and system-related service functions and innovations that far exceed their accustomed spectrum. They are no longer expected to supply simply high-performance machine systems especially designed for concrete applications. Rather, the expectation is to provide comprehensive and innovative solutions which are oriented to concrete application problems in production technology and also in production organization. Thus, depending on the rationalization strategy pursued by the user, differing demands are made on the manufacturers.

The first question is what effects the rationalization strategies of mass producers have had on the development of technological systems. In these kinds of user companies, the most important impulses for systemic rationalization originate at high levels of the company hierarchy, in the administration or data processing departments, whose interest it is to implement planning and control systems which facilitate centralized control of the entire production process. Toward this end, cooperation with large software and hardware houses plays a large role, since, in many cases, these firms have already been involved in the installation of commercial data processing systems in the administrative departments. The influence and potential involvement of machine manufacturers in determining the basic course of integration pursued by these user companies is relatively small. For this reason, the user companies generally fall back on the kind of manufacturers who provide technology which supports centralized processes. This means that the technological systems required by the users and developed by the technology manufacturers have to fulfil two basic requirements. First, they have to bring about a cost-effective increase in performance, which is primarily brought about by networking the various processing, handling and transport functions and equipping them with efficient control systems. Secondly, and of greater importance, they have to be capable of being integrated into centralized planning and control systems. What this basically means is that the new machinery and their control systems have to be extensively adjusted to meet the configurations of data processing and control systems which are already in place in the company. Thus, in many cases, the machine manufacturers have to integrate systems developed by other technology suppliers into their machines and facilities. There is also the possibility that, from the start, only machine manufacturers who have built up long term relationships with the respective system suppliers are considered. In such cases, the manufacturers and system suppliers support each other in promoting their machine and machine control systems and the planning systems and organizational structures that go along with them.

The result is as follows: the demands made by the users and the technological steps carried out in accordance with the machine and control systems' manufacturers lead to the development and at least partial implementation of integrated, comprehensive technological and organizational systems with orientations toward centralization which are generally brought onto the market by large, market-dominating data processing system suppliers or by machine manufacturers working in close cooperation with them.

Quite different are the demands made on machine manufacturers by user companies who base their rationalization on flexible production strategies. In these cases, the innovations for achieving flexibility are linked directly to technological and organizational changes at the manufacturing level; without the early involvement of the machine manufacturer, their realization is not possible. Also, for processes involving the conceptualization and implementation of systemic forms of rationalization based on changes at the production level, the machine manufacturers are involved at an early stage and in a comprehensive manner.

The measures mentioned above for rationalizing production processes and making them flexible are characterized by the fact that, from the very start, they link the organizational and technological aspects of process innovation to each other. As a result, the companies seek solutions in which organizational changes are realized in the context of particular types of technological systems. Manufacturers are expected to develop technological systems which combine (in a new way) various process functions, and moreover to integrate all of the functions. In this way, the user seeks technical solutions which enable him to implement new organizational structures. Conventional forms of technology, which make it possible to detach technology and organization, are abandoned in favour of a concept in which the technological design of the system establishes certain organizational choices. This means that the manufacturer, since he is developing his product for specific user applications, has to consider given organizational structures and the changes the user hopes to achieve. Also, as integration proceeds, the manufacturer has to create the machine and control technology necessary for a reorganization of the user's entire production process in accordance with the processes of integration.

The development of new technologies aimed at mastering the production tasks and problems of the respective users, as well as realizing particular organizational structures by means of the specific arrangement and linkage of various processing, handling and transport functions on the basis of new data processing technologies, makes it necessary for the manufacturer and the user to overcome the usual, market-mediated buyer/seller relationship in order to cooperate as closely as possible. In this way, systemic solutions can be developed and implemented that respond to the concrete flexibility needs of the user as well as to his special manufacturing and procedural requirements while taking the various structural conditions of the user company into consideration.

However, the increasing linkage and integration of production functions both within and between individual production areas, the new delineation of entire functional areas and the increase in the number and complexity of functions to be integrated into technological systems all make new demands on the manufacturers. These can only be fulfiled on a long-term basis when the manufacturers themselves are prepared to set up cooperative relationships with each other. In other words, there is a need for a new and close connection between the manufacturers of machine and control systems on the one hand, and the manufacturers of diverse machines and machine components on the other. This leads to a growing concentration among the manufacturers and thus a fundamental process of restructuring on the manufacturing market.

When flexible NC-technologies were first implemented, many medium-sized machine manufacturers from various sectors of the machine building industry and small, up-and-coming control systems manufacturers participated in the development and customizing of machine and control technology systems. They generally cooperated as equal partners, despite occasional problems and conflicts. The processes of manufacturer concentration, discussed above, however, change these relationships. For example, either financially powerful and technologically advanced machine manufacturers build up their own capacity to develop control systems, and develop the necessary control technologies for their systems themselves; or they take over independent software companies and adapt them to their own purposes. At present, there is a clear trend for these manufacturers to use their accumulated know-how of control technology to develop comprehensive planning and control systems. In doing so, they try not only to extend their range of products and service functions on the market, but also to improve the sales chances of their machines and equipment.

Under such conditions technological solutions take on a more and more systemic character, yet they are still strongly oriented to production problems. In principle, the opportunity remains open for control and intervention at the manufacturing level or for intervention from the production planning departments. The extent to which these possibilities are made use of largely depends on the interests and tactics pursued by the departments and groups advocating the implementation of these systems in the user company. In contrast to mass production technology users, these groups or departments tend to be closely connected to production, such as production planning.

At the same time, market leaders among data processing and control systems' suppliers press for cooperation with machine manufacturers to have their systems integrated into the machines and facilities. This gives them the opportunity to offer comprehensive integrated (machine and data processing) systems to users that are connected with existing systems for planning and control of production and administrative processes that have already been installed at the user companies. As pointed out above, there is a strong relationship between the trend towards centralized concepts and the growth of large suppliers' influence.

Processes of concentration are also becoming apparent in the relationship between the manufacturers of machines and machine components due to the fact that the traditional areas dividing up technology application have been broken up and newly structured through new distributions and clustering of functions. This has made it necessary for the manufacturers to concern themselves with integrated machine structures involving a variety of machining processes that were previously run separately. Thus the techniques and technologies which, until now, were produced by different groups of manufacturers, are now being developed and sold by one manufacturer. The integration and networking of various functions, such as handling and transport, in addition to machining, encourages the tendency for supply to come from a single-source as well as increasing the demand for it. This tendency is encouraged by users for reasons of compatibility, service, liability, calculability of performance, etc. Manufacturers, too, increasingly base much of their sales strategy on single source supply because it facilitates the implementation of comprehensive control systems, solves adaptation and co-ordination problems and helps prevent system disturbances, etc. Cooperation between manufacturers can take on a variety of forms: direct company acquisition or fusion, contractual regulation of the cooperative relationship, or the establishment of jointly owned, but autonomously acting, firms which take over the organizational development and planning services that the users require and charge for them separately.

Within the general context of the development of marketable technology, these processes have had various consequences. First of all, since they involve the synthesis of technical and organizational solutions, they have increasingly taken on a comprehensive and systemic character. Secondly, since they are developed on the basis of close cooperative relations, they have become more and more related to the specific needs and technical and organizational problems of given users. This means that they can only be optimally applied in the companies for which they were originally developed. At the same time, the groups of market-dominating and influential manufacturers, which have formed as the result of increasing concentration, push forward the general dissemination of these types of systems and thus their diffusion into user companies for which they were not developed.

EFFECTS AND CONSEQUENCES

The development presented above results in a two-way polarization of the technology market:

1 The first polarization consists of machine manufacturers who are in a position to develop and offer systemic solutions and those who are not, the latter then continuing to develop and offer simple technologies and components. It is our contention that the companies who cannot themselves develop integrated systems either enter into cooperative associations with other manufacturers or merge with them in order to succeed in developing systemic solutions. Otherwise they are forced into the role of component suppliers or have to retreat from a particular market niche due to the competitive price strategies of foreign manufacturers of standard technologies and may even disappear from the market altogether.

2 The *second* polarization consists, on the one side, of machine manufacturers who cooperate with large scale suppliers of production-related data processing systems in a more or less dependent relationship and who deliver the building blocks for manufacturing technologies and processes. At the other extreme are manufacturers who use their accumulated know-how of control systems to extend their range of products and service functions toward planning and control systems and in doing so succeed in developing global integration concepts.

Both of these forms of polarization entail two differing consequences: the first is product-related in that systemic solutions, which have been developed with the specific needs of certain users in mind, become the predominant form of technology on the market. The second is related to the structure of the manufacturer market and the changing market position of the users in that ever larger and more powerful manufacturers and manufacturing groups supply a number of medium-sized customers. This puts the users in a comparatively weak position in relation to the manufacturers, with regard to market position, technological capability, the ability to implement innovations and limited material resources. The chances that this type of customer's particular wishes will be taken into consideration by the manufacturers are extremely low, leading to a situation of almost complete manufacturer dominance.

The manufacturers dominance is only 'almost complete' because they have regarded many of these medium-sized companies from the very beginning as potential customers for their products and have thus always hoped that their technological solutions would cover the needs of these companies. At the same time, however, they dominate in the sense that they install products which were developed mainly for other user companies.

Because these medium-sized companies have limited options with regard to market strategy, they are highly exposed to the demands of the market. Therefore, they have to make use of the machinery and equipment developed to solve cost and flexibility problems, and of course, the organizational structures associated with them, even if these systems do not correspond to their operating conditions and the companies cannot exert the pressure to get the technology adjusted to optimally meet their specifications. The question is which concrete problems arise for medium-sized companies as a result of this situation.

Adopting a given technology in the form of a 'systemic solution' means that the flow of material and information in manufacturing subprocesses, in areas preceding and following up manufacturing, and in management and planning departments, are all fixed to a great extent. Thus, the company's work organization, personnel structure and training conditions are all affected. The consequences this entails are difficult to foresee, both for the companies and for the in-plant employee representatives. Yet, changes after the fact are problematic and involve much effort and time, thereby restricting potential reactions.

Another problem is that the manufacturers try to transfer the hardware and software structures developed for systemic solutions to other uses; this results at best in modifications in terms of some kind of modular design. Due to the costs involved, fundamental changes are rarely possible, meaning that programmes originally developed for other fields and conditions of application are implemented, making it impossible, or at least extremely difficult, to achieve an optimal adaptation for specific user conditions.

Systemic solutions incorporate a level of production, planning and control technology, and organizational know-how to a far greater extent than was ever the case with traditional technologies. This becomes an integral aspect of the respective manufacturer's product range and service functions. The result is that design knowledge passes out of the hands of user companies into those of the manufacturers, meaning that the users gradually lose the kind of competence needed to plan new kinds of production systems, especially integrated ones. Furthermore, once it has been lost, companies are no longer able adequately to build up this capacity and know-how again – a factor which is of great significance for middle-sized companies. All of this leads to a restriction of available options for system design and ignorance of potential alternatives.

The growing necessity to utilize and implement systemic solutions as the definitive technology plunges many user companies into a hopeless dilemma; on the one hand, they are no longer able to implement traditional procedures and approaches, yet, at the same time, they lack the prerequisites and conditions for integrative and comprehensive computerized production systems.

In the old days, the process of technological development was carried out in small steps; new technologies were mainly confined to single areas, especially in critical areas of production, and the technology employed was insular in character in that it only entailed isolated machines or facilities. The integration of these machines or facilities into the total production process took place within the context of work organization measures, but in many cases was also the result of a worker/machine relationship that developed slowly and over a period of time. This meant that technology did not determine the measures and alternatives for organizational design. This was and still is of particular significance to medium-sized companies because:

1 their limited financial scope does not allow them to carry out comprehensive forms of innovation entailing the long and difficult implementation phases that are usually involved when complex systems have to be adapted to specific company conditions;
2 due to their limited technical and technological know-how and their skills structure, they are generally dependent on the gradual introduction of new forms of technology into the production process; only in this way can they develop or obtain the necessary skills; and
3 process and system-specific skills generally exist to a sufficient degree in these companies. Up to now, this guaranteed a comparatively high degree of flexibility and made it possible to keep open design possibilities that lie outside the realm of technology. However, drastically different forms of technological innovation and integrated systems which are quickly implemented endanger these types of skills. Such systems require a different set of skill which can only be built up with the help of comprehensive and expensive training measures. However, the capacity for this sort of retraining does not always exist. This plunges the companies into a further dilemma. On the one hand, skills that exist and have worked well in the past are losing their significance, but new skills are not yet sufficiently available. Disadvantages such as these often cause medium-sized companies to choose centralized solutions.

The growing integrative character of technical solutions not only makes it more difficult to implement them in small, successive steps, but in time makes it more difficult to restrict their utilization to particular functions and certain areas of application since they usually bring about the implementation of a comprehensive computer-aided information flow system. A rationalization of subprocesses, which would be suitable to the financial and personnel capacity of these companies, becomes difficult, if not impossible. Thus, the companies are constantly being pushed to the limits of their capacity. As a result, their dependence on manufacturers, who have to assume training, financing and other services in addition to their usual technological and organizational services, grows.

This dependency enables the manufacturers to force the user companies into adopting systems that correspond best to the manufacturer's own production, development and financing conditions. This does not mean to say that such systems are ineffective as a means of solving flexibility and cost problems, for they are developed with precisely these ends in mind; however, it cannot be denied that the user companies have to pursue their rationalization strategies with techniques and technologies which do not fully correspond to their conditions and needs and which, above all, prevent them from trying out other alternatives. Moreover, the difficulties and problems that the user companies experience in attempting to master these suboptimal solutions increase their dependency on the manufacturers, due to the continual need for their intervention when difficulties arise and the assumption of almost all service functions by the manufacturer. This leads to a further weakening of the user company's position, for example in terms of their skill potential.

As a result of this state of manufacturer dominance, systemic solutions are now being implemented in user companies at an increasing rate. Companies which do not possess or which are unable to create the necessary conditions for systemic solutions, or which are unable adequately to cope with the outcomes, have little chance of long-term survival.

7 Thinking Machines, Dreaming Engineers? Towards Applied Expert System Technology[1]

Manfred Moldaschl

When dealing with expert systems, researchers, engineers and manufacturers tend to stress the advantages of new technologies whereas social scientists often concentrate on the risks. This paper too will concentrate on risks, my position being that the potential benefits of these new technologies can only be exploited to their fullest extent when we are aware of the risks and develop ways of avoiding them. My central thesis is that these risks are not so much the direct result of AI-technologies, but are far more the outcome of the overestimation of their performance capacity. This leads to inadequate forms of application which neither make the best use of the technology's potential nor of the personnel involved.

INTRODUCTION

> Every social scientist is faced with the task of acquiring knowledge in order to master the present and of creating ignorance in order to design the future.
>
> J. Kuczynski

> Knowledge which has not been acquired by the senses can create no other truth than one which will be harmful.
>
> Galileo

These two statements were not made with 'artificial intelligence' in mind and yet they touch on two aspects of a central problem: the significance of practical and sensory experience and the role of human ignorance in the further development of human knowledge. In other words, the central problem in the use of expert systems (XPS) is the production and reproduction of human knowledge and human experience.

In the following my basic premise is that the above-mentioned problem can be posed in very different ways, depending on company objectives and expectations in the use of XPS and the possible forms of utilization which are opted for. It is important to clarify this view, because in the current discussion on XPS the potential effects are often directly regarded as stemming from technology 'per se'; this holds true for such negative effects as the loss of empirical knowledge, as well as for positive effects including that of creating a 'more human society' (see, for example, Schank 1984).

Moreover, even in those cases where it is accepted that the effects can only be interpreted in the context of the conditions and objectives of technology utilization, the discussion usually tends to focus on the direct effects and consequences of an intended form of utilization. It is a common experience, however, also within companies, that things often go quite differently from the original intentions of a given company, or what technology assessment has anticipated. In my opinion, this is because the unintentional and the systemic effects are often underestimated and in fact are so diverse, that it is impossible to anticipate them all.

In the following considerations I will first specify some of the basic conditions of XPS and subsequently will outline some of the unintentional effects and feedback mechanisms. The effects of XPS are felt at every level: society, company, individual employee. This chapter concentrates on the employee level, specifically the risks associated with the introduction of XPS.

CONTEXT AND OBJECTIVES OF COMPANY UTILIZATION

The risks of jeopardizing the generation and preservation of human expert competence, are likely to vary according to the type of company problem which is dealt with by the use of XPS. These include:

1 problems of company performance capacity: coping with increasing flexibility requirements and complexity (for example, the rising number of product variants, greater integration, JIT);
2 increased cost pressure in certain areas: for example, when a disproportionately large number of employees are to be found in planning and service departments (while personnel in the manufacturing departments is shrinking); and
3 problems of having experts available for important tasks (shortage of experts).

It is understandable that in the first case XPS would be used more for solving technical and organizational problems. It remains open, however, as to whether this would be brought about by supporting the experts or by establishing the greatest degree of automation possible. In the second and third cases, problems of personnel management are in the forefront. In the third case it is uncertain whether the solution will lie in systems supporting the experts or in replacing them by less qualified users of 'smart' systems, the latter is likely to be the objective in the second case scenario.[2]

The manner in which the rationalization potential of AI-applications is utilized also depends on general patterns of organizing and utilizing human labour. For the last few years 'paradigms' or 'philosophies' of work design were often referred to. The range of existing rationalization strategies can be grouped around two opposite poles: the approach determined by Taylorism and the Human-Centred Approach.

'COMPUTER-AIDED TAYLORISM' AND 'COMPUTER-AIDED CRAFTSMAN'

What is referred to as 'Taylorism' has been the dominant pattern of work design in the past few decades in Europe. This 'traditional' pattern of rationalization is characterized by three basic features: the first is the process of simplifying tasks by establishing a maximum degree of division of labour, the characteristic most often identified with Taylorism. The second, more important feature, is the transformation of work planning into scientific terms: this means extracting the empirical knowledge held by workers, eliminating its subjective component and feeding it back to the workers in the form of standardized work instructions (compare Volpert 1985). The primacy of technology has emerged as the third feature, a point of view which regards man as a basically unreliable, subjective source of mistakes to be replaced to the greatest extent possible by the use of exact algorithms and technology. It should be pointed out that these basic features of Tayloristic approaches may very well continue to exist even when Taylorism is declared to have been 'overcome' in the course of a partial reduction in the division of labour.

There is a great deal of evidence that the Tayloristic rationalization approach is particularly predominant in those areas where mass production and a low level of employee skills prevail (see Part IV). Company size also plays a role (see Part VI).

In those enterprises where high quality products are manufactured in small and medium-sized series and a sufficient number of skilled employees are available, Tayloristic principles have not been able to establish themselves to a very large extent. The German machine building industry is a good example of this. The supply of skilled workers is provided by the 'dual system' of vocational training,[3] a system which is a prerequisite for the success of machine building as well as other branches of West German industry.[4] The type of work organization found under these conditions is a low degree of division of labour in terms of skills and hierarchical structure and a high degree of employee participation (see chapters 10 and 11). On the management side, there is a pronounced awareness of the central significance of skilled personnel. This type of work organization has been termed 'professional organization' (Kerr 1954). The work design focuses on human labour and its tasks, and selects technology according to its capacity for supporting human skills in the best possible way (as evidenced by the concept of the 'Computer-aided Craftsman', Jones 1984). In practice, this approach continues to be an exception. At present the so-called 'Human-Centred Approach' is more a target concept, or a basic 'design philosophy' (see Rosenbrock 1983 and Corbett 1986) than a reality.

All evidence points to the continued prevalence of the Tayloristic approach in the ideas, expectations, objectives and concepts expressed by manufacturers and potential users of XPS application at present. This is clear from the following commonly voiced objectives:

1 the reduction of the dependency on experts by 'conserving' their knowledge;
2 the imminent potential for accessing, using, and distributing expert knowledge;
3 the relief or replacement of experts by automating the processes of judgement and decision making;
4 the elimination of 'inconsistencies' between expert opinions;
5 the reduction of training expenditures through the use of XPS technology (the so-called explanatory component) and by reducing the number of experts required.

It is likely that this approach of objectification and standardization will continue to prevail due to the fact that it is just as deep-seated in engineering circles as the 'Man-is-a-Machine-View' is among AI-experts. The fundamental concept behind this approach is that human thought and action are basically identical to computer procedures and can therefore be automated. 'Skills' are regarded as a huge accumulation of facts and rules, and 'learning' as the progressive amassing of facts and rule bits.

A REAL AND CURRENT RISK: THE OVERESTIMATION OF AI

In our opinion the main danger in the application of XPS lies in overestimating their performance capacity which is a result of the above-mentioned overly mechanistic outlook. Expert systems extend the limits of what is technically feasible, yet they are confined by the general limits of formalizing human competence and the fundamental limits of the feasibility of planning all company processes and sequences (Moldaschl 1988; Lutz, Moldaschl 1989).

The first limit which is systematically underestimated is the extent to which human competence can be formalized. 'Professional Qualification' is not the same thing as knowledge, and 'knowledge' is by no means identical to knowing elementary facts and rules. In that case it would suffice to read a book or undergo theoretical instruction in order to be able to ski, drive a car or learn how to operate a printing machine. The capacity for thought is not restricted to the application of knowledge formulated as rules and you will find that AI-experts rarely have much to say about the individual's scope for taking action. It is not my intention here to criticize AI at a psychological level. However, it should be pointed out that the 'mastery' achieved by experts is based on practical experience, on the daily sensorial experience of dealing with materials, machines and human beings. It is only through taking practical action and through sensorial experience that knowledge becomes that which the German word *Können* denotes, a term which is somewhat inadequately translated as skill and which refers to a certain achievement of mastery. The 'feeling for material' or 'intuitive feeling' for taking the right course of action which is based on a certain degree of mastery can, for the time being, not be reproduced by technological means because it is corporeal, experience incorporated in the human body (see chapter 8). Gilbert Ryle, the almost forgotten critic of rationalism, introduced the distinction between 'Knowing That' and 'Knowing How' to accentuate the role of the body, although it is so often misunderstood as merely a distinction between kinds of knowledge (Ryle 1949).

With these thoughts in mind, try and think of something you are particularly capable of doing, such as giving a critical assessment of a scientific paper or driving a cross-country motorcycle over a difficult course. Could you imagine that just by talking to someone you could explain these things in such a way that the other person would be equally capable of doing them? Or what about writing down some of the relevant rules on a sheet of paper? How could the inexperienced person understand this knowledge which has been extracted from its empirical framework, in a way that would make it possible to act in an effective manner? This is indeed the situation of the XPS user who does not happen to be an expert in the particular area.

A second, fundamental restriction which the methods of artificial intelligence are confronted with, is the limit to which company operational sequences can be planned. Open systems and complex processes which are typical for modern companies, have a multitude of potential sources of disturbance which are impossible to anticipate fully and thus possess a high degree of incalculability.[5]

There is no means of planning that is perfect or system-complex enough to anticipate fully all of these factors. Also, the more complex systems become, the less scope they offer for flexible control and the more prone they become to failure through technical defects, coding mistakes (two 'Mariner' probes, for example, were lost because a period instead of a comma was written in the program: Valk 1987), or also due to human error (such as with the Exocet incident in the Falkland War where the British system could not identify the attacking missile as unfriendly because it was coded as belonging to the NATO arsenal).

However, the very nature of human intelligence includes the capability of dealing with unclear and contradictory information and to cope with open, uncertain and new situations. The capacity of human intelligence for handling ambiguous content and gaps in knowledge is based on our innate potential for learning. This means people can transform deficits in information into knowledge by taking action and by experimenting. Thus lack of knowledge can be used as a source of acquiring new knowledge.

Human competence remains indispensable for complex technical systems. But the question remains open as to how it will be integrated. Instead of developing software which would be able to give human beings the best possible support for doing what they can do better than computers, current developments are directed towards achieving a technical simulation of such competence, in order to replace[6] those persons holding expert knowledge. And this is happening at a time when the individual within the company is playing an increasingly central role: fewer and fewer workers are confronted with increasingly complex machinery. Therefore, overestimating the self-regulation capacity of technology poses a serious danger to productivity, process safety, and mental health. It is the technology-oriented, 'Tayloristic' approach which tends to succumb particularly strongly to an overestimation of technology's potential.

INTENTIONS AND UNINTENTIONAL EFFECTS

So far we have pointed out two sources of risk that have to be considered with the implementation of XPS. The first is the effects resulting from the performance and rationalization potential of XPS; the other is the effects stemming from the overestimation of technological potential. In the following I will briefly outline some of the possible unintentional effects which result from an incorrect assessment of human competence and the overestimation of technical performance capacity. In doing so I will restrict myself to a description of direct and counterproductive effects at the company level.

1 Companies want to become independent of experts yet make themselves dependent on a form of technology which is only capable of handling

routine operations. At the same time, as automation increases, special situations and disturbances become more and more common.

2 Companies want to overcome the shortage of experts yet they are exacerbating the shortage by dispensing with the training or recruitment of human experts.
3 Companies seek to reduce their training expenditures, yet are generating a lack of skill, due to the fact that the 'explanatory component' of XPS is not capable of replacing either practical experience or vocational training; it is also not capable of providing explanations in the true sense of the word because the computer has no genuine 'understanding' of the field for which it is being used.
4 Companies would like to eliminate the inconsistencies between different expert opinions, yet they invariably induce inconsistencies between model and reality – inconsistencies which users are subsequently confronted with and cannot handle adequately.
5 Companies are trying to avoid mistakes and disturbances by objectifying human knowledge and in doing so are inviting what they would like to avoid because the system is only capable of partially reproducing the complex interrelationships.

There are, of course, less technology-centred approaches which are in danger of losing the potential benefits of XPS through an inadequate use of technology and personnel; for example, when they:

1 try to take an excessive work load off the shoulders of experts and at the same time expect them to make more decisions in less time and additionally monitor the conclusions the system has arrived at,
2 try and take the 'burden' of routine work off their hands and at the same time cut off their source of potential practical experience – such as the case of a supervisor in a power plant who is only supposed to intervene in the case of an emergency, but who does not know how the system reacts to intervention.

These are only a few of the problematic effects which may directly result from the use of XPS unintentionally. The question of what other possible indirect effects, both intentional and unintentional, could arise (for example on departments and employees not at all involved with the use of XPS), should be considered. And finally it would be very interesting to investigate how companies are reacting to these effects. Before going on to the question of how companies might react to the problems looming on the horizon, the risks associated with XPS which touch the employee in particular, such as those affecting qualification (and acceptance), should be mentioned:

1 The present job held by an expert is subdivided into that of a knowledge engineer and one or several data suppliers for the system or system user, associated with a loss of status and dequalification of experts – or an overburdening of poorly qualified users.
2 The expropriation of expert knowledge takes place. Individual knowledge does not actually disappear altogether, but it no longer belongs to the individual and that means an increased danger of loss of status and control, or of job loss.
3 Expert knowledge is lost through lack of use in the case of extensive automation ('use it or lose it').
4 Expert knowledge stagnates due to the loss of possibilities for gathering experience and learning (also due to the necessity of having to adhere to system guidelines under time pressure).
5 Qualification is reduced to man-machine 'interaction', associated with a loss of communication and social support in learning situations and the danger of deficient and solely machine-based qualification.

These risks relating to employee skills are associated with severe stress factors for personnel:

1 Loss of status and control, or the fear of such loss both cause emotional stress, as does overstrain.
2 So-called users are left alone with the system during the learning period, resulting in overstrain and a feeling of insecurity.
3 Although skills waste away, stagnate and are no longer specifically developed, they are still called for in emergency situations in order to avert serious consequences of disturbances. (Stress is caused by contradictory work requirements – emergency dilemma, or Harrisburg syndrome; see Hirschhorn 1984).
4 Although neither experts nor lesser qualified personnel are able to verify data provided by the system under time pressure, they are nevertheless responsible for the results if bad comes to worse (anxiety and stress caused by the dilemma of responsibility).

While these effects are hypothetical with regard to XPS utilization, they are by no means new. They are typical for highly automated systems, although not XPS-specific. By increasing the degree of automation and technical mediation they serve to intensify problems. In our investigations of FMS (Flexible Manufacturing Systems) or of PPS (Production Planning Systems) we frequently encountered discrepancies such as those between the anticipated effects and reality as well as between the demands made on employees and their scope for taking effective action. Thus, these discrepancies are by no means the never-ceasing pessimistic apprehensions of social scientists, but are widespread facts which affect productivity and cause 'inexplicable' failures, or failures which are explained away as being 'human'.

CORRECTING THE MISCALCULATION – OR 'CORRECTING' REALITY?

Up to now our main concern has been the risks which arise through an overestimation of AI-technologies, on the grounds that the effects have less to do with the technology itself than with the way in which it is used and the objectives which are pursued. If it is true that overestimation of this kind causes such grave problems one might anticipate that the expectations would soon be corrected by practical experience. In agreement with Mertens *et al.* (1988) one should assume that the expectations held by the companies and society concerning this new technology may follow a quasi natural curve: overestimation – frustration – realism.

Investigations concerned with other applications of information technology in the manufacturing and office sector (Moldaschl 1989; Weltz 1988) have shown, however, that this expectation curve is not necessarily an inevitable consequence. An attitude of overestimation may remain despite frustration and may lead to the attempt to solve the problems through an increased use of technology. The reasons for the frustration are either looked for in the wrong area or are not recognized at all.

Some of the reasons for this blindness to the above-mentioned problems and risks could be:

1 the compensatory efforts of employees by which the discrepancies between formal requirements and actual conditions are covered up as far as possible at the cost of considerable additional effort, personal risk and strain (such efforts help to sustain further the illusion of complete control);
2 the reversal of the cause and effect principle, for example when the lack of employee motivation or competence is held responsible for the discrepancies between planning and execution.
3 when the Tayloristic approach has become so firmly entrenched in engineering structures that it is no longer perceived as a paradigm, that is as one of a number of possible approaches.

A curious, but typical, example of the last mechanism is exemplified by the following incident, encountered in one of our field studies (see Moldaschl 1990). A software engineer built up a highly centralized and automated control system for an FMS. The system never worked because of the new problems – whether organizational, technical or social – which continuously arose. So the engineer revised the system again and again, until it was a completely new and in fact worker-oriented system. But he did not seem to be particularly aware of the fact for he told us: 'The system was perfect, but the circumstances were lousy.' Meaning there was nothing wrong with the model, it was reality that was flawed.

This implies that all the pleas in favour of the expert and assertions of not wanting to replace him, are not worth very much in practical terms, because we are lacking the ways and means of pursuing a better course of action. It also means that a supervisor, for example, can derive no feeling of safety from 'knowing' that he is basically not replaceable, (as Boden 1988 points out). Nor should we rely on the Mertens-curve. The image of an oscillation being continuously re-excited might be more apt. Moreover, the current discussion on Parallel Processing and Neuronal Networks is adding fuel to the fire to the concept of omnipotence harboured in AI-circles.

Whichever way the level of expectations may turn the present discussion gives us more reason to fear that the company's reality will have to adapt to the restrictions set by expert systems, rather than hope that XPS-technology will be adapted to the reality of work, to human potential and thus to a humane, low-risk and future-oriented organization of work. This is indicated by the highly optimistic forecasts voiced by promoters, manufacturers and potential users of AI-products who rarely mention risks or brush them aside as a minor issue (see Schefe 1988).

Deeply concerned and worried, yet not without hope, two influential AI-critics concluded a paper with the warning:

> Should calculative rationality triumph, no one will notice that something is missing.
>
> (Dreyfus, Dreyfus 1986: 206)

POLITICAL OPTIONS

There are two basic requirements for political control measures and research policies: first is the need for policies which will promote technical and organizational alternatives to the predominant Tayloristic course of development. There have been a number of practical efforts in this direction in several companies. Similar efforts in other areas in which information technology is applied have been receiving government support in the Federal Republic.

Explosive government support of technology development in the AI-sector must be augmented by a massive promotion of theoretical and methodological alternatives such as the 'Human-Centred Approach' (for example Corbett 1986) and 'Contrastive Work Analysis' (Volpert 1987). 'Contrastive Analysis' means: see what people need to do their work and what man can do better than the computer. Ask in what way these functions and human competence could be granted optimal support by system design. These concepts also form the foundation of a 'forward assessment' with regards to desirable and undesirable future developments.

Second there is a need to control and substantiate the effects of promotional measures by evaluating AI-systems where such measures were employed and those which did not receive such support ('backward assessment'). The discussion has reached a point where further policies cannot be developed without feedback from empirical assessments.

SUMMARY – SIX THESES

1 It is impossible to speak of 'the' effects of the use of expert systems. It is only possible to investigate their effects together with the company's basic conditions and the rationalization strategies directed towards these specific conditions. Therefore scenarios concerning the effects of expert systems must include the above-cited factors as central components.
2 Rationalization paths and scenarios display a range of variance between the two poles of 'Computer-aided Taylorism' and the 'Human Centred Approach'. So far Tayloristic solutions have predominated.
3 The Tayloristic paradigm (production knowledge is expropriated, formulated in scientific terms, and 'returned' to the manufacturing area as highly standardized instructions; human labour is regarded as a disturbance factor to be excluded, and top priority is given to technology) is likely to prevail in companies when AI-systems are employed, as it conforms with the mentality usually found in AI-circles. Hence XPS act as 'trend amplifiers'.
4 Expert systems increase the gap between the reality of planning and the 'actual reality'. The discrepancy between the formal planning models and the restricted calculability of complex manufacturing processes – as evidenced by the use of information technology to date – is certain to increase with the spread of formal models. Therefore XPS act as 'failure-amplifiers'.
5 The greatest danger in the use of expert systems is the fact that their performance capacity is overestimated and the discrepancies are not recognized. It is very likely that expert systems will not be primarily adapted to complex reality, but that reality will have to adapt to fit the expert systems. Reality will be reduced to that which can be formalized and fit into the expert system.

6 Consequences: The limitation of the generation and reproduction of knowledge and skills must not only be determined for the individual employee, but also for the company and society. This is particularly necessary due to the fact that the significance of the individual employee grows along with the rising degree of automation and decreasing number of persons involved in the ongoing manufacturing process.

8 Computerized Manufacturing and Sensory Perception – New Demands on the Analysis of Work[1]

Fritz Böhle, Brigitte Milkau, Helmuth Rose

INTRODUCTION

This chapter introduces an approach which broadens the scope of sensory perception analysis and its significance for the work process. We discuss phenomena such as the individual's feeling for certain materials and the way workers allow themselves to be guided by the sound of certain machines. The concept of 'subjectifying action' is the central focus and will be explained in depth using the tasks performed by skilled workers in mechanical engineering industries. This chapter also deals with changes in work which occur in connection with the application of new technologies and which have received little systematic examination up to now. We also probe the causes of new forms of mental and nervous stress as well as factors which endanger the continued maintenance of workers' skills and qualifications.

THE LIMITS OF EXISTING CONCEPTS

The idea that the application of new technologies involves changes in sensory perception during the work process is generally accepted among social scientists. The changes come mainly from the increasing technological intervention in the relationship between workers and their work tasks on the one hand and production processes on the other. The element of technical mediation is by no means a new one, yet with the utilization of new technologies it is attaining new quantitative (more fields of application) and qualitative (data processing, progressive automation) dimensions. Existing work analysis concepts, however, are inadequate for grasping, let alone assessing, the resulting changes occurring in sensory perception. To date, the sensorial components involved in the work process have only been considered in terms of their implications for imposing physical strain on the individual (noise, heat, demands made on physical endurance). When sensory perception is seen in connection with its effect on the work process, then concepts come to the fore which are primarily oriented to the cognitive, rational model of receiving and processing information. A characteristic of this model is that without cognitive, rational interpretation, whatever is perceived sensorially remains at the level of simple, physiological 'stimulus-response processes' and more or less automatic motor sequences. This is particularly evident in the category of 'sensory motor skills'. The 'theory of action regulation' (Hacker, Volpert) places this category at the lowest level of action regulation that is shaped and controlled by cognitive, rational processes. Without rational thought, action takes place purely on the basis of primary physiological and mechanical processes or the remnants of what used to be governed by cognitive, rational processes (habitualization).[2] Such an interpretation is in keeping with the theories of sensory perception developed and accepted in general (psychological and physiological) research on perception, particularly since the turn around to the domination of cognitive psychology has taken place.[3] With this approach, however, the tendency is to greatly belittle the practical significance of sensory perception for actions and to describe it in biased terms.

If one examines actual work actions, one runs into a number of phenomena that are difficult to fit into the currently accepted analysis of sensory perception. Examples of this are, for instance, the frequently mentioned 'feeling for the material' or such work practices as being guided by the sound of a machine during its monitoring and inspection. Similarly, there are reports of skilled workers engaged in the monitoring of highly automated facilities having (or indeed requiring) a 'sixth sense' that anticipates what the instruments are going to indicate or do not indicate at all. Important decisions are often made – particularly by skilled workers – on the basis of feelings, without rational grounds.

Another example is 'technical sensitivity',[4] a category with a firm tradition in industrial sociology. More recent discussions refer to skills of this type in such terms as 'tacit skills' and 'empirical knowledge' (Wood 1986; Polanyi 1985; Malsch 1987). However, here too, one finds only a rather vague reference without a precise explanation of what is really meant.[5]

In view of the increasing utilization of computer-based information, control and communication technologies, it is becoming increasingly relevant to subject such phenomena within the work process to a systematic analysis especially since these phenomena seem to be affected by changes in the work process that are currently emerging. Moreover, they play an important part in debates about the differences between man and the computer and about the limits of artificial intelligence.[6]

In the following, we shall present our findings on this topic. Our studies are aimed at broadening the analysis of sensory perception. This means that sensory perception is not viewed in isolation but in its relations with other components of work-related actions.[7] Within the framework of industrial sociological research there are connections to research approaches in which 'subjectivity' is discussed as a constitutive factor in the human capacity for work. Within the framework of research into the psychology and science of work there are links to approaches aimed at a 'contrastive analysis of work'.[8]

We shall proceed by describing some theoretical concepts for an extended analysis of sensory perception; then, we shall present the findings of our empirical studies and, finally, point out some consequences for further research.

SENSORY PERCEPTION AND SUBJECTIFYING ACTION

For our investigation it is expedient to distinguish between objectifying and subjectifying action. The category of 'objectifying action' refers to forms of action which – in correspondence with the predominant consensus – are performed according to rationally grounded action (see, for example, Habermas 1981). The basis for this form of action is a specific shaping of sensory perception, of the relationship to the (social and physical) environment and in the dealings with it, and the role played by feelings. It is characterized by a distance between the subject and the environment which is seen as being different from and independent of the subject doing the action. Moreover, there is the assumption that the environment displays generally valid, and in this sense, objectifiable characteristics and properties. To recognize these characteristics and properties and use them in practical action is the key aim of objectifying action. Thus sensory perception of this kind occurs based on the exact and reliable collection of (objective) information. Furthermore, it is a sensory perception which leads to a far more adequate understanding of the environment and appropriate forms of action if it is guided and interpreted by the intellect (comprehended, categorized, etc.). Knowledge that derives from scientific method and action based upon it are prototypical for such objectifying action (at least in theory). This coincides with the separation of sensory perception from subjective feeling. From this perspective, feelings are primarily an intra-psychic occurrence in the sense of individual states of experience (well-being/ill-being, happiness/unhappiness). They are thus wrapped in an aura of 'subjective introspection'.[9] Feelings either precede or succeed practical action; however, in the course of action itself they tend to 'disturb' and cause subjective distortions.[10] The predominant concepts of sensory perception and the analysis of work are both oriented primarily towards objectifying action.

By contrast, the concept of 'subjectifying action' describes forms of action in which not only sensory perception, but also feelings, the relationship to the environment and the way one deals with it attain a different quality and significance in terms of practical action. Using the concept of 'subjectifying action', an attempt will be made to show the various components of an action, e.g. intuitive action, sensory perception, and associative and intuitive thought, in their interactive context as elements of specific forms of practical action. In doing so, we shall adopt findings from a wide range of interdisciplinary research and apply them to this analysis.[11]

We start with forms of sensory perception that can be described as a perceptiveness based on feelings and senses. The basis is a sort of participatory perception in which the subject experiences the environment not as something separate, but rather as a part or even an extension of himself. This is, for instance, like a hearing which 'listens to itself'. This use of the senses is commonly referred to as 'sensing' or 'feeling'. This sort of sensory perception does not take place in a fragmented and isolated way via individual sensory organs; it is all-embracing and is connected to the entire body.[12]

The basis for this kind of sensory perception is a relationship with the environment which can be termed 'sympathetic'. The subject doing the acting experiences the environment (things and people) not as unfamiliar, external objects, but as belonging to himself or herself; in the process of taking action they form a 'unity'. While in action the person seeks to synchronize his/her action with the environment. In this way, non-human things are made quasi 'human'. This does not mean that human qualities are projected onto them. The crux is that in performing an action, the subject identifies with characteristics and properties of the environment so that the person acting and the object being acted upon draw closer together and become more similar. In this sense, the environment becomes a subject.

Also characteristic of sensory perception are ways of acting in which the fundamental elements are empathy and subjective involvement in the sense of 'getting involved'. The mimetic, identificational reproduction of motions as well as dialogical ways of acting in which the subject neither influences nor reacts to his environment one-sidedly, but rather strives for a certain result through a process of mutual exchange, and 'shared endeavour' are vitally important. A constitutive element is the receptiveness to the possibilities inherent within and offered by both objects and persons which only come about or are developed by coming into contact with them. Action is therefore characterized by the union and simultaneousness of action and reaction; the effects of one's own action are experienced directly and at the same time have the effect of directing action. Planned action is no hindrance to proceeding in such a way, provided enough space is left for the unity of active and reactive action.

Sensory perception and feeling cannot be separated in the ways of acting described here. The relationship of feeling to the object is fundamentally important. It is the medium in which the connection between subject and environment gets produced. In conjunction with sensory perception, feeling is an important factor in the recognition of properties and qualities in the environment as well as for the regulation of action. This kind of intuitive recognition has its counterpart in an intuitive, empathic way of thinking in which feeling, thinking and sensory perception are various aspects of the same process. Such intuitive recognition is based essentially on experiences which in this context should be understood as 'getting involved' and not merely as practice versus theory.

As described here, subjectifying and objectifying action are not hierarchically related, nor can they be reduced to nor substitute one another because each does something different. The fundamental point is the proposition that both subjectifying and objectifying action can focus on gaining knowledge about the environment which is relevant for action and on organizing that action in a suitable way.[13] The relevant insights and rules governing subjectifying action could be 'supra-individual' – that is, shared and effective at the collective level. What is crucial is that they are integrated into concrete actions and can only be experienced, imparted and learned in this way; they cannot be isolated and in this sense 'objectifying'.

We do not, therefore, consider it a correct scientific premise to attribute subjectifying action only to certain social areas, e.g. to the so-called cultural sphere or processes of interpersonal interaction and communication. Rather, it is our thesis that subjectifying action is indeed significant in the work process, particularly in dealing with things, i.e. work materials, tools and machines. Such a thesis cannot, of course, be grounded on only theories and concepts (although in this area, too, a number of further studies are both possible and necessary). Rather, an empirical base is also necessary. In our opinion, two things need to be accomplished in this area: first to demonstrate how the forms of sensory perception and ways of acting as defined by the concept of subjectifying action are essential aspects of practical work actions. Second – and this was our initial point of departure – it has to be shown that important changes in the work process after the introduction of information, communication and control technology can only be identified and analyzed when the concept of subjectifying action is adopted. In the following, we shall present examples of empirical findings which fulfil this twofold objective using a selected area of industrial production. The focus will be on skilled workers' activities at conventional and CNC-machine tools in the mechanical engineering industry.[14]

AN EMPIRICAL ILLUSTRATION OF SUBJECTIFYING WORK-RELATED ACTION – SKILLED WORK AT CONVENTIONAL MACHINES

Taking the example of skilled workers' activities at conventional machines, we shall demonstrate that working practices such as going by the sound of a machine or the feeling for materials are based essentially on a subjectifying type of action.[15] In conventional manufacturing processes this kind of work action is an important part of the worker's skill and the way he does his work. The following section discusses the essential characteristics of work-related action.[16]

Sensory perception mainly takes place via several senses simultaneously, i.e. the eye, the ear, hand, and the body and movements of the body. Manual movements, such as turning a handle or mounting a workpiece are not isolated, individual manipulations that exist by themselves. Seeing and hearing, too, are directly connected to movements of the body which change angles and distances. If one takes a closer look at the use of the senses, it becomes clear that an important part is played by precisely the perception of those circumstances that cannot be identified and defined objectively, rationally and unambiguously. It is, for instance, imperative that skilled workers check and identify defects and disturbances using the sound of the machine and the various processing operations. What it is exactly that skilled workers hear and how they recognize whether 'everything is running smoothly' during operation, 'cannot be precisely described and measured', to quote one skilled worker. It is clear that 'feeling' plays a particularly important role. This becomes evident when workers describe how they can recognize faults by the sound of the machine. Comments are made such as: 'You have to hear, to feel if its running smoothly'. The same is also true for tactile contact to work equipment, for example, in the use of the hand: 'You feel it in your hand. The hand recognizes something. To mount a work piece, you need that feeling in your hands. You couldn't do anything with just a gauge. A gauge is only important in confirming what your hands feel and tell you.' And in the case of visual perception: besides having to read off measured values and scales, you have to have the 'right eye' for material properties and tool wear and tear. Here, too, it is emphasized that, 'only a skilled worker with a feeling for it would see that.'

Sensory perception is closely connected to a specific *relationship* to the machine and the materials. Skilled workers have not only an objective and impersonal relationship with their machines, but also an extremely 'personal' one. One hears of skilled workers being 'fused with' or 'married to' their machines and that 'man and machine make up a unit'. Neither the skilled workers themselves nor their superiors see this merely as an expression of some kind of personal need. They stress quite emphatically that this kind of relationship to the machine is necessary because each machine has its 'bugs', and even two identical machines each have their own peculiarities. For this reason it is necessary to have an 'intimate' knowledge of the machine.

It is characteristic for the skilled worker's *handling* of machines and of the way he works that he does not work 'on' the machine inasmuch as he sets certain work processes in motion or operates the machine. Instead, he works 'with' the machine as if it were a tool. For this way of working it is essential for the skilled worker to believe that the machine only executes the desired processes because he is operating and controlling it. The object of the work process is not the machine but the particular workpiece being processed by the machine. Correspondingly, the processes being executed by the machine become a part of the worker's own operations and are directly connected with them – the same as with a tool. Here are some typical statements to corroborate this: 'The machine is like a tool to me. After all, on their own, conventional machines cannot do anything at all, so the machine is like a tool'. Tools in this sense are seen as something belonging to one's person, representing a sort of extension of one's own 'organs'. In this connection it is essential to be 'able to handle' the machine. An important basis for this is manual control, i.e. control via knobs and handles. As one skilled worker expressed it: 'With the knob in your hand, you have the machine under control. I can feel what I'm doing in my hands'. By proceeding in this way, the worker is adding a subjective dimension to his execution of the machine operations. This means he has to 'get fully involved' with his machine – or in the words of one skilled worker: 'That is, so to speak, getting into the machine. It hurts when the machine is running badly'. Even though work with the machine proceeds according to a strict schedule, workers execute the individual machining operations step by step. Typical is a way of working which displays the characteristic dialogical or interactive action: each step in a process builds on the one before it by taking into account the preceding results. This means that in a sense the material's 'response' to a processing step is taken into account. Experimenting with the machine and thus gradually finding out its full capacity for performance plays an important part in this regard.

Feeling plays a part in this way of working not merely in the sense of affective satisfaction or job motivation. Instead, it is an important basis for perceiving, recognizing and assessing the properties of materials and machining operations and for regulating the execution of jobs. This is particularly clear in the case of the sensory perceptions described above. What is perceived and the way in which it is interpreted takes place to a large extent on the basis of feelings. Examples of this are to be found in such statements as: 'What you hear or don't hear depends on whether you have a feeling for it or not.' Similar statements are heard in the case of visual, optical perception: 'Everyone sees things differently and if someone doesn't have a feel for it, then he simply can't see it; it's a matter of feeling.' What is most significant here is the fact that 'intuitive perception and judgement are not in any way seen as being 'just a feeling' which would lead to uncertainty and only tentative assessments. On the contrary: it is precisely the intuitive assessments which lend certainty and confidence in handling both machines and materials. Therefore, feelings are assessed as a necessary prerequisite for carrying out a task in the same way as theoretical and expert knowledge. This becomes particularly clear when theoretical knowledge is compared with feelings. Seen from the point of view of the skilled worker, technicians and engineers often lack the right feeling: 'An engineer knows an awful lot, but he has no feeling for it. Only the skilled worker has that' – is one of many typical statements.

We should not jump to the conclusion that such a way of working can be explained by the skilled workers' inability to proceed according to the criteria and findings of technical, scientific rationality – in the sense that they lack the qualifications. On the contrary, it is clear that if they concentrate exclusively on such criteria, they would not be able to fulfil the tasks and challenges that face them. Points that should be stressed in particular are: peculiarities of the material, i.e. different material properties that are influenced by a number of factors which cannot each be identified and measured individually; variations within basically similar work processes as well as deviations from the norm; peculiarities of individual machines ('bugs') and the necessity to improvise and take into account the unpredictable, particularly with new products and when correcting errors.

Our empirical findings suggest that such ways of working are far more than a preliminary stage on the way towards scientific analytical thinking, or indeed, basically inferior to such thinking. They tend, rather, to be an independent and – in this sense – qualitatively different form of 'grasping' reality which is needed to complement the technical, scientific way of getting to the bottom of the production process.

USING NEW TECHNOLOGIES

The following sections will show in what way those elements of work based on subjectifying actions are effected when computer-based information and control technologies are introduced.

Our studies centred on activities at CNC-machines. According to our investigations, the most important technical and organizational changes when using CNC-machines as opposed to conventional production technology are:

1 Control over the machines ensues from a computer program by which the actual production runs and individual machining steps on the machines are directed. 'Manual control' is also changing as the result of electronic control equipment.
2 The creation of programs can be organized along various lines. Centralized programming in the planning or technical departments and the so-called 'shop-floor' programming directly at the machine roughly represent the two poles of work organization solutions.
3 Individual machines are more complex, i.e. several machining operations are being performed on single machines (especially machining centres).
4 Machines are increasingly enclosed.
5 As a rule the use of CNC-machines goes hand in hand with flexible assignment of personnel and a proliferation of shift work.

Our studies concentrated on activities that can be termed skilled work. For the most part the creation of programs in this area takes place centrally (i.e. not at the machine), but the skilled worker is still left with important functions to perform, such as: optimization of programs, i.e. their adaptation to the actual conditions of the machines; running in of the programs; inspecting and monitoring the performance of machining operations; correcting malfunctions. This is why activities of this type are also referred to as skilled work tasks with 'de facto competence to act'. Studies show that activities of this kind currently predominate in the production of small and medium-sized series.[17]

In existing studies in industrial sociology, these developments are assessed mainly positively, especially when viewed in contrast to a lowering of skill requirements toward semi-skilled production.[18] However, our investigations reveal that the subjectifying coming to terms with job requirements is affected in several – partially contradictory – ways by these developments. It is under these working conditions where you will find some major reasons for a number of new problems and risks in the work process. There are signs in particular that:

1 The subjectifying execution of job requirements is being repressed, undermined and impeded, yet at the same time

2 new requirements are arising for subjectifying work activities, but the necessary technical and organizational foundations are lacking.

These two developments are interactive in the actual work situation. After they are described briefly, attention will be drawn to the associated stresses and risks for workers.

Repression of subjectifying work-related action

The key to this change does not lie in particular factors, but is more of a syndrome, in which changes in the control system and in the outer appearance of the machine (notably its casing), the flexible assignment of personnel and greater integration of on-the-machine activity within the overall production sequence are mutually reinforcing. These developments are leading to a complex change in work overall.

Skilled workers and their superiors share more or less the same opinion with regard to the relationship to machines: 'Of course, there isn't the same kind of relationship with the new machines as there was with the conventional ones.' At the same time, it is always stressed that: 'Intimate knowledge of the machine isn't important anymore. You don't need to be intimate with the machine any more.' Some superiors and management representatives no longer think it desirable for skilled workers to have such a close relationship to the machine, because: 'It must be possible for the CNC-worker to rely fully on the proper functioning of the machine . . . What's important for the person at the machine is what comes out of the machine. He doesn't need to be interested in the machine's inner workings.' Flexible assignment of personnel is yet another obstacle to the development of a machine–worker relationship of the type described above.

Use of electronic controls is resulting in an important change for the handling of machines (way of working). The manual control that remains is restricted essentially to pressing a button or activating a switch to trigger the desired operations. The possibility of regulating machining operations directly (as the worker experiences it) is thus limited. Skilled workers emphasize that they have to rely on 'the machine doing the right thing', without being able to exert a direct influence on it. A typical comment in this regard is: 'You used to have the machine under control, especially when moving the spindle up to the workpiece. This used to be done by hand. With the CNC-machine the skilled worker has to rely on the machine moving up to the workpiece at high speed and stopping at the right point.' This is why infinitely variable speed control is considered a major advance in the design of control systems. The man at the machine, so it is stressed, 'has nothing left in his hand'. These changes are often illustrated by the example of driving a car. You have to imagine driving a car that is controlled by just buttons and switches and no steering-wheel. Then imagine the situation where you are sitting in such a car approaching a house at a speed of 100 km/h, and you have to rely on it stopping two meters in front of the house or reducing speed in good time to drive around a bend – whichever is programmed. Everything has to be planned in advance. In the actual execution of machining operations you have to 'trust' both the technology and the programs. On a CNC-machine it is no longer possible to perform the various machining operations 'step by step' and with each step succeeding from the last as with conventional machines. For workers the consequences are: 'It used to be possible to watch and make decisions during the machining operation with the intention of delaying tool wear by lowering the cutting rate. You can't do this any more. You have to decide and plan everything in advance. Direct reactions to changes in the work process are impossible. The most you can do is switch off prematurely.'

The change in sensory perception in light of these developments makes it clear that the loss of direct 'graspable' control and influence of the machine is a central point. One of the main differences between a CNC-machine and a conventional machine is emphasized thus: 'There aren't any handwheels any longer on a CNC-machine.' The important thing here is that the skilled worker used to be able to 'sense in his hand' what he was putting into effect. Even the visual perception of the machining operations is being obstructed and impeded. The main factors here are: higher speed; complexity of the machining operations, making it impossible to view everything 'simultaneously'; encasement of the machines, obstructing access to the machine and impairing visual contact. These changes are mutually dependent. High speeds, for example, necessitate the use of coolant, and lead – in the event of tool breakage – to a higher risk of accident. Therefore, the machines must be increasingly encased. But: 'One's view of the machining operation is reduced by the encasement.' In an extreme case this means: 'If you can't see anything any longer, you have to rely on the program, and that's very unsafe.' And finally, there is also impairment of the worker's acoustic perception. Certainly, the greater protection against noise and dirt provided by the encasement is viewed positively. But this also 'makes it more difficult to set up and run-in the machine, because you can no longer hear and see as well as you used to'.

For skilled workers, a major factor in the impairment of visual and acoustic perception is the (poor) accessibility to the machine. It is no longer possible, or at least not to the same extent as it was with a conventional machine, for a worker to arrange and alter his distance and angle of vision to the machine – to suit requirements – by moving the whole body. To use the skilled workers' own words, you can no longer 'go straight up to' or 'walk around the machine'.

Finally, the worker's intuitive judgement and discernment are impeded in various ways by these developments. Far more than on conventional machines it is necessary to set up the 'empirical processes' in objectifiable scientific and technical data and in (mathematical) relationships. The development of intuitive judgements is getting impaired to an even greater extent. This finds clear expression in a skilled worker's assertion that: 'In the long run, CNC-workers are losing their skills by working on machining centres. They no longer come into contact with the workpieces in these situations. Part of the worker's skill, namely his sense of precision, is vanishing.'

The need for subjectifying work-related action remains and is developing new forms

CNC-machines place requirements on workers, which – according to our analyses – they can only fulfil through subjectifying work action. In spite of the changes mentioned above, companies stress that the most crucial factor in the optimization of programs and the monitoring of machining operations is the skilled worker's special knowledge of the practical situation.[19] A machine company owner's comment is typical in this regard: 'The skilled worker continues to be important because of the unpredictables, it isn't possible to plan for everything. Nothing can be planned 100 per cent.' And this tallies with a skilled worker's assessments: 'The skilled worker's contribution is his knowledge of practicalities in the CNC field. The skilled worker contributes his practical experience as a supplement to the program during optimization. This is the point where the skilled worker is vital.' Skilled workers and their superiors also share the opinion that it is as important now as it ever was for the skilled worker to be able to 'hear' tool wear – when monitoring the machining operations – so as to be able to intervene in time or at least prevent consequential damage. The resultant situation is thus a contradictory one for skilled workers. On the one hand, the technical and organizational bases for subjectifying interaction with his work are impaired, but on the other hand, these actions are necessary at CNC-machines. In the actual work situation this means that skilled workers try to engage in subjectifying work actions – in a sense – against the flow of changes dismantling such actions. A few examples of this follow.

1 Skilled workers try to acquire 'trust in technology', even at CNC-machines. Here it is important – as is partially the case with some newer

control systems – for the speed to be directly regulatable when running in the machine and when optimizing the program. The following comment is typical in this regard: 'You feel safer with a switch you can keep pressed down than with one you just turn on and off. I know that what I'm doing at that moment, the machine is doing, too.' If this is so, you have the feeling – at least to a limited extent – of 'having a grip' on the machine, even if it is a CNC-machine.

2 To improve visual perception – especially during optimization and running in of programs – the worker will try to open the machine casing or enter into the encased machining operation.

3 One response to impeded visual monitoring is to shift to the sense of 'hearing'. In the words of one skilled worker: 'When you can't see and can't get a look inside, you've no option but to listen.' An assertion typical in this regard is: 'I monitor 60 per cent by ear and at most 30 per cent by sight. For the remaining 10 per cent you can't do anything.'

4 Skilled workers on CNC-machines develop an ability that may be termed 'abstract perceptiveness'. What this means is creating a mental image of processes and operations, independent from direct sensory perception. It is important when operating buttons and switches, for example, to imagine which operations are being triggered on the machines and what the machine is doing. To quote a skilled worker: 'Only when you have an idea of what the machine does will you be able to reconstruct the action.' Skilled workers describe it in these terms: 'You used to feel it in your hand when turning the crank. Today you have to sense whether the machine is doing something, without touching anything.' The same is done when checking the program: 'With complicated parts I drive the program step by step. I imagine in my mind what the machine is doing.'

5 Finally, skilled workers, when monitoring machining operations – in the so-called 'down times' – try to behave in a way that is not just 'passive and reactive'. Even when they seem to have 'nothing to do', they are trying to reconstruct the machining operations and 'stay on the ball'. Secondary activities such as reading the newspaper, doing crossword puzzles or chatting with colleagues are not a contradiction, but actually make possible or support the work. These secondary activities, in fact, create a situation in which the workers can 'immerse' themselves in their work and – as on conventional machines – 'be on the ball with all five senses'. A foreman's comment was: 'Reading a newspaper doesn't have a negative effect on work. On the contrary. People want to have something to do to stay attentive. It's funny and strange – but the work doesn't suffer from it.'

New types of strain

Skilled workers need to employ their 'old' skills on CNC-machines, too, but – as has been shown – their attempts to do so are simultaneously obstructed and impeded in numerous ways by changes in technology and work organization. Our findings show that it is precisely this contradictory situation which contains important catalysts for a number of new sources of strain and risk for workers.[20] Here, these can only be summarized briefly as follows:

1 Skilled workers feel overtaxed and under pressure, because on the one hand they are responsible for the machine, but on the other they 'don't have it under their control any longer'. It is notable that demands on the workers' responsibility are tending to increase rather than decrease due to the higher costs of the machine, increasing down-times and defective piece production. The following summary drawn by a foreman can be taken as typical of this situation: 'The greater stress arises because the skilled worker has more responsibility for the products than before, but is less and less able to influence and control the machine.'

2 Skilled workers are subjected to mental stress because they have to concentrate more and always keep their minds on their work. This stress does not arise because the workers are unable to concentrate or think in mathematical terms and programming rules. They are caused more by the fact that the skilled worker's task – as opposed, say, to a programmer – is to consider the practical conditions on the machine at the same time as the theoretical ones. For them it is not only important that a program be correct. It must also prove successful in actual practice on the machine. Even a perfect program can prove deficient if certain particulars of the machine and the material were not taken into account. Yet precisely this is what is difficult to calculate in advance, and is, in any case, possible only to a limited degree. The skilled worker is under pressure, therefore, always to think of 'what might happen.' and 'whether they've thought of everything.' He is never certain of himself because his questions cannot be answered until the machine is running. And in most cases it is no longer possible to prevent the defects that then arise. This uncertainty means that the skilled worker is unable to 'switch off' in his spare time. To quote one skilled worker: 'You have to have everything in your head. And if you're having to think all the time whether everything's o.k., whether you've made a mistake or forgotten something, you simply can't switch off after work. You keep running mentally in top gear – and that's stress.'

If skilled workers try to transfer the working methods customary on conventional machines to a CNC-machine, they will feel more secure on the one hand, but they will also end up with new problems. If the casing is opened when the machine is being run, or if the workers enter the machining space, there is a greater risk of accident. If they try to be guided by the noise of the machine, this is all the more difficult because of the overall higher noise level – not to mention the effect of removing the machine casing – and requires a higher level of 'concentration'. Attempts to listen to the 'right' noise results in a higher sensitivity to other sources of noise and thus to higher stress on the psyche and nerves.

If skilled workers try to fill their down times with secondary activities, they find themselves in conflict with job discipline. There is also the risk that these secondary activities will be used as a reason to give them extra jobs to do (especially operating several machines). 'Side jobs' of this type, however, make it difficult to have the necessary attentiveness. It is possible simultaneously to monitor a machine by ear and read a newspaper, but not both to monitor one machine and optimize another. Workers are forced, therefore, either to forgo secondary activities, such as reading a newspaper, or else to keep them secret. In the words of one skilled worker: 'The skilled worker used to be on his toes. Today he just needs to pretend to be on his toes.

Problems arise even when skilled workers adapt to the new situation. For workers this means they have to reconcile themselves to no longer being able to influence and control the new machines in the same way as conventional machines. This means you just have to 'grin and bear the increasing problems'. More than anything else this entails 'having to become thicker skinned and trying "not to be so sensitive about the work any more"' – in either the positive or negative sense. But not only does this make the work less interesting and satisfying, skilled workers also run the risk of being considered unreliable, irresponsible, disinterested and unskilled. In other words, they are putting into question precisely those qualifications and working practices on which a company's interest in the use of skilled workers is essentially based.

Finally, there is the risk when working on CNC-machines of forgetting or of no longer being able to learn and develop that special 'empirical knowledge' of the skilled worker. The prevailing opinion is: 'You can't learn these essential basics on a CNC-machine. If you had only CNC-machines – it wouldn't work out.' In the skilled worker's eyes, 'Work on conventional machines is important for getting a feel for what a machine does and can do.' It is, above all, the experience which is acquired on conventional machines and transferred to the CNC-machine that plays an important part in precisely this capacity for 'abstract perceptiveness'. But how is such 'empirical knowledge' to be acquired in future if production –

particularly in conjunction with a more extensive interlinkage of data systems (CIM etc.) – is converted increasingly to CNC-machines. This is a problem that is just beginning to emerge, and one which appears to be intensifying.

CONSEQUENCES FOR FUTURE ANALYSES OF WORK AND OPEN QUESTIONS

It is a common opinion that the problem outlined in this paper is only a question of short-term adaptation and transition. Our findings, however, do not speak for such an assessment. More investigations have to be carried out to see to what extent such developments occur in other areas of production and whether the results of this study can be generally applied to the use of information and control technologies.

On the basis of our results up to now it is clear that the repression of subjectifying action as well as the contradictory demands made on workers have to be investigated more closely. The question involves developments which cannot merely be limited to dequalification as in the case of Tayloristic work forms. The greater weight placed on objectifying actions at work could result in a maintenance, and even the creation of 'qualified tasks', as well as an expansion in the scope of job discretion and decision-making for workers. The main issue here, however, is the qualitative changes in the qualifications necessary in the work process. Therefore, it does not make any sense to put these changes in categories such as 'more – less' or 'higher – lower'. For future investigations it is necessary to clarify in what ways the skilled worker tasks described here appear in other areas of industrial production, especially in process industries.

The repression of subjectifying actions does not lose any of its power just because in practice it turns out that such work actions are still necessary. On the contrary, as our findings show, even though 'empirical knowledge' is expected from workers as before, the foundations and preconditions for the necessary (subjectifying) work actions are endangered. The design and development of the technical system are overwhelmingly oriented to the objectification of work actions.

Our investigations demonstrate that subjectifying actions at work are an important part of what the worker can contribute to the work process. A central question surrounding this issue is what effects, both within and outside of the company, can be expected if the development and application of subjectifying work actions are endangered. Future studies will have to clarify not only how this leads to new problems in the production process, but also what deep-seated changes could occur in dealing with life outside of the workplace.

In terms of the design of technology and work organization, we are not trying to portray work forms for the skilled worker on conventional machines nostalgically. It is impossible to ignore the existing strains and restrictions that have characterized conventional tasks in industrial production up to now. The question is not one of a return to 'old' technology. The decisive point is to discover whether it would be possible and which practical approaches might exist to set up future developments in technology and organization systematically so as to facilitate the use of subjectifying action at work.

Social scientists have the challenge to take the analyses of work outlined here and develop them further both at a theoretical and conceptual level as well as at an empirical one. There is no doubt that this research issue is one which extends beyond the boundaries of industrial sociology. This topic can only be dealt with 'collectively' through interdisciplinary research. We hope that the analysis presented here will act as a stimulus for this future research.

Part III

CIM and Organizational Change in the Mechanical Engineering Industries

9 Diffusion of CIM-Technologies – Dynamic Dissemination and Alternative Paths of Innovation[1]

Rainer Schultz-Wild

INTRODUCTION

The CIM trend appears to be well under way. Discussions and debates on the 'factory of the future' have given rise to speculations concerning the present use of computer-integrated automation technology which are, in many respects, exaggerated and overdrawn. A survey in the West German capital goods industry reveals that there is still a long way to go towards the realization of fully automated and computer-integrated manufacturing. However, we are currently experiencing a dynamic diffusion of CIM-components, but without the emergence of a uniform model of future factory and work structures. Therefore the effects accompanying the application of this type of technology reported by companies are varied and partly contradictory.

TOWARDS THE FACTORY OF THE FUTURE: A NEW TYPE OF RATIONALIZATION

Discussions on the 'factory of the future' refer almost exclusively to the new technologies of computer-integrated manufacturing (CIM). One major approach within the field of engineering sciences is that material technology and information technology will merge to form a new type of production technology, namely 'an information machine oriented to production technology, or a production machine with integrated information technology' (Spur 1986: 7). Such perspectives still reflect the old ideal upheld by engineering sciences of the fully automated factory. It might appear that there were few alternatives to this development.

In the 1970s initial attempts were made by pioneering companies to extend the integration of computer and control systems which had been utilized up until then in a more selective and insular manner. In the early 1980s such projects not only gained significance in engineering science circles, but also in industrial practice. In view of the success in innovation and rationalization being reported on all sides,[2] one might wonder whether a company without CIM-technologies could survive.

Examples of this can be seen in the introduction of DNC systems or the integration of CNC-machines with transport and information systems to form flexible manufacturing cells and flexible manufacturing systems.[3] But above all it is the functional expansion of production planning and control systems (PPC) as well as computer integration processes, starting with the design sector and developing in the direction of process planning and machine control (CAD/CAM), which seem to be playing an increasingly important role (see figure 9.1).[4]

Thus the outlines of a new kind of rationalization process are currently being observed. This pattern can be termed 'systemic rationalization' (see chapter 4) because the implementation of CIM components and CIM systems encompass individual job shops and departments and tend to be oriented to the company as a whole.

In view of the significance which these new innovation and rationalization patterns are thought to hold for economic and industrial development, it is surprising how little is actually known about the ways in which the companies are approaching computer integration, at what speed it is being accomplished and what the effects are on the design and structure of production tasks and work organization.

CIM-COMPONENTS LESS WIDESPREAD THAN EXPECTED

The results of our survey of about 1,100 companies of the West German capital goods industry (with the machine-building, electrical and automobile branches as its largest subgroups) provide some information on the following issues: the present diffusion of computer systems used in manufacturing processes and related services such as planning, control and monitoring; the extent to which company functions have been integrated by computers; and the speed of diffusion processes in this field of innovation.[5]

The use of modern, computer-based technology is not as widespread as one might assume from the ongoing discussion among engineering scientists or the information provided by hardware and software manufacturers, for example at trade fairs (see figure 9.2).

1 While most companies utilize computers in one form or another, they are only widespread (more than 70 per cent) in administrative areas such as financial and general accounting and payroll accounting, i.e. fields in which computer systems have been developed and available on the market for a long time.

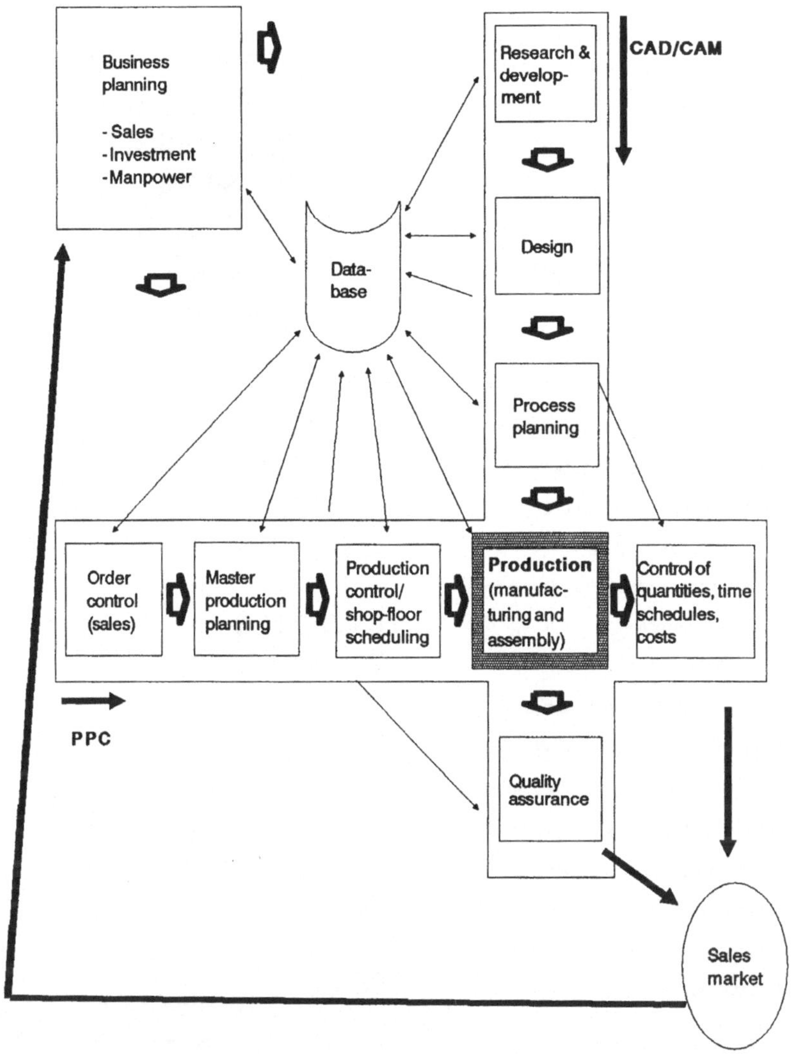

Figure 9.1 Computer-integrated manufacturing (CIM). ISF 1991

2 CNC-machine tools and other numerically controlled machines are fairly widespread (in about half of all companies and in approximately two-thirds of the machine-building companies). However, less than 4 per cent of companies have installed more modern and complex machine and control systems, namely flexible manufacturing cells (FMC), flexible manufacturing systems (FMS) or DNC-systems.

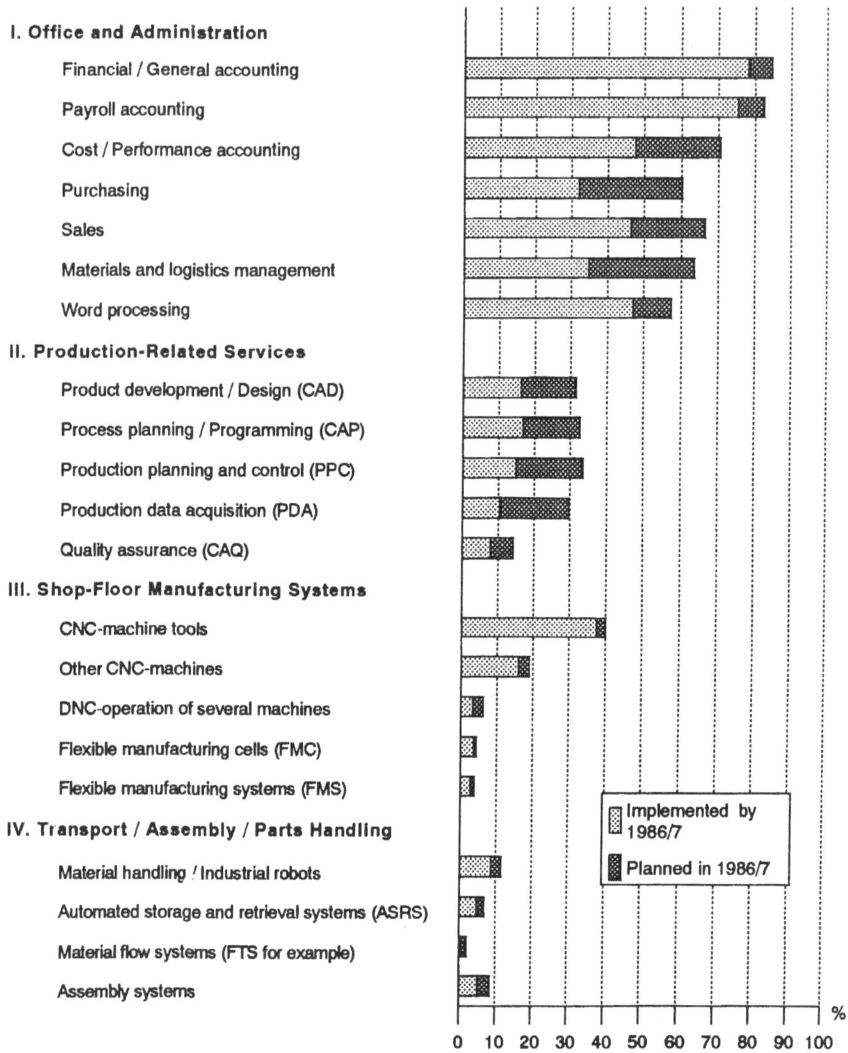

Figure 9.2 Computer-based technologies – 1986/7 implemented or planned – utilization rates in the West German capital goods industry (N = 1,096). ISF 1991

3 Other CIM-components, such as industrial robots/automatic material handling systems (8.6 per cent), computer-based assembly systems (5.1 per cent), storage (4.8 per cent) and material flow systems (0.7 per cent) still have rather low diffusion rates (in terms of the application of at least one system per company).

Less than one-fifth of the companies currently use computers in areas which can be termed collectively as production-related services, or for company functions such as planning, control and monitoring, which in many cases are regarded as being the key functions of future CIM-structures. A total of 15 per cent to 17 per cent of all companies employ PPC-,[6] CAP-[7] and CAD-[8]systems, while PDA-[9]systems are used in 10 per cent of the companies and only 8 per cent use computer-aided quality assurance (CAQ).

However, the application of computers and CIM-components is increasing in some administrative areas as well as in production-related services: according to the planning intentions of the companies, diffusion rates in areas such as design, process planning, manufacturing scheduling etc. will approximately double by 1990, which means that about one-third of the companies can be expected to fulfil these functions with the help of computers by the beginning of the 1990s.

Regarding all functions, computer applications are more widespread in large companies than in small ones. This implies that relatively larger numbers of employees are potentially affected by these new technologies than would be indicated by the company diffusion rates. However, an increasing number of medium- and even small-sized companies are about to introduce new technologies such as PPC-systems. Viewed from this perspective, there will be an increase in the use of computer technology starting with large companies, and extending through the medium-sized companies right down to small companies. Thus, for example, there is a remarkably high rate of medium-sized companies (with 100 to 500 employees) planning to venture into computer-based process planning/programming (CAP, see figure 9.3a) or production planning and control (PPC, see figure 9.3b) from 1988 to 1990.

Computer integration, in the sense of on-line computer networks connecting various functional areas of a given company, is just beginning to spread. By 1986–7 only 9 per cent of the companies had established at least one integration link between different functions. An increased pace of development can be expected in this area also. More than one-fifth (23 per cent) of the companies are engaged in establishing computer integration links such as CAP/PPC, CAD/PPC and PPC/CAM (see figure 9.4).

DYNAMIC DIFFUSION RATES IN THE MACHINE-BUILDING SECTOR

Although discussions on computer integrated manufacturing are by no means restricted to the machine-building industry, this sector does occupy a central position with regards to development and considerations

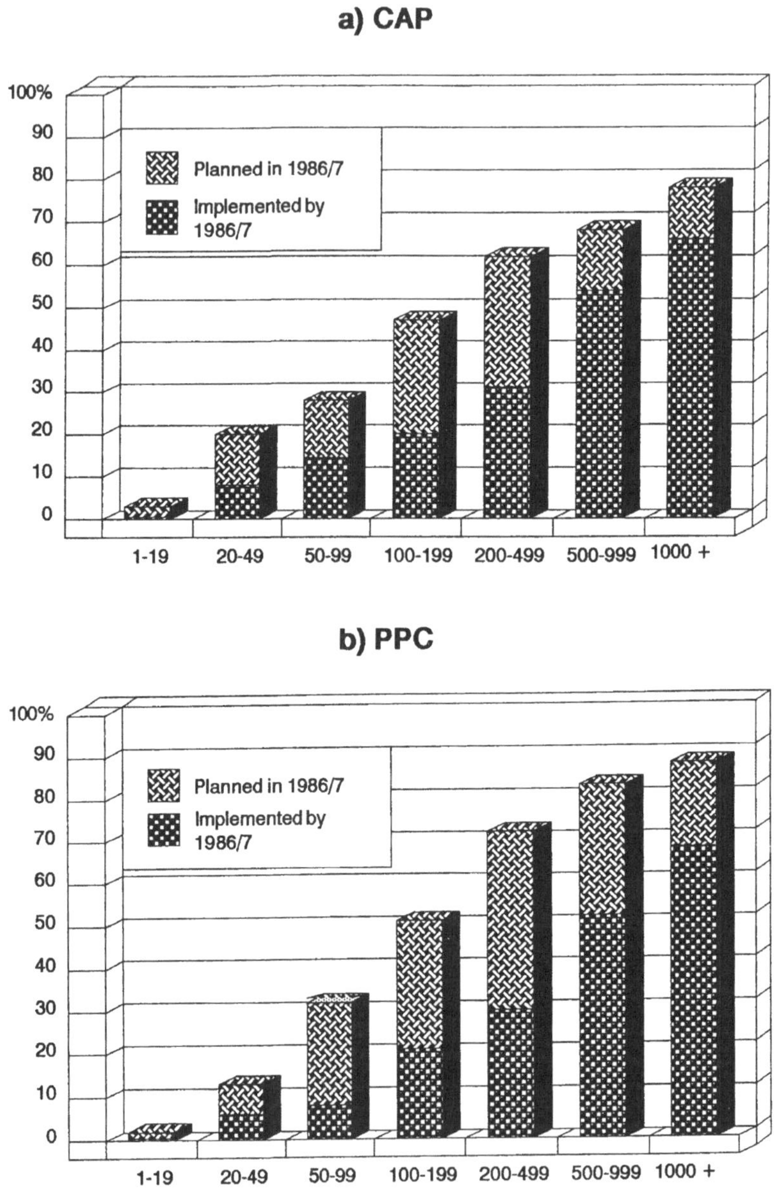

Figure 9.3 CAP and PPC by company size (number of workers) – utilization rates in the West German capital goods industry (N = 1,096). ISF 1991

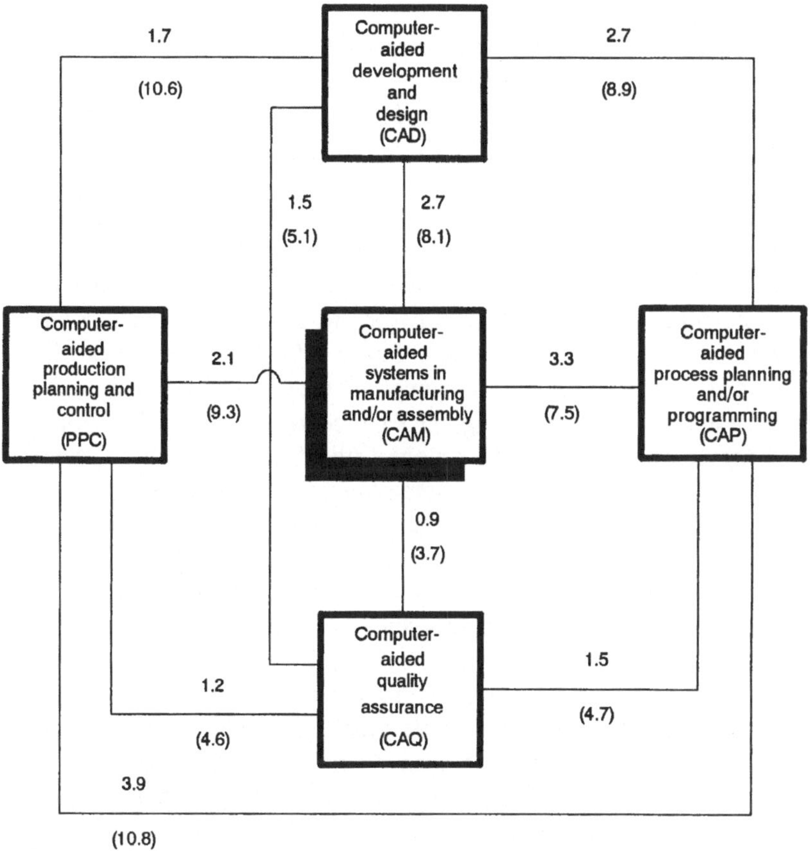

1986/7 at least one computer link implemented: 9.0% of companies.
1986/7 at least one computer link implemented or planned: 22.9% of companies.

The figures refer to the number of companies (in %) with computer integration between two functional areas, independent of whatever other integration lines may exist; the percentages of companies planning such integration is stated in brackets.

Figure 9.4 Computer integration – in 1986/7 implemented or planned – West German capital goods industry (N = 1,096). ISF 1991

concerning future production structures. Apart from the significance of this sector within German industry, this is also because the demand made for flexibility in scheduling and production, and in manufacturing quality, is particularly high. At the same time this sector is significant in terms of labour policies because production structures which require a high percentage of skilled workers still play an important role within the often smaller or medium-sized machine-building companies.[10] By contrast, big companies in the automotive and electrical industry are more strongly oriented towards large-scale or mass production and employ larger shares of unskilled or semi-skilled labour.

To what extent do diffusion rates of computer and control technologies in mechanical engineering differ from the capital goods industry as a whole?

The general pattern of computer diffusion is similar to that of the capital goods industry: mainly administrative functions are being fulfiled with the help of computer technology. A smaller number of companies have computerized production related services or are utilizing modern, complex machine and transport systems (see figure 9.5). Thus more than three-quarters of machine-building companies are currently using computers for financial and general accounting (77.7 per cent) or for payroll accounting (79.3 per cent), while only about one-fifth of the companies are using computers for process planning/programming (20.6 per cent), engineering/design (18.3 per cent) and/or production planning and control (18 per cent).

The use of CNC-machine tools and other numerical-controlled machines is far more widespread in mechanical engineering than in the capital goods industry in general: approximately two-thirds of the machine-building companies (as opposed to less than half of all the other companies) have already had some experience with this technology; rising growth rates through new first-time users cannot be expected in this sector.[11]

While in 1986–7 the diffusion rates of computer systems in production-related services in the mechanical engineering industry were only slightly higher than the overall rates, the number of companies in this branch planning to implement such systems for the first time is well above average. According to company plans, there will be relatively more users in this sector than in the rest of the capital goods industry within a short period of time, particularly in the areas of process planning/programming (CAP – approximately 43 per cent), production planning and control (PPC – 43 per cent), design and engineering (CAD – about 39 per cent) and production data acquisition (PDA – approximately 34 per cent). It would appear that machine-building companies expect that the new products for these types of control technologies or control systems will provide attractive solutions to their problems.

Machine-building companies are increasingly attracted by computer-based systems in production planning and control functions. This is shown by an analysis of diffusion according to company size. Provided the planned projects are carried out, about nine out of ten machine-building companies with 500 employees upwards will be using computer systems in the areas of CAD, CAP and PPC by the beginning of the 1990s (see figures

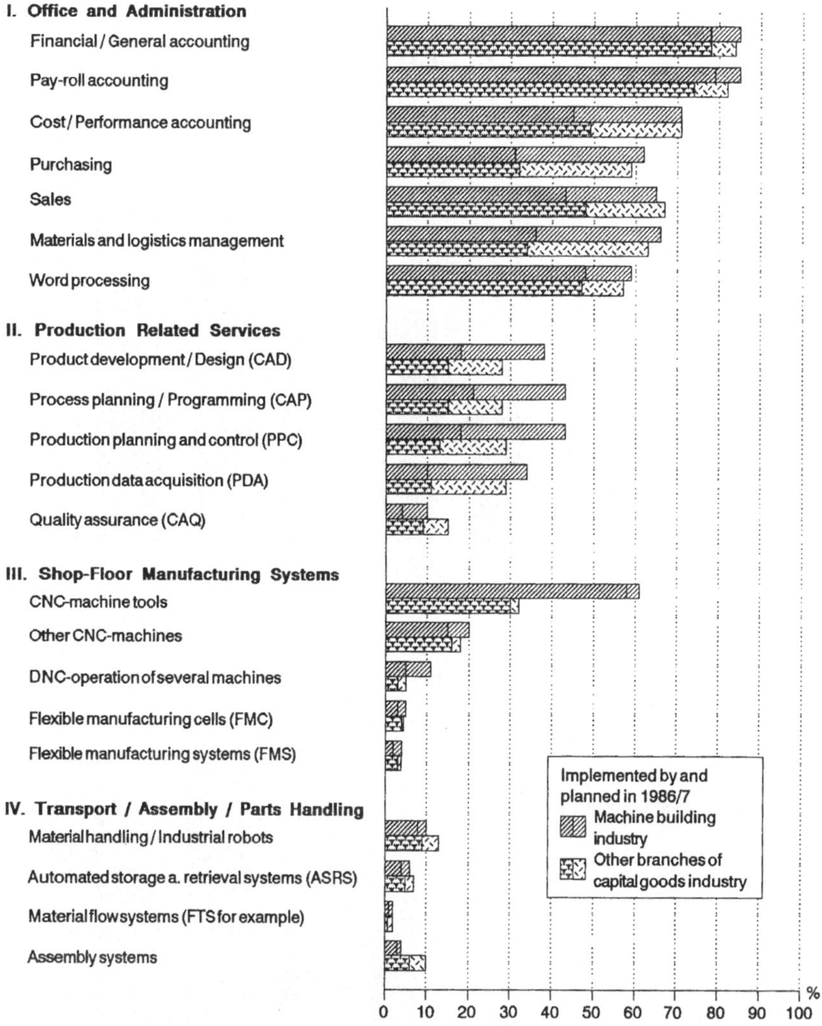

Figure 9.5 Utilization rates of computer-based technologies in machine-building industry (N = 572) and other branches of the capital goods industry (N = 524). ISF 1991

9.6 and 9.7(a)). In companies with 100 to 500 employees, two-thirds are expected to use computers in the PPC area in the near future. The same also holds for CAD and CAP systems as well as for the application of PDA. Even in companies with less than 100 employees, increases of between 20 and 40 per cent in the rates of computer application are anticipated. However, the introduction of computer-based quality assurance systems

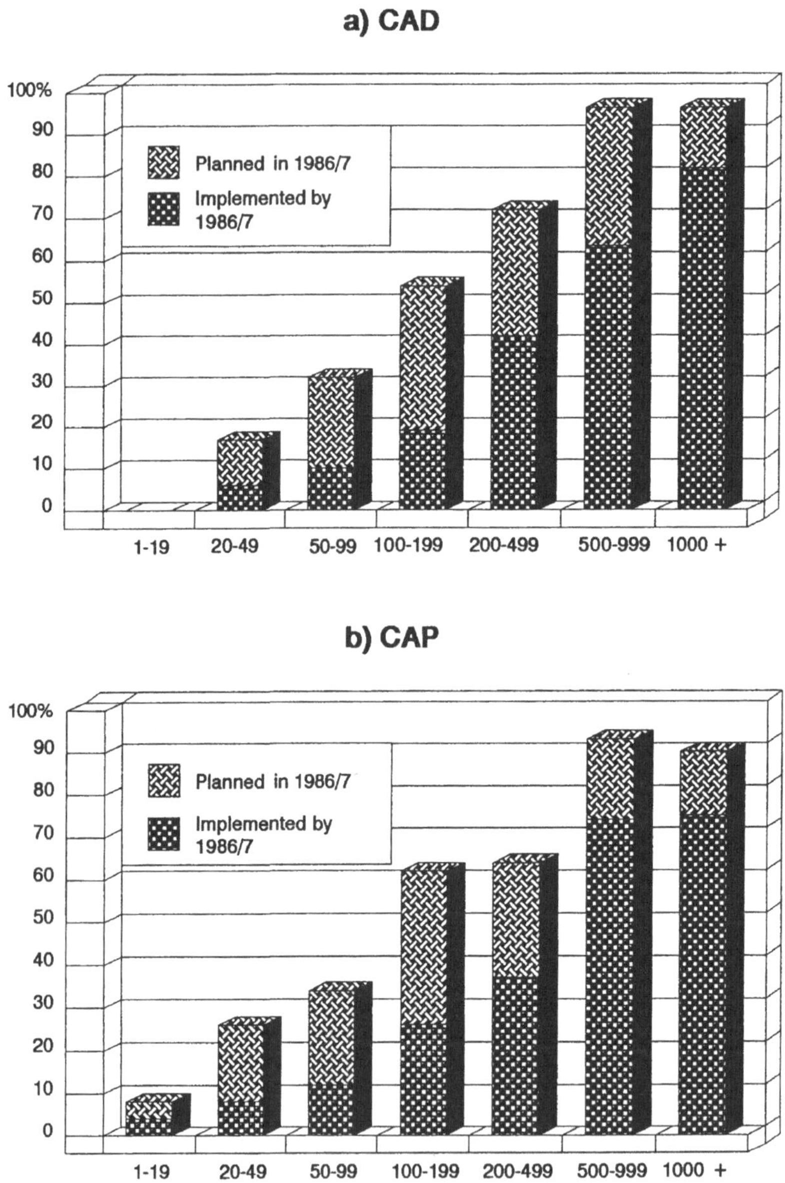

Figure 9.6 CAD and CAP by company size (number of workers) utilization rates in the West German machine-building industry (N = 572). ISF 1991

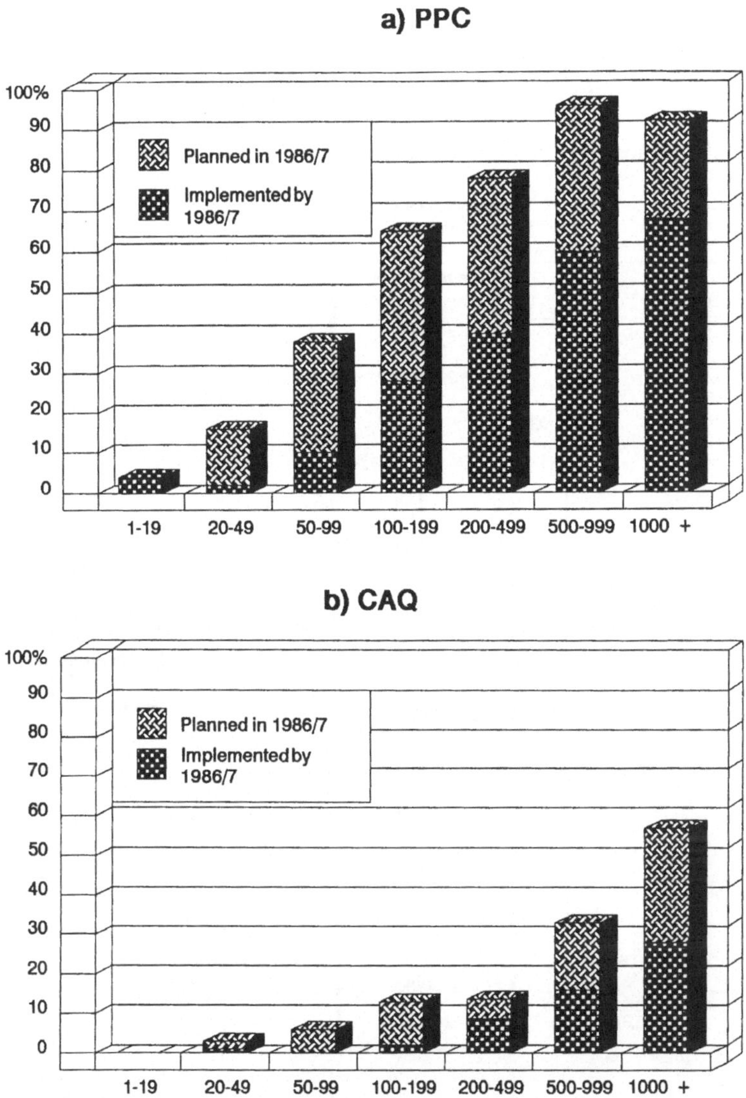

Figure 9.7 PPC and CAQ by company size (number of workers) – utilization rates in the West German machine-building industry (N = 572). ISF 1991

(CAQ, see figure 9.7(b)) is apparently of lesser interest for the majority of machine-building companies.[12]

An above average rate of computer integration of production related functions is notable in mechanical engineering.[13] A total of 12 per cent have realized at least one internal network (as opposed to the 9 per cent average in the capital goods industry), whereas 31 per cent are planning *or* realizing an on-line network (as opposed to an average of 23 per cent). Also many machine-building companies are investing in these computer-integration technologies for the first time (compare Schultz-Wild *et al.* 1989: 69).

The types of integration lines which have been established in mechanical engineering so far are mainly between the functions CAP/CAM

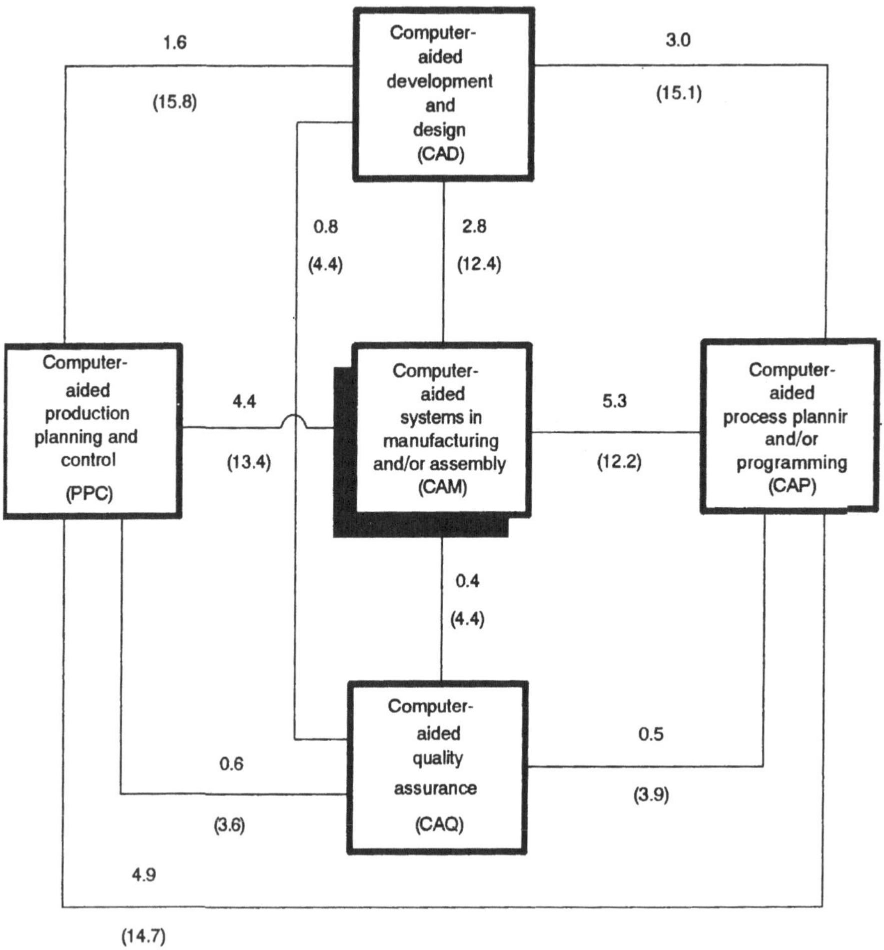

1986/7 at least one computer link implemented: 12.1% of machine building companies.
1986/7 at least one computer link implemented or planned: 31.2% of machine building companies.

The figures refer to the number of companies (in %) with computer integration between two functional areas, independent of whatever other integration lines may exist; the percentages of companies planning such integration is stated in brackets.

Figure 9.8 Computer integration – in 1986/7 implemented or planned – West German machine-building industry (N = 572). ISF 1991

(5.3 per cent), PPC/CAP (4.9 per cent), PPC/CAM (4.4 per cent), CAD/CAP (3 per cent) and CAD/CAM (2.8 per cent, see figure 9.8). As far as the types of networks planned are concerned, the most favoured are CAD/PPC (15 per cent), CAD/CAP (15 per cent) and PPC/CAP (14.7 per cent). Vertical and horizontal types of networks are planned equally often whereas the integration of quality assurance systems still lags a long way behind.

These survey findings indicate clearly that, for the entire capital goods industry as well as for mechanical engineering, while the process of factory automation being discussed under the term CIM is still in its infancy, there are changes under way in precisely those areas which are being focused on in labour policy debates concerning 'new production concepts' (Kern, Schumann 1984; chapter 1 of this volume), i.e. (re-)skilling of industrial work, etc.

ALTERNATIVE PATHS OF INNOVATION

More comprehensive and detailed analyses of company innovation processes reveal, contrary to the impression often gained from engineering science circles, that in mechanical manufacturing of small- and medium-sized series, no uniform model of technology application and computer-based integration is evident (Hirsch-Kreinsen, Schultz-Wild 1990, 1990a). Technical CIM components or systems and developments observed in user companies can be differentiated according to the main points of departure of innovation policies and the work concepts inherent in technology.

The predominant points of departure and prime objectives of introducing computer-based integration technologies differ according to criteria, such as:

1 more office-oriented versus workshop-oriented concepts of technology and innovation;
2 the predominance of 'horizontal' integration strategies centred on PPC-functions versus the option of 'vertical' integration oriented toward CAD/CAM-networks;
3 the prevalence of technological data processing versus business data processing in the company implementation and rationalization process; and
4 the scope and degree of the desired computer-networking within and beyond the company.

The influence technological systems available on the market exert on the design of company and work organization, and the use of skills, particularly on the shop-floor, differ according to two fundamental concepts in technology:

1 Most technology systems and components contain implicitly defined 'Tayloristic'-oriented design rules which are 'ready wired' and exert a strong influence in the direction of centralized planning, controlling and supervising functions, increased division of labour and – at least in the long run – an erosion of skilled production work on the shop-floor.
2 Apart from these standard products on the market, there are technological concepts which cannot be said to 'contain' alternative, skill-oriented and more comprehensive forms of work, but can be implemented with such a degree of versatility in job design and work organization, that they allow the user company to either pursue the course of computer-based Taylorism or to form structures for skilled production work.

Differences of this type have long been observed and discussed in the context of NC technology (compare chapter 5). They become even more important with the development of computer integrated systems. Various CIM-components can serve as examples for these alternative concepts of technology[14] (see figure 9.9).

Systems of *production planning and control* (PPC) directed at an overall regulation of complex manufacturing processes in terms of time schedules, production flow, etc., can be designed in different ways.

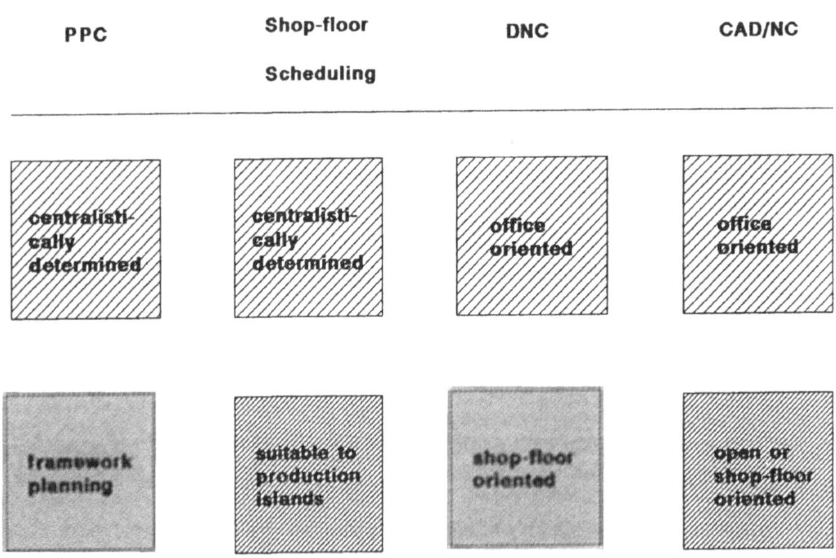

Figure 9.9 Alternative organizational concepts inherent in CIM-components. ISF 1991

Many well established PPC-systems are designed to achieve detailed *total planning* of all work processes. On the basis of complex algorithms and comprehensive production data acquisition, such systems are oriented towards establishing the most exact planning, control and monitoring of the work process possible. In such cases there remains very little scope for making decisions on scheduling, job sequences and capacity on the shop-floor. Within such PPC-systems, production workers are supposed to refrain from handling any orders which have not been pre-planned in detail and which cannot be controlled and monitored from a superordinate position. The functions of planning and coordination which were formerly carried out by personnel, have now been rendered into an 'objective' form within the computer system. Such centralistic PPC-systems have long been offered on the technology market, as well as used in many companies. Primarily by transforming certain components into modular units, such systems are being further developed toward achieving a greater degree of flexibility, while retaining their inherent concept of work organization.

By comparison, PPC concepts which only provide for *framework planning* (Manske 1987) are far better suited for supporting skill-based forms of production work. Systems of this kind dispense with precise and comprehensive pre-planning; instead they intentionally provide 'loopholes' for planning and scheduling at later stages of the process. In this case the system remains restricted to roughly pre-calculating and stating certain essential and intermediate dates for defined steps of the manufacturing process, and also periodically assembles a pool of orders for defined production areas. Detailed planning, i.e. the exact scheduling of order sequences, the coordination between different machining stations, service functions etc., can be carried out in a number of different ways by shop-floor workers. Thus the design of the system technology fulfils the essential precondition for integrating important planning functions such as scheduling, capacity planning etc., into shop-floor operations, which in turn forms a crucial aspect of skill-based production work.

In the Federal Republic, such alternative PPC concepts with a modular design have been developed for some years – not least of which is due to government support. These concepts display a high degree of flexibility in regard to work organization and are suitable for various forms of framework planning.

In the last years an increased number of systems for *shop-floor scheduling* have been making their appearance on the technology market – either as independent systems or as modules of PPC-systems. The inherent concepts of such components are strongly oriented to centralistic PPC systems in terms of their hardware and software design, so that they represent little more than modules within total planning systems which have been made somewhat more flexible. Due to their complex hardware, a range of functions which are to a large extent automated, and oriented to man–machine interfaces which in many cases do not meet shop-floor workers' needs, the use of these systems depends on specialized personnel in an office-like environment. Particularly in connection with comprehensive and detailed production data acquisition systems, these systems amount to a *centralistic type* of shop-floor control. Compared with traditional systems of total planning, they offer considerable advantages in terms of flexibility, and thus efficiency, yet allow shop-floor personnel little scope for further planning.

As opposed to these concepts, there are recently systems on the market and partially in use which support decentralized decision making. They display a type of operator logic similar to the conventional planning board (Jackson, Browne 1989; Hirsch-Kreinsen *et al.* 1990a). This type of 'shop-floor control centre' (*Leitstand*) is well suited for use in job shops and specially in production islands due to simple hardware, its limited automation of planning functions and its potential for interactive planning along a number of criteria. Particularly due to their specific, object-oriented man–machine interface, these control centres can be used by shop-floor workers as computer-based information aids in decisions concerning scheduling, work sequences, machine-operation planning etc.

DNC-systems, which essentially allow the computer-based control of several NC-machines, also display some widely differing concepts. There are a number of systems based on the integration of *office-oriented* NC-programming systems with a DNC computer and NC-machine tools on the shop-floor. Together with a more or less automatic and often integrated machine or production data-acquisition system (PDA), these systems aim to to make centralistic NC-organization more efficient to cope with some of the shortcomings it has in many cases. Therefore these systems would oppose the formal transfer of programming functions to the shop-floor. Moreover, by achieving more rigid formal organization and improving pre-programming, these systems seek to reduce the necessity for improvization on the shop-floor, thus restricting the scope for planning at this level.

For some years there have been a number of DNC-systems developed which are open with regard to work organization and therefore are compatible with a retention or increase of *shop-floor programming*. Their main design characteristic consists of a NC-programming system which is neither dependent on an office environment nor on the qualifications of specialized technicians. Here we find programming systems in various versions with forms of operator logic and hardware design which, associated with the appropriate configuration of other system components, allows shop-floor operation. Together with the interaction with CNC-machines which can be programmed on the shop-floor, these concepts form the technical preconditions for flexible, shop-floor-oriented DNC operations.

Finally, the interlinkage of *CAD-sytems* in the design department and NC-programming systems to form CAD/NC-*systems*, is recently a rapidly spreading CIM component. As far as can be presently estimated, almost all CAD-NC concepts on the market today are *office-oriented* and amount to increased division of labour between planning and programming on the one hand, while restricting shop-floor workers to operative functions on the other. The reason for this is that the possibilities of interlinking CAD-systems in various forms generally only encompass NC-programming systems, which suggest office oriented operation for reasons of system costs as well as those of information technology. These systems tend to transfer programming functions to the design departments.

There are a number of *open concepts* emerging, which avoid such effects and also allow *shop-floor programming* under the conditions of CAD/NC integration. First of all there is an increase in systems being offered which are simpler in terms of hardware and functions. These CAD/NC-systems are available in the form of compact work-stations which can be utilized in close vicinity of machine operation on the shop-floor, even in small companies. Secondly, there are concepts presently in the phase of development which involve linking CAD-systems directly to CNC-machine tools (Beck *et al.* 1990; Hirsch-Kreinsen *et al.* 1990a). On the basis of the input of geometrical data from the CAD-system into a machine control unit programming will be carried out exclusively on the shop-floor. The successful development of such CAD/CNC-systems to the stage where they are ready to go into operation would thus represent a major technical precondition for a shop-floor-oriented utilization of CAD data.

While the various types of CIM components available on the technology market and in operation in user companies do not necessarily *determine* the actual forms of work organization and work design, they must be regarded as important factors which may either facilitate or obstruct certain forms of work organization and concepts of production work.[15]

A MULTITUDE OF RESULTING EFFECTS

In view of the many different technological solutions on the market and the varying objectives and forms of innovation policies pursued by companies, it is no surprise that management reports manifold and diverse effects on company and work organization, as well as on manpower issues. The five points below summarize managers' judgements regarding existing or expected effects associated with the introduction of computer-based technologies in the capital goods industry. The percentages (see figure 9.10) refer to number of companies who responded positively to the appearance or anticipated appearance of each effect. According to the survey responses, most companies which use computer-based technologies experience considerable change: the percentage witnessing no effects (6 per cent) is relatively low.

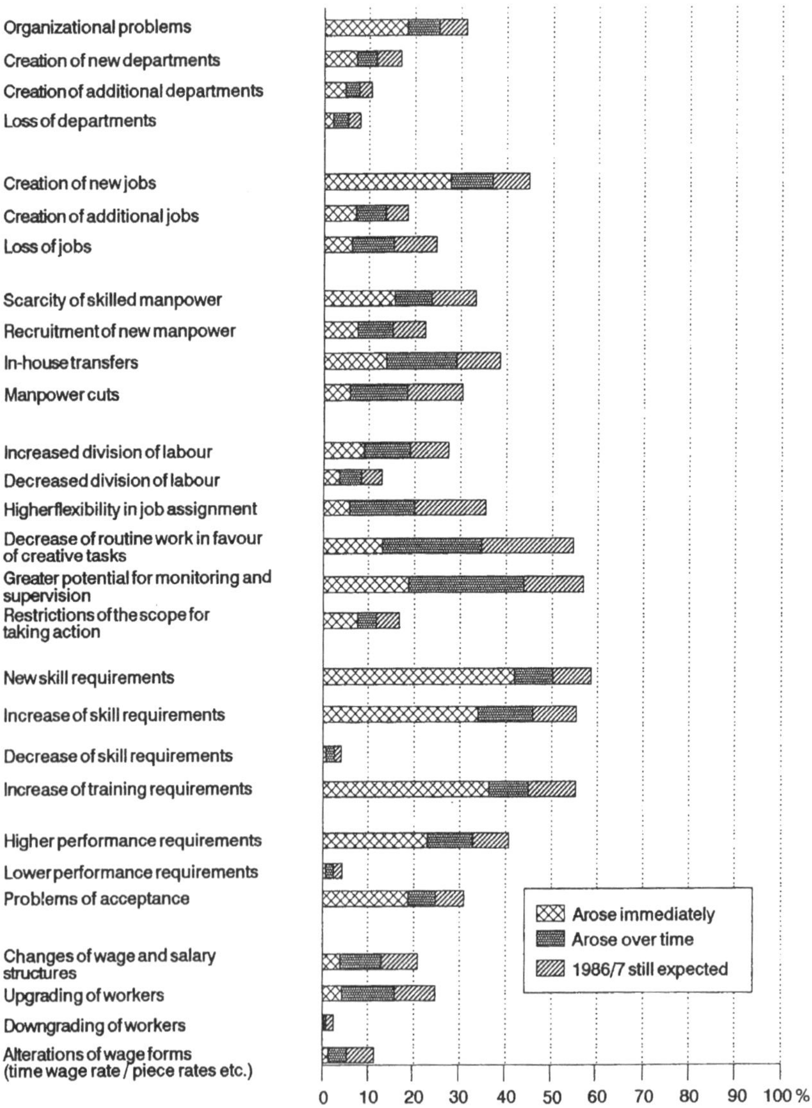

Figure 9.10 Effects of the implementation of computer-based technologies – West German capital goods industry (N = 1,096). ISF 1991

1 The company's organization often needs to be altered. Approximately one-third of all companies experience organizational problems, while a similar number report the creation of new or additional departments. In nearly one out of ten companies, former departments are dispensed with.
2 In many companies individual jobs are changed; in many cases there are reports of new (45 per cent) or additional (18 per cent) jobs being created, where as job losses occur in approximately one quarter of cases.
3 Division of labour has increased in 27 per cent of the companies, whereas 13 per cent have recorded a reduction.
4 The effects on employees also vary. The percentage of companies in which routine work has been replaced or is expected to be replaced by more creative tasks as a result of implementing new technologies (55 per cent) is about as high as the percentage of companies in which the potential for monitoring and supervision has increased (57 per cent).
5 Changes in skill requirements are apparent in many cases, with reports of new or increased skill requirements (59 per cent and 56 per cent, respectively), far outnumbering those of reduced requirements (2.5 per cent).

These few examples suggest that the changes taking place in the context of CIM implementation cannot be adequately grasped by concentrating exclusively on the technological-managerial problems of investment, and the choice of hardware and software. The emerging, dynamic dissemination of new technology, especially in the economically important machine building industry, makes it necessary to arrive at a better understanding of the related risks and opportunities for industrial work.[16]

These results illustrate that there is at present no uniform structural pattern of computer-integrated manufacturing and no genuinely predominant image of the 'factory of the future' emerging either from the few solutions realized thus far or the larger number of planned ventures into computer-based integration. By the end of the 1980s, the process of application of CIM-systems in West German industries had not yet progressed very far. But a remarkable tendency towards dynamic diffusion, particularly in medium and small sized metalworking companies, is being observed. As far as the restructuring of industrial work goes, there is also no clearly pre-dominant concept. Efforts to make traditional Tayloristic forms of work organization more efficient by means of new computer-based technologies is the path still taken by many companies, while others are seeking ways and means of maintaining or revitalizing forms of skilled production work on the shop-floor.

10 Technological Innovation – Organizational Conservatism?[1]

Christoph Köhler, Klaus Schmierl

INTRODUCTION

During the post-war period, German industry has been dominated by Tayloristic rationalization strategies. However, this rationalization strategy has not asserted itself equally in all branches of industry. In the field of machine-building, for instance, complex products, small batch sizes and complicated production sequences, together with constantly changing market requirements for company flexibility and quality, have created obstacles to imposing an extreme separation of conception and execution. Nevertheless, even in this industry the last 40 years can be described as a process of progressive standardization and specialization of functions and skills.

This chapter investigates the ways in which work organization in the West German capital goods industry is changing with the utilization and integration of CIM-components. Our central question is: will conservative Tayloristic rationalization strategies continue to dominate in central branches of West German industry? Or are new ways of integrating functions and tasks being tested on a broad scale?

THE CAPITAL GOODS INDUSTRY

The capital goods industry plays a decisive role in the economy of West Germany. In 1988 it employed approximately 3.8 million employees (about 18 per cent of all West German employees in the FRG) and it had a turnover of about 713 billion DM (Statistisches Bundesamt 1989). The capital goods industry is a very heterogeneous part of the economy, including, among other branches, machine-building, the automobile industry, shipbuilding, the electrical industry, office machine manufacturers and the aerospace industry.

Within this industry, machine-building is the industrial branch which employs the most workers, followed by the electrical industry. These industries are superseded in turnover by the automobile industry, which, however, takes third place in the number of workers employed. Thus, somewhat more than three-quarters of all workers employed in the capital goods industry are employed in these three branches. The other branches follow only at a vastly inferior magnitude regarding both the number of employees and turnover.

In terms of product complexity, series size and the dominant forms of production organization, the capital goods industry can be characterized as displaying a continuum in which the machine-building industry represents the one extreme (high product complexity, small series and job shop manufacturing of discrete parts) and the production of iron, sheet metal and other metal goods represents the other extreme (low product complexity, high-volume mass production and sequential line production).

DIFFERENT WORK SYSTEMS

There are roughly *three groups of functions* distinguishable in the manufacture of durable goods in the capital goods industry. The first comprises production tasks in a strict sense. The dominant technology can be stand-alone machine tools (e.g. NC/CNC-boring or milling machines, lathes, etc.) or more complex and interconnected facilities such as transfer lines, flexible manufacturing cells and systems, and processing centres, etc. Tasks directly related to this first group of functions are machine-setting, loading/unloading, and monitoring. The second group of functions includes service tasks such as pre-setting of tools, maintenance and repair, quality assurance, as well as cleaning and transport. The third group of functions comprises all aspects of production planning and control, shop-floor scheduling, process planning, and NC-programming (see figure 10.1).

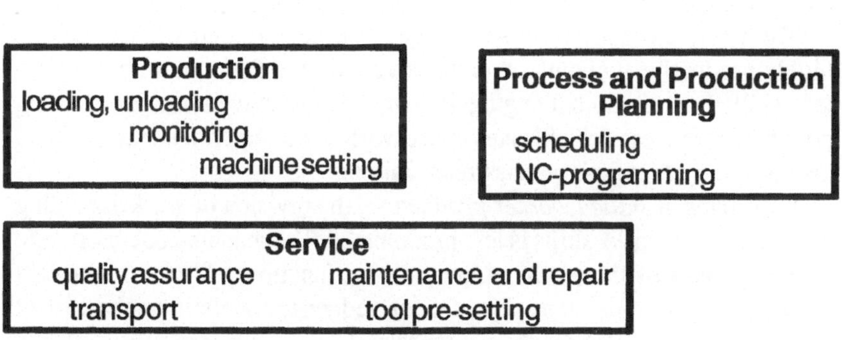

Figure 10.1 Functions and tasks. ISF 1991

The way in which these functions and tasks are divided decides the job structure of a given plant. The term 'functional division of labour' describes the differentiation of functions such as process planning, quality control etc. into independent organizational units. The term 'division of labour according to tasks (skills)' means the extent to which tasks are divided up within the organizational units both in a horizontal and a vertical (hierarchical) dimension. Skill requirements for individual jobs are directly linked to the complexity of the demands of a job. This complexity is in turn dependent upon the way in which the division of labour is organized according to functions and tasks.

Job structures are tied to forms of recruitment, training and remuneration and therefore constitute an integral part of work systems. Two ideal-types of work systems can be distinguished in the the capital goods industry of West Germany today (see figure 10.2): skilled worker production and semi-skilled worker production.

The job structures in systems of *semi-skilled worker production* are marked by a high degree of both functional and task specialization (see chapter 11; Köhler, Grüner 1990). The technical offices dealing with process and production planning as well as the service departments for maintenance and quality assurance, etc. play the decisive role in the planning, control and supervision of the production process. The majority of production workers are mainly semi-skilled machine-operators whose skills are not only specialized in specific production techniques (such as drilling, turning and milling), but also in certain types of machines and a narrowly limited spectrum of parts. This horizontal specialization is supplemented by a clearly defined vertical division of labour, ranging from unskilled labourers to machine-operators of various levels of skill and experience, to reliefmen, job setters, group leaders and foremen. The division of labour between job setters and operators is not clearly delineated. In routine jobs experienced operators may set up their machines themselves, whereas more complicated tasks are carried out by job setters.

The work is organized in such a way that newly hired workers can be highly productive without comprehensive training. Paths of mobility from less skilled to more challenging jobs makes gradual on-the-job training possible, a process which takes place with little support in the form of instruction by superiors or classroom training.

In systems of *skilled worker production*, the division of work according to functions and tasks/skills is less pronounced. The machine operators take on small tasks of the planning and service functions. Thus, to a certain extent they determine themselves the job sequences on their machine, read process plans and decide about processing steps, and can carry out less complicated maintenance tasks. Also the vertical and horizontal division of

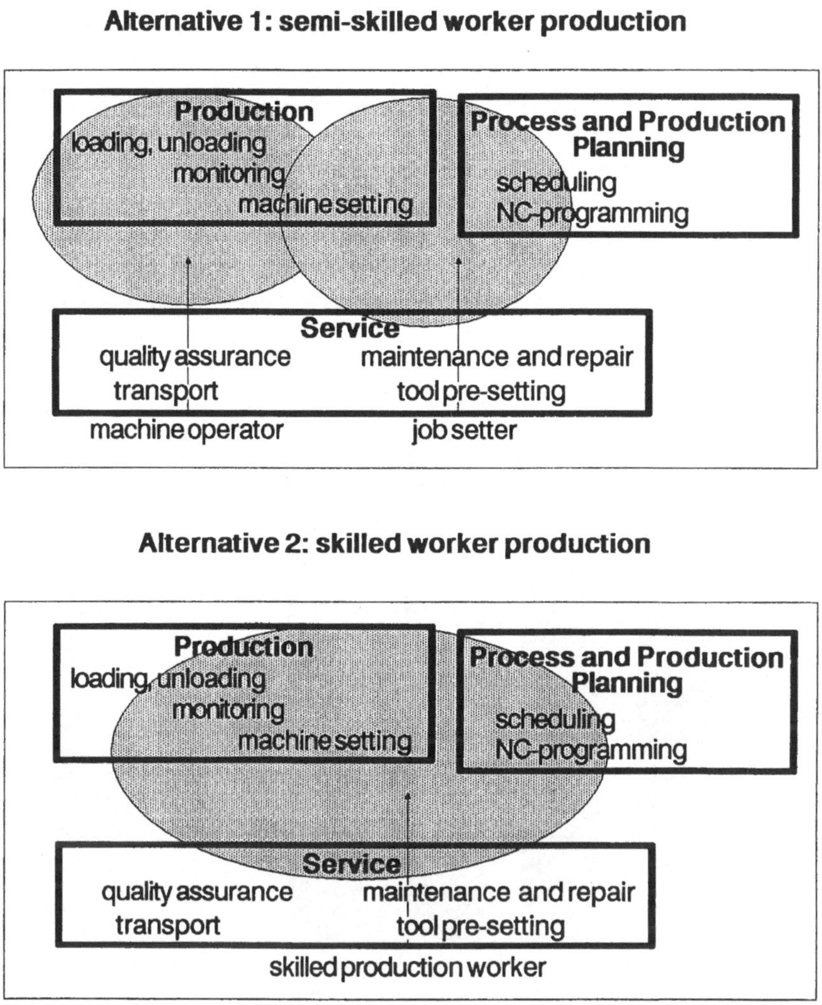

Figure 10.2 Job design in alternative work systems. ISF 1991

labour according to tasks/skills within the production departments is less pronounced, compared to systems of semi-skilled worker production. Skilled workers (*Facharbeiter*) carry out all necessary production tasks themselves – including as a rule the job setting for complicated parts and machines. They are often capable of working with a variety of machines involving a number of different processing methods.

The two systems of work are related to a certain extent. Often, they can be found side by side in one enterprise which has differentiated production processes according to series size and product complexity. Systems of skilled worker production dominate in custom and small batch production; systems of semi-skilled worker production dominate in series production.

This tendency towards a product group related, area specific differentiation of systems of work extends up to the industry wide level in accordance with dominant product and manufacturing structures. Figure 10.3 shows a close relation between the individual branch and the dominance of a particular work system. 'Skilled worker plants' are defined as companies in which the share of skilled workers out of all blue-collar workers reaches at least 60 per cent. 'Semi-skilled worker plants' are those

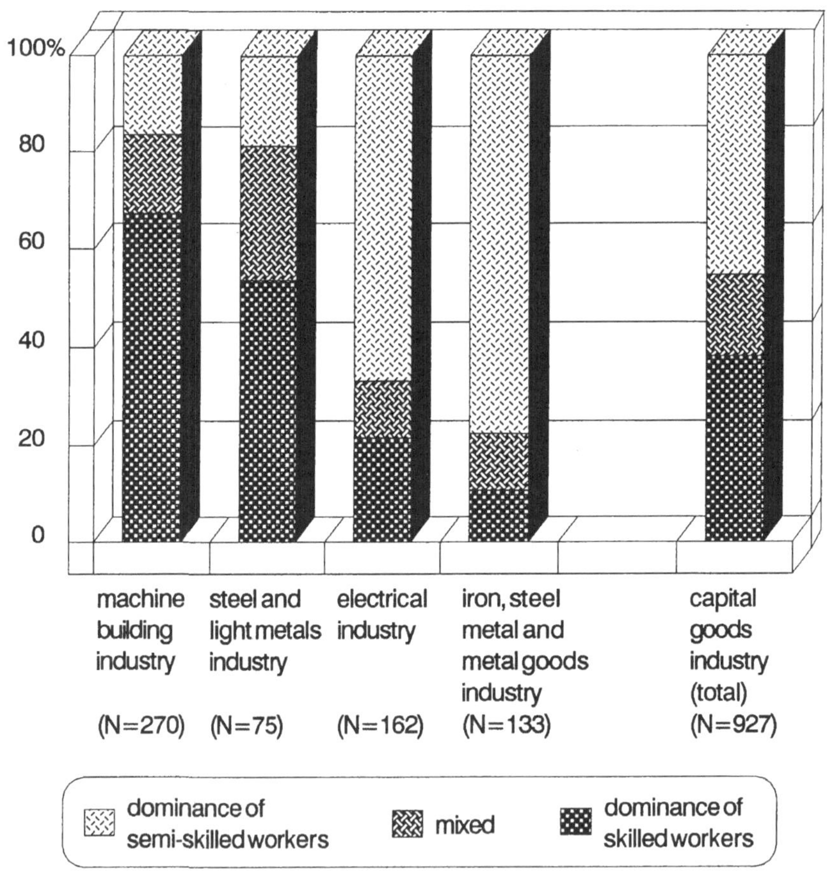

Figure 10.3 Work systems in selected branches of the capital goods industry. ISF 1991

in which the share of semi-skilled and unskilled workers among all blue-collar workers is at least 60 per cent. 'Mixed plants' show no clear dominance of either of the two dichotomous qualification forms. The skilled work system clearly dominates in the machine-building industry and in the steel and light metal industry, in which the product range consists of mainly complex and small batch products (this applies in roughly one half to two-thirds of such plants). At the other end of the spectrum, where semi-skilled worker plants dominate, we find the iron, sheet metal and metal goods branch with its large production series and rather simple standard products as well as the – in terms of the product structure – extremely heterogeneous electrical industry.

Both systems of work are relatively new from a historical point of view (see chapter 3). Their history can be described as a process of a growing differentiation of functions and tasks and an increasing narrowing of skill profiles in production work. This is definitely true for systems of semi-skilled worker production which were generalized after the Second World War (in contrast to the development in the United States). It is also true for the systems employing predominantly skilled workers whose jobs just a few decades ago, had included a wide variety of tasks of the planning and service functions next to production tasks in the narrow sense, and at this time could be characterized as traditional forms of craft production.

This development is an expression of company strategies to separate conception and execution, differentiate functions and tasks, and eliminate human labour to the greatest extent possible through technology and automation. We call such strategies Tayloristic in accordance with the definitions by the field of business administration (mainly related to functional differentiation, see Scheer 1988) and the social sciences (mainly related to task specialization, see chapter 3).

TRENDS IN WORK ORGANIZATION – BETWEEN CONSERVATISM AND INNOVATION

The problems that have arisen in the German capital goods industry (market saturation trends, sensitivity to economic cycles, strong international competition) combined with drastically sinking prices for computer technology have created a push to rationalize which is just in its first stages. The results of our study on the dissemination of CIM technologies (see chapter 9) demonstrate that German industry has only just embarked on its journey towards the 'factory of the future', especially in the fields of flexible automation and the integration of computer systems. High rates of distribution have, however, already been reached in the case of certain CIM-components (such as CAD-systems, PPS-systems, and CNC-machines), and high rates of growth are to be expected in the integration of computer systems. Moreover, the process of rationalization encompasses more and more the company as a whole, giving it the character of a 'systemic rationalization' (see chapter 4).

If one follows the international discussion about the future of work, one gets the impression that the application and the integration of new technologies are linked to a rejection of Tayloristic rationalization strategies on a large scale. Our question was whether such statements hold up to scrutiny for a central area of German industry.

In order to investigate these questions 60 case studies in companies with a high rate of computer utilization and integration were carried out.[2] Following the trends in the division of labour and the dominant rationalization strategies, we can distinguish three types of companies (figure 10.4):

1 Companies with a *conservative strategy* seek to stabilize or indeed extend the existing division of labour according to functions and skills. In this sense Tayloristic rationalization strategies predominate.
2 Companies with an *experimental strategy* operate similar to their conservative counterparts with Tayloristic rationalization principles. However, in some areas of production they are prepared to experiment with structures integrating functions and tasks.

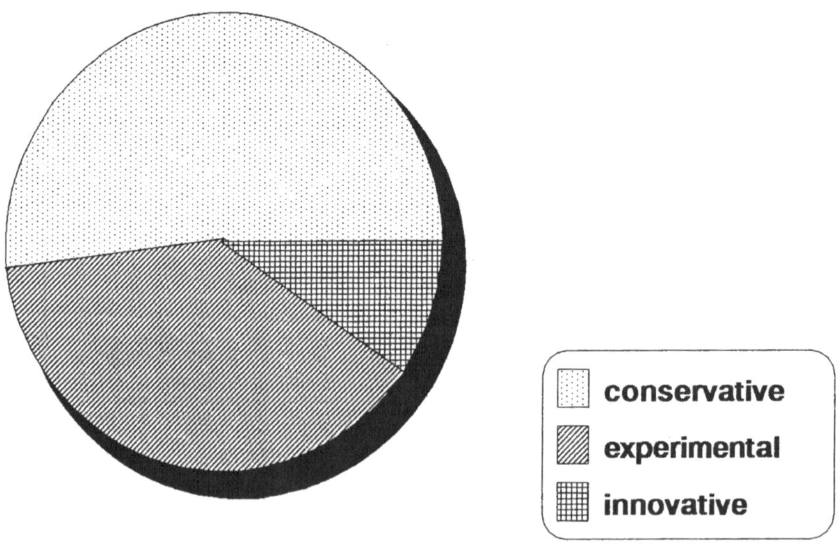

Figure 10.4 Companies according to their dominant organizational strategy (60 cases out of West German capital goods industry). ISF 1991

3 Companies with an *innovative strategy* deviate completely from the Tayloristic rationalization paradigm in central areas of their production and work organization.

Conservative strategy

At least half of the 60 cases can be classified as the *conservative* type of company. Tayloristic strategies of the differentiation of functions and tasks predominate. In these companies the highly differentiated and centralistic organizational structures in place are maintained or indeed further consolidated after new technologies are introduced. These forms of work and production organization, in turn, influence the selection and utilization of CIM-components. Centralistic CIM-structures back up the more or less marked degree of division of labour and, part, reinforce it.[3]

An example of this is a company in the iron-, sheet metal and other metal goods branch whose entire spectrum of products consists of one-piece or simply structured steel and plastic parts, manufactured in mass production with batch sizes of between 500 and several million pieces. The production structure is reflected in the work system. The company has 200 employees. The highly differentiated work organization can be roughly subdivided into two areas: mainly unskilled or semi-skilled workers are engaged in the manufacturing process (pressroom, injection moulding shop), supported by a few skilled workers for more complex tasks; service departments, e.g. toolmaking, mouldmaking or quality control are filled almost 100 per cent by skilled workers.

Management's major concern is the transparency of all company operations. For this reason, great importance is attached to integration by means of Production Planning and Control Systems (PPC). For the past two years, the company has been operating two CNC-milling machines and ten robots for simpler insertion and sorting activities. There are plans to install CAD, CAP, CAQ and PDA systems within the next two or three years. These kinds of automation technology all lead to an explicit centralization strategy for all planning and control activities. The purpose is to be less dependent on the skills and discretion of the machine operators. The already insignificant NC-programming on the shop-floor will therefore be shifted to the process engineering department when the CAD system is introduced. With the aid of a PDA system, existing knowledge of breakdown susceptibility in the manufacturing process and individual operation times will be input into computer systems for future reference in work scheduling. This strategy results in a further, deliberate development of the functional division of labour supported by CIM-components, while maintaining a high degree of division of labour according to tasks and skills. It also increases the potential for supervision and displacement of workers.

In our sample, conservative rationalization strategies could also be found in machine building companies with small and medium-sized batch production. One enterprise in this category manufactures cylindrical and rectangular moulded parts with a large number of variants and adaptations tailored to specific customer requirements in sequential line production (1400 employees). The firm has been expanding steadily for many years and utilizes a broad spectrum of CIM-components. Flexible manufacturing systems are utilized almost throughout; even transport between the systems is to a large degree automated. Process control and shop-floor scheduling are handled by a highly differentiated and centralistically structured computer hierarchy. The NC-programs for controlling the machines are compiled in the office and transmitted to the machines via DNC.

Production and work organization is based on a high degree of division of labour, although the company can rely on a well-trained body of skilled workers. Thus, not only planning and control, but also maintenance, tool presetting, etc. are all carried out by specialized departments. Moreover, there is a marked degree of division of labour within the FMS. Semi-skilled workers handle the mounting of parts. Specialized skilled workers are responsible for setting the pallet fixtures or individual machines. Spare men serve to bridge personnel gaps. A next step in automation will probably do away with the palletizers' jobs. There is no discussion of reintegrating the highly differentiated tasks or for introducing forms of group work.

Experimental strategy

Tayloristic methods of differentiation, specialization and control also play a predominant role in at least another third of the companies: those characterized as *experimental*. In contrast to the conservative type of company, this group of companies tends to experiment with alternative structures. These experiments can go into the direction of integrating functions and tasks in new forms of skilled production work ('computer-aided Drucker'; see chapter 3).[4] They can also mean the decentralization of functions to the shop-floor accompanied by an increased polarization of skills ('computer-aided Huxley'; chapter 3). However, the search for alternatives is limited to individual departments or lines of technology.

Experimenting companies usually initiate the changes of production and work organization in manufacturing areas with special requirements; for instance, on the shop-floor where either extreme precision or speed and fast reactions are required of the work-force. These experiments are frequently so successful that they are gradually extended to other parts of the manufacturing process.

The presence of differing concepts of work design within one and the same company is particularly clear in the case of two German plants of a large European producer of electronics. In plant A, with 1,200 mostly semi-skilled employees, the detailed planning and shop-floor scheduling by technical offices reflects a strong centralistic philosophy. However, in the same plant, on a highly automated system for assembling circuit boards, a new type of work organization has been realized. In order to facilitate the re-integration of repair, maintenance and set-up jobs into group work, a hybrid type of skill has been created in which the knowledge of mechanics, electronics, hydraulics and pneumatics has been combined. By contrast, an identical assembly system in plant B functions with a fairly high degree of division of labour.

A further example of the experimenting group is a machine-building company with 700 employees in which complex machines for the textile industry are manufactured by a skilled work-force in small batches with variations to fit customers' specific requirements. Frequent, rapid changes in market requirements force this company to attempt to reduce throughput times and to meet short-term deadlines. Toward this purpose, decentralized shop-floor control centres are equipped with interactive terminals so that the schedule supervisors can call up workshop orders and assign them to the machine operators themselves.

In order to increase flexibility in several areas of production (e.g. in CNC-sheet metal processing, in assembly and in a flexible manufacturing cell), a start has been made to restructure work organization towards semi-autonomous groups. These groups are given the opportunity to schedule their work independently within a certain time frame. Also, quality circles are planned for the near future. In other areas within the company, however, there is a tendency towards centralizing areas of authority: for example, the Production Planning and Control System (PPC)) stipulates detailed sequences of production orders for the shop-floor.

Innovative strategy

The remaining tenth of the cases studied belong to the innovative type of company. These are companies which not only experiment with alternative forms of work organization in individual departments, areas or functions, but who are restructuring the entire manufacturing area towards non-Tayloristic, skill enhancing forms of work ('computer-aided Drucker'; see chapter 3). Production work is reskilled as a consequence of a (re-)integration of planning and other functions which were in the past performed by the technical offices. Furthermore, it involves the utilization of shop-floor and skilled worker-oriented CIM-technologies such as those which are already being offered in the area of CNC-programming or interactive scheduling (see chapter 9).

Examples of these kinds of reorganization measures can be found particularly in cases where planning and decision-making powers have been decentralized to production departments. In this respect, companies with generalized structures of 'manufacturing islands' or 'cellular manufacturing' (sometimes discussed as the 'factory within the factory') are exemplary (see Brödner 1987). In these cases, changes in the organization of production, such as the transition to 'group technology', are combined with experiments in work organization towards skilled group work. Reductions in divisions of labour were also found in companies with more classical forms of job shop and sequential line production. The impulse for such 'radical' changes often stems from dramatic crises in a corporation due, for example, to excessive overheads.

A typical case in this regard is an electronics company, a manufacturer of special high-performance semiconductors and complex components, covering a market niche, where a few years ago the economic situation was so disastrous that without comprehensive changes there was no hope of survival. The manufacturing process in this company with 600 employees is characterized by two different sections: the manufacture of chips in a continuous production process is highly automated; subsequently, the chips are integrated into the corresponding components in an assembly line with mostly manual work.

There are two reasons why the company cannot achieve a high degree of automation in its assembly process: on the one hand, the automation technology necessary is at present too expensive and inefficient and is likely to remain so in the foreseeable future; on the other, it is frequently necessary to fulfil customers' specifications. In order to be able to meet such market requirements, the firm, whose labour system might best be characterized as semi-skilled production, relies mainly on the human potential for flexibility. In accordance with this philosophy, the company pursues a training and restructuring policy whereby the – mainly unskilled female – assembly workers are given systematic training with the goal of establishing a workforce with a homogeneous level of skill. Thus the female workers are to be given the opportunity and the skills necessary for a relatively free hand in the distribution of tasks among themselves. They will also be able to carry out additional tasks on their own which hitherto had been delegated to other departments, such as the setting of measuring systems, transport jobs, quality assurance and maintenance work.

A similar case to the one just described is that of a manufacturer of complex machines for road construction with 550 mostly skilled workers. Prior to the reorganization of manufacturing it was suffering a severe economic crisis mainly due to high overheads and almost landed in bankruptcy court. Under the guidance of a charismatic and decisive managing director, production of complex programme products for road construction, manufactured in small batches, was fundamentally reorganized with the purpose of reintegrating conception and execution and achieving a flexible utilization of manpower. Due to high overheads, the production planning departments and technical offices were reduced and some of their functions shifted to the shop-floor. By 'pulling out the plug from one day to the next' the deterministic PPC-system was dispensed with, since, according to our interview partner's statements, it had never adequately reflected company reality and had produced only 'rubbish'.

Functions such as NC-programming, tool-presetting, servicing, quality control and process planning, which had previously been removed from the shop-floor, were partially reintegrated into the production departments. In future an interactive graphic NC-programming system will aid shop-floor personnel by transferring geometric data from the CAD system. Restructuring led to a reduction in the division of labour. The more repetitive auxiliary functions such as deburring and transport are integrated in machine operators' scope of duties. The same is true for NC-programming, tool setting, minor repairs etc. The workers were given such broad training that within a very short time they could be transferred and deployed in many areas of production and even in assembly operations.

A LARGE NUMBER OF CONTEXTUAL FACTORS

A large number of external and internal contextual conditions contribute the direction that technological and organizational change takes in a company. Since there are several, frequently contradictory, constellations, it is not possible to demonstrate unequivocal causal relationships between particular conditions and a certain type of rationalization strategy. However, to a limited degree one can indicate the direction in which a given variable takes effect.

Developments on the *product market* can be important in two senses. Taken over a longer period of time a given company's product market will be reflected in the structures of its manufacturing process. Product range and product structure suggest corresponding production technologies and organizational solutions. If product market requirements begin to shift toward greater product diversity and shorter delivery times, incongruities may arise. Increasing process complexity may induce a re-evaluation of the flexibility potential of the work-force as a productive resource and encourage experiments in the area of job design and manpower management.

Product market developments are also significant in terms of a corporation's profit situation. Profitability crises may cause shifts within in-house power constellations – strengthening top management – and provoke new ways of designing the organization of production and work. Thus a characteristic of all the innovative companies in our sample was the fact that a more or less dramatic crisis situation was what initiated the company's reorganization process.

A further significant factor is the labour supply[5] of the *internal labour market*. In companies with a predominance of semi-skilled workers, the transition to new work structures generally implies intensive and comprehensive training processes. In contrast, in companies with predominantly skilled workers, the skills and motivation required for reintegrating the conception and execution of certain tasks are more likely to be available. The predominance of either semi-skilled or skilled workers corresponds closely to the industry the company belongs to and its inherent complexity of products and processes.

The labour supply on the *external labour market* has a similar influence. When skilled workers are available there is a greater chance of reorganizing manufacturing operations towards alternative work forms, while a lack of skilled workers is more likely to reinforce trends to conservative procedures. Two electronics plants we studied, for example, both manufacturing comparable products with almost identical technological facilities, set out in opposite directions to restructure their work organization. The decisive factor for the realization of skilled group work established in one plant was the particularly favourable situation on the regional labour market.

The role played by the *level of automation* cannot be ignored. The case studies show a relationship between the degree to which technology is utilized and the reduction or increase in division of labour. There are several reasons for this. The majority of CIM-components and their system architecture are oriented towards centralization and are utilized accordingly. Also the vision of the 'unmanned factory', prevailing in some of the highly automated companies, may play a role in the pursuit of Tayloristic rationalization strategies. It is striking that three of the five innovative companies pursue a slow and cautious CIM-strategy. This corresponds directly to managements declared strategy of achieving flexibility and productivity by means of a sensible and efficient combination of human labour and technology and not by eliminating the human factor from the production process.

The *size of the company* also appears to play a part. Interestingly, most firms with experimental or innovative strategies are to be found among companies with 500 to 1,000 employees, while they are more seldom seen in companies with fewer than 500 or more than 1,000 employees. This trend is all the more significant because four of the five innovative companies are within this medium-sized range. It may be that such companies are sufficiently large to develop and implement new concepts; on the other hand, they are not too large in the sense that the centralization interests of middle management (e.g. production and process planning which is traditionally geared toward controlling the manufacturing process) could effectively block alternatives.

Dependence on large customers or corporation headquarters can also influence the development of rationalization strategies. In our sample we found that in some individual cases, the decision to purchase technical systems was enforced on the grounds of demands made by large customers, yet against the interest of the company's management. To what degree such stipulations concerning specific technology or a certain system also have implications for work organization depends, among other things, to what degree the technology concerned favours more centralistic solutions with a high degree of division of labour, or instead shows a preference for shop floor or user oriented solutions. If such systems are geared to structures with a high degree of division of labour, as a rule centralistic organizational concepts find their way into the companies concerned.

IMPLEMENTATION PROCESSES OF PARTICULAR IMPORTANCE

The stated set of external and internal context variables must not be interpreted in a deterministic way. These conditions and factors should instead be seen as impulses, urging in one direction or another, but which can be contravened by other sets of circumstances. A crucial point is that the different strategies found in conservative, experimenting, and innovative companies cannot be satisfactorily explained by the internal and external contextual conditions alone. In the cases of comparable constellations of context variables, one frequently finds entirely different solutions for problematic situations within companies. Undoubtedly, implementation processes of CIM-components and -systems play an important, and indeed quite autonomous, role in determining the direction of the company's rationalization strategy and thus also the development of production and work organization (Hirsch-Kreinsen, Schultz-Wild 1990). Most significant seem to be the following dimensions: planning concepts (technology, organization, personnel), planning agents (production, planning departments, top management, external advisers); planning resources (skills, time, money); and participation (rank and file, shop stewards, shop committee).

With regard to these dimensions, certain sets of conditions can be identified in conservative, experimenting and innovative companies. Not every firm has all, but every company possesses some of the characteristics named below.

In *conservative companies* there is a high incidence of the following constellation of characteristics:

1 the formation of project teams, particularly from the staff of the technical offices and the process engineering departments, for whom the decentralization of functions into the manufacturing area could be disadvantageous (loss of personnel, power, process control, etc.);
2 the project team members consist mainly of representatives of middle management who frequently have to handle the work parallel to day-to-day operations and whose limited resources force them to plan and implement along conventional lines;
3 the predominance of purely technical considerations when selecting and installing new technologies; a total concentration on such questions as the technically optimal coordination of the components, hardware and software design and the problems of interfaces, etc.;
4 particularly in the case of small companies, there is often a corresponding predominance of a large, all-powerful computer system manufacturer, whose concepts are basically geared to large-scale companies and with structures involving a high degree of division of labour;
5 strong priority of the following objectives: high level of automation, central process control and the selection of corresponding technologies;
6 traditional methods of cost/benefit-analyses, excluding factors which are only indirectly quantifiable such as flexibility and skills, which therefore favour structures with a high degree of division of labour and are oriented to cutting costs of production personnel;
7 the employees concerned, their immediate superiors, shop stewards and members of works councils are as a rule excluded from the planning process.

In the case of *experimenting companies*, different constellations of characteristics were found:

1 project teams with a broad representation of all departments affected directly or indirectly by the project, and with more or less comprehensive, but at least, as a rule, sufficient resources in terms of technology and time;
2 problems of work organization included in planning concepts from the outset and various alternative solutions considered;
3 system manufacturers and users cooperate on the basis of detailed performance and program specifications;

4 the main emphasis in the utilization of CIM-components is in the manufacturing area, sometimes with explicitly shop-floor-oriented computer concepts;
5 the predominant goals are: 'meeting narrow deadlines' and an 'increase of flexibility';
6 the more or less marked participation of employees concerned and their representatives in the planning and implementation process.

Innovative companies as a rule demonstrate additional characteristics over and above those already mentioned for the experimenting group. The most important are the following:

1 the continuous participation and dominance of top management in the implementation process, which in some cases has to be pushed through against the opposition of other in-house interest groups;
2 explicit strategies of holding open options, particularly by means of planned improvements in the company's skills and flexibility potential;
3 anticipation of long-range problems (the lack of skilled workers, for example) and initiated early counter-measures;
4 full involvement of workers affected by changes, their representatives and superiors.

THE CONTRADICTIONS OF POST-TAYLORISTIC RATIONALIZATION CREATE POLITICAL OPTIONS

By categorizing the companies surveyed, we came to the following findings: the majority of the companies continue to pursue traditional Tayloristic rationalization strategies in order to consolidate or indeed further develop the division of labour according to functions and tasks and simultaneously seek to automate to the greatest extent possible. The next largest group of companies is experimenting with new possibilities in certain areas and only a small minority has implemented alternative structures throughout production, or indeed the company as a whole.

A closer analysis of our sample shows that there is a large number of factors effecting the development and design of the company rationalization strategies. These include the external conditions of a company (sales and labour markets, for example) as well as in-house structures (product and process complexity, level of automation etc.). It seems, however, that different strategies of technological and organizational change cannot be adequately explained by the *internal and external contextual conditions*. Thus we find quite different strategies applied to solve company problems in cases showing a similar set of contextual conditions. There can be no doubt that the *implementation processes* of CIM-components and -systems play an independent role in determining the direction that technological and organizational change takes in companies. Of importance are planning concepts and strategies of the change agents, as well as power constellations and resources.

These findings are somewhat surprising. Labour economics as well as industrial sociology have provided a multitude of evidence over the last thirty years of how macro- and micro-economic forces determine the structural development of systems of work. The classical variables typically applied are the sales market, company size and series size, the level of automation, and the company's internal and external labour market.

Our *assumption* is that the shift from economic to 'political' factors is grounded in the erosion of the 'Tayloristic syndrome' characteristic of the post-war period (see chapter 3). This syndrome was determined by the following macro- and micro-economic forces: the expansion of sales markets, economies of scale, standardization and automation of production, and a corresponding technology market as well as a large labour force reserve. Tayloristic strategies were an expression, as well as a component, of this constellation. They were adapted to the different economic, technical, and social structures of particular branches and types of plants and thus created distinctive patterns of work systems like the ones described above as skilled worker production and semi-skilled worker production.

However, this constellation seems to have dissolved during the 1970s and 1980s. Now markets are tending towards product variety and smaller batch sizes, while at the same time, the pressure to increase productivity through processes of automation is increasing. The increasing capital intensity demands process continuity, yet production's susceptibility to technical failure is growing in a causal relationship to the increasing complexity of technical systems. The availability of skilled and motivated workers prepared to take on industrial production jobs is becoming a problem. Micro- and macro-economic forces which were moving together in the past, developed in the 1970s and 1980s into a multidimensional force field, whose vectors are directed in various and partially contradictory directions. A new equilibrium is not yet in sight. Strategies, coalitions, and activities on the part of management and employee fractions can be of decisive influence in this situation. The current disunity in the development of rationalization processes can only be understood in this context.

If the conservative, (neo-)Tayloristic strategy succeeds in becoming dominant, the destruction of the remaining skill-based systems of work in Germany can be expected in the long run. (see chapter 3; Hirsch-Kreinsen *et al.* 1990). The distinction between skilled and semi-skilled worker production, which still exists today, would probably dissolve in favour of a sharper polarization of skills within both systems. If a generalization of innovative strategies emerges, a new form of skill-based systems of work will develop. Which of the strategies will be dominant in the long run remains at present unclear.

These results differ in two ways from the more optimistic statements made by the proponents of the 'new production concepts' (Kern, Schumann 1984, 1987, 1989): our empirical analysis shows the conservative forces as being substantially stronger; our prognosis is that various paths of development are possible.

Different and divergent patterns and strategies have been revealed in case studies. An analysis of the forces influencing the development of work systems shows a high degree of openness. This leads to the conclusion that various outcomes are possible in the process of restructuring industrial work in the capital goods industry of West Germany. Competing and contradictory economic and social forces are creating opportunities for a more humane and socially acceptable design of work systems and for political action.

11 Flexible Manufacturing Systems and Work Organization[1]

Marhild von Behr, Hartmut Hirsch-Kreinsen, Christoph Köhler, Christoph Nuber, Rainer Schultz-Wild

INTRODUCTION

Flexible manufacturing systems (FMS) can be seen as a new stage in the automation of small batch production. The prevailing manpower approach to these systems is characterized by concepts of a hierarchical and differentiated job structure. An alternative is to develop new forms of skilled group work. The latter approach increases problems and costs of implementation processes; however, it may be in the interest of companies because of higher utilization times, more flexibility, and better quality. It also meets worker interests in challenging and rewarding work. This chapter analyses experiences with new forms of skilled group work in a flexible manufacturing system.

SETTING THE SCENE: THE COMPANY

The FMS was developed and installed in the main plant of a company specializing in the production of transmission equipment for the automotive industries. The plant is located in the south of the Federal Republic of Germany and has approximately 7,000 employees, the majority of whom are semi-skilled production workers.

The plant mainly produces gearboxes for trucks, buses and special vehicles in medium- and large-scale series. Therefore, it is dependent on the ordering behaviour of its customers, most of them being large-scale companies from the heavy vehicle sector, and is thus sensitive to business fluctuations. For this reason, the company is under considerable pressure in the following areas:

1 First, product development has to be continually pursued in order to satisfy specific demands and growing requirements and to minimize the danger of customers undertaking the production of aggregates themselves.

2 Second, there is constant pressure to automate and rationalize in order to remain competitive, not only with other suppliers, but also with customer self production.
3 Third, there is considerable pressure to keep the production system flexible because the ability to deliver a relatively small number of products with different specifications has to be maintained at all times, along with the ability to change specifications (for example, specific auxiliary drives for transmissions) up to two weeks before delivery.

Production is divided according to product groups and production methods and thus mainly organized along the principle of job shop manufacturing. In spite of the variety of different products, not to mention their parts and components, the company has tried to utilize the advantages of standard production in large series and lots to the greatest extent. In some fields the employment of transfer lines has proved to be economical. However, stand-alone machine-tools (some of them equipped with robots or parts magazines) that can be adapted to differing demands still represent the most important means of production. Comparatively long production throughput times (of up to six weeks) from the raw material to the finished part ready for assembly have to be tolerated, as well as the respective costs for storage, intermediate storage and capital tied up in semi-finished products.

The job structure in production is characterized by a relatively strong division of labour according to functions and tasks, typical for semi-skilled work systems in West Germany (see chapter 10). Figure 11.1 shows the wage and occupational structure of a typical gear production department. As a rule, the workers continuously operate one and the same machine tool and help out at two or three other machines of the same type. Reliefmen take over in case of illness or holiday; they are usually responsible for equipment of the same method (e.g. turning) in one workshop. At one level higher are the job setters, who are also normally responsible for only one group of machine tools. Above this are the group leaders and foremen (*Meister*). Labourers or material handlers, who transport workpieces, load and unload magazines and fulfil other auxiliary tasks, are located at the bottom of this hierarchy.

Because of the high functional division of labour this staff alone is not in a position to carry out all necessary tasks. The technical services of other units of the plant have to intervene to a greater or lesser extent, especially with regard to programming, maintenance, repairing of breakdowns, presetting tools, quality assurance, etc.

About one third of the company's blue-collar workers are classified as 'skilled'. The majority of them have completed apprenticeships in the metal trades and/or have long-term experience on jobs with high skill demands.

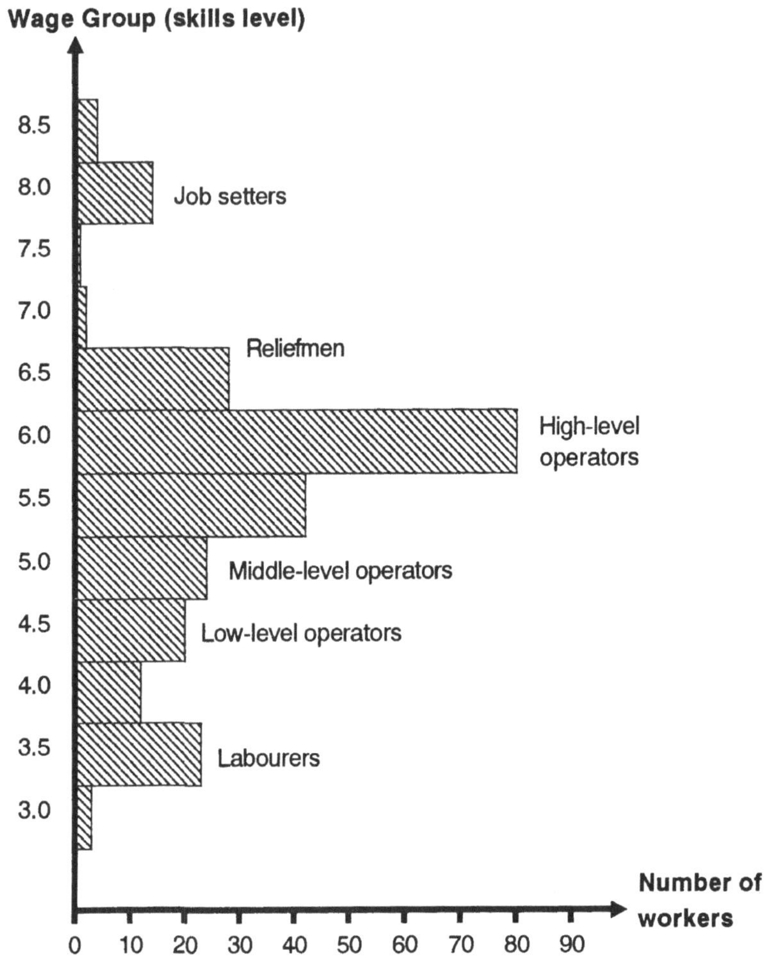

Figure 11.1 Wage structure in a gear production department by occupation (1982). IFS 1991

They work mostly in typical skilled trades departments such as tool making, maintenance and repair, etc. A smaller proportion of the skilled workers are assigned to complex production jobs in the capacity of reliefmen, job setters and foremen.

Two-thirds of the blue-collar workers are classified as unskilled or semi-skilled. Many of them have received vocational training outside of the metal trades (to become butchers, barbers, bakers, etc.), and being unable to find attractive jobs, have ended up as factory workers. Most of them work on jobs in production and assembly, whereas a minority are assigned to service departments such as quality assurance, transport, etc.

The social structure of the work-force has changed dramatically over the past 30 years. The proportion of foreign workers has increased from almost zero in the early 1960s, to about one-third of all blue-collar workers in the 1980s. They are under-represented in the skilled trades departments and over-represented in production departments where, in many cases, they comprise more than 50 per cent of the work-force and have penetrated all levels of the job hierarchy (see Köhler, Grüner 1990).

Women represent less than 10 per cent of the blue-collar work-force. In production departments their percentage is even lower, and they work mostly as labourers, helpers and machine operators in low skilled jobs involving lightweight pieces.

SETTING OBJECTIVES FOR TECHNOLOGICAL INNOVATION

During the 1970s, the market for heavy vehicles became increasingly turbulent and the company came under considerable pressure. One of their reactions was to increase investments during the second half of the decade. The decision to design and install a flexible manufacturing system was part of this increased effort, whereby several objectives were pursued:

1 Production optimization was to play a major role. This meant productivity gains through automation as well as high flexibility. It was felt that a reduction of throughput times would increase delivery capacity while helping reduce storage costs and capital requirements for in-house inventories.
2 There was also an interest in developing a new line of products. The design of manufacturing equipment and handling technology that could be applied to other plants was to help investigate the chances for a venture into a new market sector.

After a two-year preliminary planning phase, the design and construction of the FMS was started in 1977 with a grant supplied by the Federal Ministry of Research and Technology (BMFT). In 1981, the first machine tools were installed; 1982 saw the first test runs of parts in the system, and from the autumn of 1984 on, the system began to attain normal running conditions.

Investment appraisals carried out at an early stage of the project showed that the innovation could not be legitimated with conventional methods alone, because the costs saved on staff and stock reduction were countered by high investments in machine technology, handling technology and control engineering. In the second half of the 1970s there were no established means of evaluating factors such as flexibility and scheduling effectiveness due to short throughput times, etc.

In spite of these problems, the decision to carry out the project was taken by the managing board. Strategic considerations concerning the possibility of taking the first step towards flexible automation played a major role in this decision, because the company wanted to start gaining experience with new forms of technology at an early stage. The decision was facilitated by the prospect of financial support from the Ministry for Research and Technology, and by plans eventually to market the results of the project.

The system was conceived for the metal cutting of gears from three families of parts in small to medium-sized batches. It was intended to handle the entire 'soft machining' process, that is, all necessary operational phases that are carried out before the workpieces are hardened. A total of fourteen machine tools were grouped into thirteen machining cells, each containing one robot and three stations for workpiece carriers. A fourteenth cell was designated as the central loading and unloading station with component carrier stations and a robot. As far as material flow was concerned, the cells were concatenated by means of a gantry crane and central workpiece storage facilities; in terms of information flow, the cells were interlinked by a super-ordinate control unit by means of a mainframe computer (see figure 11.2).

The development of system technology did not aim for total automation, since a system of this size and complexity would demand the presence of several operators. While a reduction of staff compared with the previous production process was intended, a system team of about six persons per shift was opted for from the start.

The difference in terms of manufacturing structures between conventional production with stand-alone machines and the new system consisted primarily of the interlinkage of the processing stations by automating the transport and handling operations. The system team was left with the following work tasks:

1 Their main task was to set up the machine tools and robots, to read in and optimize NC programs, to exchange tools and adjust clamping devices.
2 A further task, which was comparatively undemanding in terms of skills yet physically strenuous, was the manual feeding of the system with workpieces and the unloading of the processed gears at the system's central loading and unloading station.
3 Finally, there was a whole series of controlling and monitoring tasks pertaining to the operation of the entire system as well as tasks involving individual machining cells.

Apart from these functions, there were also a number of additional tasks ranging from simple service jobs, such as clearing away chips in the turning cells, to more demanding jobs such as quality assurance or tool presetting.

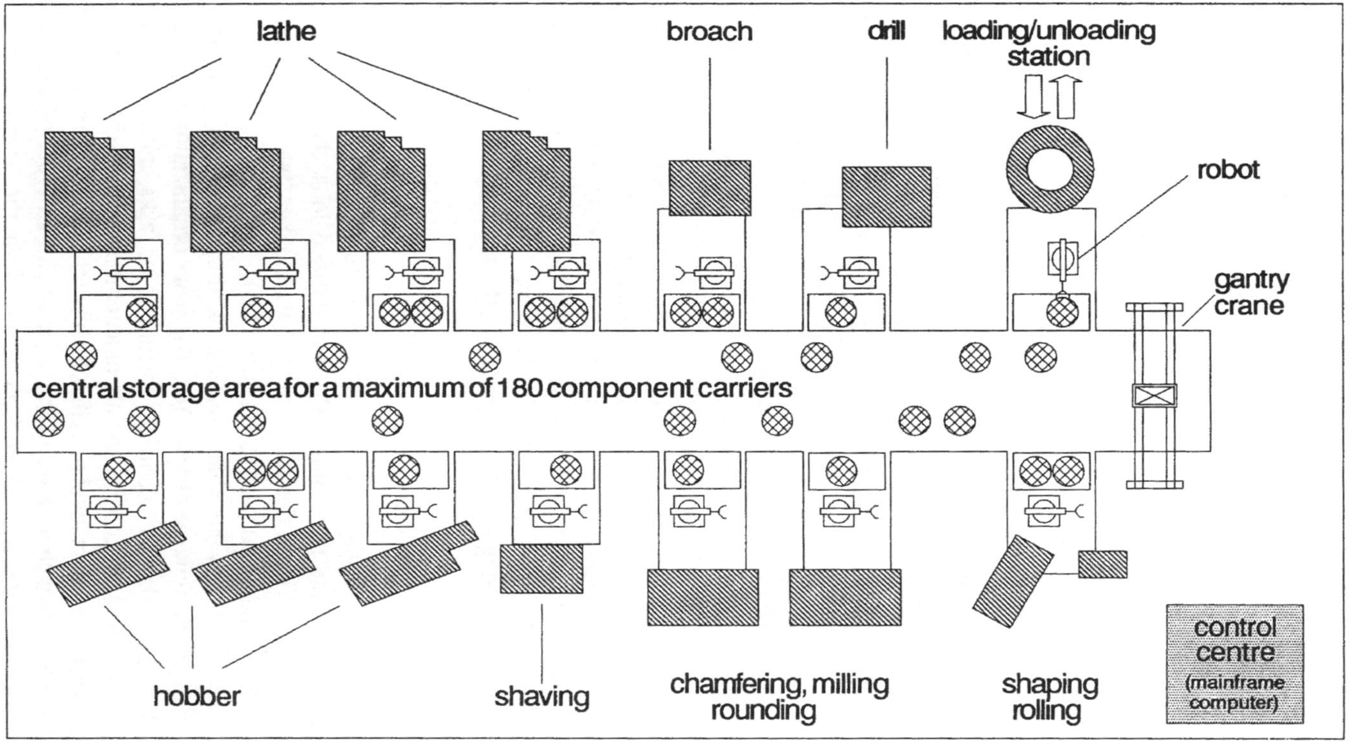

Figure 11.2 Floor plan of the FMS. IFS 1991

MANPOWER POLICY ALTERNATIVES

The discussion of manpower policy issues focused on three problem areas:

1 the employment of manpower from the company versus manpower from the external labour market;
2 the choice of a low or high degree of division of labour; and
3 special training for the respective jobs versus immediate selection of the best suited workers for the FMS team.

External versus internal recruitment

In the case of implementing production equipment that is complex and highly innovative compared with conventional production equipment, it would seem an obvious policy to recruit persons who have the necessary skills and experience from the external labour market. This course of action was not seriously considered, however. This was due partly to the influence of the works council and the union, and partly to the project's pilot character. Instead, the project management decided to staff the new jobs with suitable workers from within the company, thus allowing the staff to benefit from the planned improvement in working conditions.

For the same reasons, the possibility of simply dismissing workers from the traditional production area who were displaced by the new system was not given much consideration. For the members of staff who could not be provided with jobs within the new system, in-house transfers took place. In this way the works council was able to make use of its right of co-determination and safeguard the interests of affected workers.

High versus low degree of division of labour

In designing work organization within a flexible manufacturing system, that is combining the remaining tasks into defined jobs, there were, according to the experience that had been gathered to date, two basic courses of action that could be pursued (see figure 11.3).

The first case can be characterized as the traditional Tayloristic approach to job design (see chapter 10). The work organization is based on a high division of labour according to functions, tasks, and hierarchical levels. The system staff consists mainly of a number of specialized machine operators with a medium level of skill who have the task of monitoring and setting up several machine tools of the same or similar type. Additionally, there are one or two unskilled workers who perform the strenuous task of loading and unloading the parts as well as other auxiliary functions. There are also one or two shift leaders who have to be qualified technicians with

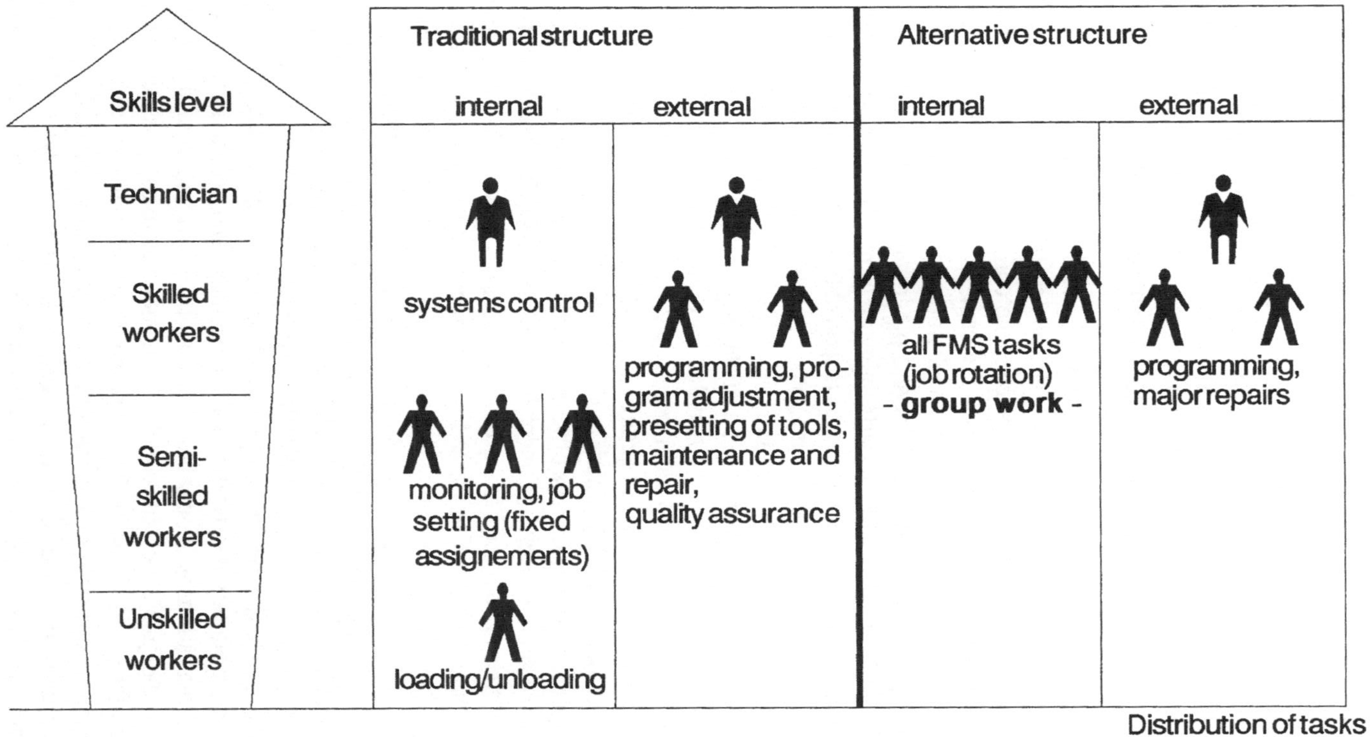

Figure 11.3 Alternative job structures in flexible manufacturing systems (FMS). IFS 1991

additional special knowledge. Their job is to control the entire system, manage the scheduling and coordination, supervise the system's workers and take care of all tasks requiring communication and interaction with other company departments. This type of system team cannot fulfil all necessary functions itself, which means that personnel from other departments have to intervene to a greater or lesser extent in order to preset tools, and carry out quality assurance, programming and program adjustment as well as repair and maintenance tasks.

The second case represents an alternative to the Tayloristic model. The work organization is based on a low division of labour with respect to functions, tasks and hierachical levels. The entire system group consists exclusively of skilled workers who are capable of performing all of the system's metal-cutting processing techniques and who have additional knowledge of electronics, robotics and programming, etc. Each of these system operators is capable of performing all or most of the relevant tasks. The operators form a group which is responsible for the system as a whole. Apart from direct production functions, personnel of this type are able to take on additional tasks – naturally in varying degrees, according to the amount of work to be performed and staffing levels given – which in the first case described have to be handled by other departments. Only complicated programming jobs and repair and maintenance tasks requiring extensive or special skills and qualifications (for example, in the electronics sector) are performed by system-external specialists.

Obviously, these two types of work organization are both extreme cases. Yet, while mixed forms are most likely to be found in actual company practice, they always tend to a greater or lesser extent towards one extreme or the other. In the case examined it was expected that the organization of work within the flexible manufacturing system would tend towards the first type of work organization based on a distinct division of labour. Turning the FMS tasks into a number of jobs with different tasks and differing skills requirements would have accorded with the hierarchical job structures prevailing in the area of conventional production. However, after considerable discussion between members of management, the works council, union representatives and external consultants and researchers, it was decided that all tasks should belong to the operating realm of the system group to the greatest extent possible. The aim was to establish the lowest degree of fixed division of labour possible; in other words, to avoid the permanent assignment of specific tasks to individual workers. This meant that the tasks that had to be dealt with at a given time were performed according to a task rotation scheme while adapting to the varying amounts of work and personnel availability.

Personnel selection versus training

The decision to use internal recruitment and skilled group work within the FMS made two alternative courses of action for filling the new jobs with suitable personnel possible:

1 The first solution amounted to minimizing training expenditure and implementation costs by selecting those members of the work-force who were best suited for the new tasks and who already possessed relevant know-how and experience.
2 The alternative solution involved giving production workers with average skills, such as those from the respective sectors of conventional production that were to be partially closed down, practical and theoretical training for the new tasks within the FMS.

Most companies pursue the 'creaming off' alternative, particularly in cases when technological innovation is introduced in small steps. Such a policy, however, is inadequate when extensive innovations with new job structures are to be carried out, and in our case, would have placed the workers under considerable stress.

In accordance with the special conditions of the project, a course of action was chosen in which comprehensive on- and off-the-job training measures played an important role. All in all, the decision ensured that the introduction of the FMS installed not only innovative technology, but also manpower policies of an equally innovative nature.

SKILLED GROUP WORK IN THE FMS

Both traditional and newly developed principles and methods were combined in designing and implementing the FMS organization of work.

Work organization and skill requirements

Compared with the traditional job structures of the plant, the job design chosen reduced the division of labour in three respects:

1 The hierarchical division of labour as traditionally given between unskilled and semi-skilled workers, machine operators, reliefmen, jobsetters, group leaders, and foremen, which is differentiated according to managerial authority and skills levels, was more or less dissolved.
2 The division of labour by task, particularly between the different machining methods and machine jobs (such as turning, milling, broaching), no longer existed.
3 Lastly, there was a reduction of the functional division of labour

between the production workshop and the technical offices and departments (such as shop-floor scheduling, presetting of tools, programming, and quality assurance).

This reduction in the division of labour corresponded to the way the technical equipment was structured and resulted in a new form of skilled group work: the tasks to be performed were not rigidly assigned to individual members of the staff, but were handled by the group of workers as a whole and were subsequently divided up according to the given work situation.

Workers' interests were well represented in this form of work organization. Jobs were created that had balanced work tasks with a high amount of diversity, requiring a certain level of skill while offering reasonable working conditions without overtaxing physical work burdens. This structure also complied with various company interests: use of manpower was able to adapt flexibly to changing requirements for intervention (for example, job setting, or loading and unloading tasks that occurred irregularly), while vacancies caused by illness, holidays, etc. could be more easily compensated for.

By the very nature of its structure and content, this work organization made new demands on the skills and qualifications of the FMS operating group, since skills of a relatively high level are called for when all group members have to perform diverse work tasks according to a rotation system. Basically, this resulted in a new profile of skills and knowledge:

1 First, there were a number of processing techniques and machines which system workers had to be able to operate. These different metal cutting production methods were also found in conventional production; there however, a machine operator was usually specialized in one type of machine and a job setter would normally have to deal with only one or two related processing methods.
2 In addition, a command of robotics was required as well as competency in a number of technologies and methods (tool presetting, programming, quality assurance, etc.), which in other systems were normally handled by specialized service departments.
3 Finally, a certain knowledge of the overall system was required, including a knowledge of the system's integration into super-ordinate company structures (such as production planning and scheduling, material and workpiece supply, repair and maintenance, for example). A grasp of such comprehensive information, which went far beyond the usual demands made on production workers in conventional production processes, was an essential precondition for the system group's capacity for taking responsibility for the overall running of the system.

Therefore, a new type of production worker was called for, one who would combine the skills common to semi-skilled machinists as well as skilled craftsmen, and moreover some of the skills and know-how held by technicians and foremen. At the time these innovations were being implemented, there was no institutionalized trade in the metal industries for production workers of this type, nor did any respective regular vocational training scheme exist in the Federal Republic of Germany. A new occupational system in the metal trades was not introduced until 1985 (see chapter 12). Given the inadequacy of the existing system of industrial occupations and vocational training, and the lack of further training measures within companies, innovative training policies were developed during the course of FMS implementation.

Personnel selection and training

The new jobs were advertised within the company and more than fifty workers subsequently applied. Two pilot groups of ten persons each were then chosen. The selection criteria were contradictory in part: on the one hand, the objective was to get the new and complex system running as efficiently as possible, thus requiring experienced and well-qualified personnel, particularly in view of the fact that there was no fixed training program. On the other hand, there was an intention (particularly on the part of the works council) to use the experience gained in FMS training for a general extension of in-house further training, meaning that production workers with average skills (semi-skilled workers) should be selected.

In the end the persons chosen for the first pilot group were predominantly workers already possessing a certain level of relevant skill and, in some cases, many years of practical experience. The second selection criterion actually played a somewhat greater part in the formation of the second pilot group which started training at a later date than the first group. Only one-third of the selected twenty workers had completed vocational training in the metal trades.

The manpower structure of the pilot groups did, however, deviate from that of the plant's average production workers, and in particular from that of machine operators in the respective area of production in conventional production. On average, the FMS workers were younger, had been employed within the company for a shorter period of time, and above all, had better training. Several foreign workers were included, but in comparison to corresponding areas of conventional production, their participation in the FMS groups was decidedly low.

Due to its status as a large-scale development project with a long planning and running-in period (and moreover being government supported and accompanied by researchers and other company external experts), the FMS became something of a learning workshop. The following elements of FMS training differed from traditional on-the-job training as well as from the regular further training measures within the company:

1 The trainees were released from work while continuing to receive full pay and thus given the comparatively favourable chance of learning without the immediate pressure of an ongoing production process. Phases of on-the-job training in the system were followed by modules of classroom courses which lasted from one to several days.
2 On- and off-the-job learning within one and the same group was continued for a long uninterrupted period. In this way cooperation and mutual support could develop.
3 The courses, some of which had been specially developed, complemented the practical knowledge and experience gained in FMS operations, thus enabling a somewhat experimental mixture of theoretical learning and learning by doing.

These relatively positive learning conditions were certainly responsible for the fact that there were no dropouts in the training program and that even the workers with average qualifications and experience (including the foreign workers) completed the course successfully.

All in all, this comprehensive FMS training scheme, which included basic theoretical knowledge, led to the same skills level as that held by skilled industrial craftsmen. The courses dealt with several production methods (not in the same depth, however, as in the apprenticeship programs for lathe operators, milling workers, etc.) and also included programming, technical drafting, basic information on control technology, etc.

The FMS training ended with a two-day examination. The certificates could not be recognized as regular trade certificates because the training program did not correspond to any classification in the metal trades at this time.

HOW THE NEW SYSTEM WORKS

In March 1984 and June 1984, the first and second pilot groups successfully completed their FMS training with an in-house examination. During the second half of the year, the groups were reduced to the planned strength of six workers per shift, and production conditions increasingly approached a normal state.

System operation

Several alterations were made concerning the originally planned range of workpieces; the demand for certain parts had dropped off significantly, so the decision was made to concentrate on manufacturing heavy workpieces with the FMS in order to relieve the workers employed in conventional production. This resulted in a partial reorganization of the system: in one cell, a machine tool was exchanged in order to adapt the capacities of the various metal-cutting methods to the altered requirements. The experience with the adaptation proved the system's modular design worthy with regard to demands on flexibility.

The actual design of the system's overall control unit (with the DNC and scheduling functions) deviated considerably from what had been originally planned. The development of the software required far more time than anticipated. The control system was initially designed to allow a high degree of automation, which, however, proved inadequate in view of the system's complexity and the great number of possible system failures. Therefore a system with a lower level of automation was realized which allowed the FMS workers more means for intervention through interactive communication with the control system (mainly for the scheduling and DNC functions).

By mid-1985, the system had reached its 'normal' capacity and was processing mostly heavy gears of different shapes (50 parts from one parts family) in lot sizes of between 50 and 500. Monthly production averaged about 12,000 pieces, which represented a high percentage of the overall capacity for this parts family but a small percentage of the company's total gear production. The system operated in two shifts. Unmanned production took place in the third shift but was limited by machine breakdowns and the need for manual set-up.

Throughput times were down from the four to six weeks necessary in conventional production to two to five days in the FMS. The short-run flexibility (indicated by the parts range that could be processed without changes in the hardware and software) was high and never fully utilized. The long-run flexibility (indicated by the feasibility of system alterations for new types of parts) was also high; due to the modular structure of the system, it was relatively easy to change machine tools and adjust to new demands.

A comparison of the FMS to conventional production with regard to the man hours required to produce the monthly average of 12,000 gears showed a productivity increase of about 70 per cent and savings in manpower of approximately 40 per cent. In conventional production, it took the capacity of about 17 direct workers and five indirect workers to produce the FMS's output of gears, whereas the FMS was operated by two groups of six workers and one foreman. The productivity gap slowly decreased over time because of the continuous modernization of the machine park in conventional production (resulting in productivity increases of 3 – 5 per cent per annum).

Several factors impeded higher capacity utilization. Some of the machine tools, most of them newly developed prototypes, proved to be highly susceptible to failure; the lack of service personnel, particularly in the electronics sector, could not be fully compensated for, and the integration of the FMS into the respective area of production was not realized for reasons of cost. Experience gained from the new system cannot be generalized without considering these factors.

Training and job structures

All in all, the training measures proved successful. The efficient running-in of the system with its many newly developed components, the level of capacity utilization achieved and the system's overall performance would not have been possible without the comprehensive training programme. The FMS workers played a considerable part in overcoming the numerous faults and breakdowns which occurred during machine testing and the instalment of the system; working instructions and operator manuals were lacking and had to be compiled, and programming and program optimization of the NC-machines had to be carried out by means of trial and error, etc.

The principle of keeping the division of labour as low as possible and preventing individual workers from strongly specializing in a restricted number of tasks within the overall system proved its worth. However, a completely open structure of assigning tasks according to immediate requirements was not established. Instead the following changes were brought about:

1 The system leaders constituted an independent job category; both groups now had their own system leader and a substitute. These workers (having been selected out of the two pilot FMS groups) were responsible for controlling the entire system by means of interactive communication with the data-processing control system; they also took over tasks at other work stations (including loading and unloading the system). The system leaders were graded one level above the rest of the group.
2 The division of labour was organized along different lines within the two system groups (figure 11.4). On the one shift three workers were deployed to each side of the system; task assignment was relatively open

and was determined by immediate requirements. After eight weeks the workers changed system sides. On the other shift, the entire system was subdivided into four areas. The responsibility for each one of these areas was held by one worker and the positions were exchanged according to a four-week rotation scheme.

3 A foreman who worked on the normal shift and who was primarily concerned with administrative functions (such as ordering materials and tools and maintaining contact with other departments, etc.) was additionally assigned to the system. This foreman, to whom the two groups had to report, had many years of experience in conventional production, but did not have any particular knowledge of the special control, handling and other technologies relevant to operating the FMS. The foreman acted primarily as an interface between the FMS and the surrounding departments.

In the operation of the FMS, the division of labour according to functions was kept low in comparison to traditional structures in the plant, but did not disappear entirely. The generation of new NC programs was relatively rare and took place, when it did, in the process engineering departments. The task of program optimization and administration was assigned to the workshop.

It was not possible to achieve the integration of a maintenance specialist, as requested by both shifts, into the group. Small disturbances were generally dealt with by members of the groups; larger ones by maintenance specialists. Quality assurance and tool presetting were almost exclusively carried out by the system groups. The system leader was responsible for scheduling in the FMS operational area.

Although a higher degree of division of labour had been established than originally intended, there were nevertheless strong differences to the far more differentiated, traditional organization of work in the area of conventional production. The FMS workers were able to step in and take over tasks when and wherever it was necessary. In the case of resetting and retooling and other tasks that arose periodically, this kind of flexible manpower utilization was of great significance for the entire system's performance level and allowed substitution problems due to illness, holiday, etc. to be dealt with more easily.

Moreover, this type of skilled group work complied with workers' interests in having diversified jobs and enabled the integration of tasks that would be unduly strenuous by themselves, such as the loading and unloading of the system. The union demand to include the system leader's tasks in the group's responsibility was mainly opposed by members of lower and middle management and would have posed wage-grading

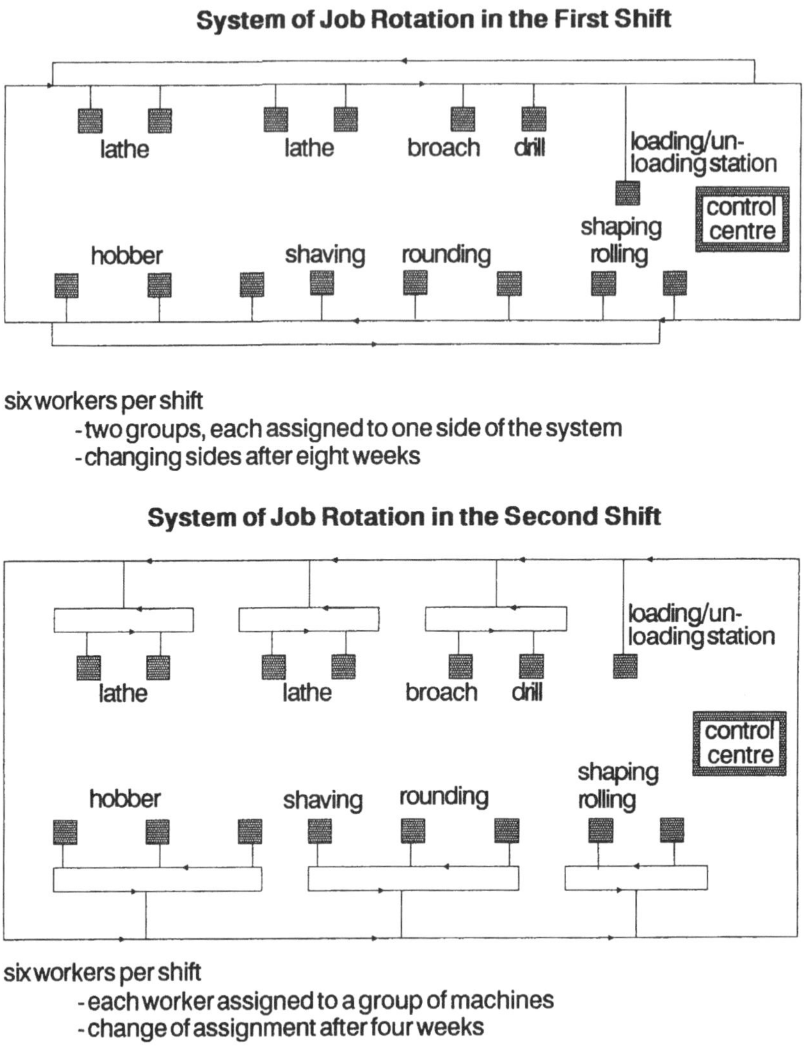

Figure 11.4 Systems of job rotation. ISF 1991

problems which would have been difficult to solve. On the other hand, a return to the specialization of individual machine tools and machine types, which appears advantageous in terms of utilizing existing know-how of certain metal-cutting methods, was not regarded as an adequate solution as it would have stood in contradiction to the system character of the production equipment implemented. The resulting organization of work was a compromise; it was capable of coping with the many-sided and complex demands made on the system while avoiding unbalanced, one-sided or unduly strenuous jobs.

PROS AND CONS OF THE NEW SYSTEM

As has been explained above, getting high and fast returns on investment in the FMS was not the main goal in setting up the project; instead, development strategies and the desire to gather experience with flexible automation processes at an early stage were the major considerations.

At the end of the 1970s and the beginning of the 1980s, high development costs had to be accepted. This not only applied to the individual components of the system (many of the machines were prototypes) and to the control systems (CNC, DNC, PDE, etc., and the overall systems control), but also to the development of work organization and manpower.

Technology

The company had both positive and negative experiences with the new technology. The consistent utilization of modular, autonomous production cells proved worthwhile although they involve more detailed planning and some redundancies. This structure makes it possible to run individual production cells in an autonomous manner, if desired, and to limit the consequences of disturbances in the system. It also facilitates adaptation to parts changes and, in the case of small-sized companies, step-by-step adoption of FMS.

FMS based on complementary machinery did not prove very successful. Flexible manufacturing systems run with complementary machines entail the basic disadvantage that the capacity utilization of individual stations (except the bottleneck station) is significantly dependent on the product mix. Also chain reactions are likely if breakdowns occur. Production carried out mainly with substituting machines enables far-reaching possibilities of capacity alignment within and between different flexible manufacturing systems. The time needed for parts transfer between the individual flexible systems is of little importance due to automated transport and storage systems.

The size of the system turned out to be disadvantageous. In spite of the modular structure, disturbances in individual components were passed on throughout the system and reduced overall availability. This is one reason why primarily small flexible manufacturing systems were installed during the 1980s (Fix-Sterz *et al.* 1990; Martin *et al.* 1987).

Development of organization and personnel

In view of spiralling investment costs – in some cases an exponential rise – connected with the integration of production and information technology, there are three variables which are decisive for the economic efficiency of alternatives of work organization: investment costs, system availability and the short and long term flexibility of the system. Other factors such as manpower costs and the costs of implementing organizational innovations also have to be considered.

The use of skilled workers helped to save planning and investment costs for the FMS since the level of automation could be kept low in problem areas and complex tasks could be left for human intervention. A prime example of this could be seen in the functioning of the overall control system (mainly for the scheduling and DNC functions). Due to their production knowledge and company experience, the workers were able to recognize the 'childhood diseases' of the new technological system and to solve these problems in cooperation with engineers and mechanics from the plant and the system manufacturers.

In view of the high costs of the FMS it was of particular importance that the group be capable of achieving a considerable reduction of the risk and duration of breakdown times. The workers themselves could prevent a high number of breakdowns by immediate intervention and deal with many system failures themselves. In other cases, they were capable of diagnosing the causes for failures and calling for the required specialists. Competence of this kind was especially valuable when maintenance workers were not immediately available such as when specialists were engaged in other departments, or outside of normal working hours when maintenance personnel was considerably reduced.

The conversion of production facilities could be performed very fast and thus changes within the planned product range (short-term flexibility); resetting did not depend on the specialized knowledge of a job setter who might be occupied elsewhere, but could be carried out by the FMS group itself when and where required. As far as long term flexibility was concerned (system alterations), the workers who were familiar with all aspects of their system were able to support the implementation of changes to a considerable extent.

In order to achieve the above-mentioned advantages, namely the reduction of investment costs and an increase in system availability and flexibility, it was necessary to plan personnel levels according to maximum requirements. Despite this, the chosen structure of skilled group work did achieve savings effects in comparison with alternative job and manpower structures. First there were savings in direct manpower. The fact that all group members could replace each other allowed lower personnel levels than would have been the case in an organization with a greater division of labour. Second, cost savings were made in indirect manpower because many of the indirect tasks were carried out by the FMS group. This resulted in a direct savings effect in technical service departments and an indirect increase in efficiency – particularly due to improved communication.

The advantages offered by the new job structure have to be seen in light of disadvantages resulting from the particular costs and problems involved in introducing and stabilizing organizational innovations that deviate significantly from the predominating structures of the plant. The decision-making processes for organizational and personnel development required a lot of time and personnel resources. The main decision making body, a joint commission of management, works council and outside experts (see below), accompanied all stages of the project. Training of the FMS groups amounted to up to 5 per cent of the total investment cost for the system. This included the wages for the two pilot groups as well as expenditures for infrastructures like training programs, manuals and teachers. The control technology of the system had to be adapted to the capabilities of production workers. In particular the overall control unit required a lot of additional programming for developing an acceptable interface. Finally frictions between the different production departments have to be pointed out. Wage levels within the FMS were higher than those for skilled jobs in other departments, which caused expectations for wage increases and thus conflicts.

In our case the short-term disadvantages resulting from implementation problems and costs arising from organizational innovations were outweighed by the long-run advantages of the system's economic efficiency. With hourly costs of more than 1,000 DM for the FMS, the costs for the development of personnel and organization seem somewhat insignificant.

In other cases, however, the immediate costs and risks involved in the implementation of organizational innovations can, depending on the initial situation of the company involved, bear more weight in management's decision-making process than the benefits gradually incurred by the new organizational structures. It is difficult for individual companies to estimate whether models of skilled group work are profitable in the final analysis. The effects on factors such as system availability and flexibility are difficult to quantify; also, the various cost factors tend to converge. In some circumstances, the advantages are confronted with strong disadvantages in the form of special costs and problems arising from the introduction and stabilization of structures that differ from predominant forms of production work. These problems have probably helped contribute to the fact that such alternative job structures are frequently discussed but rarely implemented.

THE PREREQUISITES OF INNOVATION

Although new technologies and new organizational patterns are the subject of much discussion in the Federal Republic of Germany, Tayloristic rationalization strategies still dominate in the everyday reality of company life and only a minority of companies have started to experiment with alternative structures (see chapter 10). Which factors, then, explain the far-reaching technological and organizational innovations achieved in the company examined in this study?

Studies on technological and organizational change in West German industry show that a large number of external factors, such as the sales market, manpower supply in internal and external labour markets, the degree of automation and integration of production processes, company size, etc., can influence the design and development of rationalization strategies. These factors alone, however, do not sufficiently elucidate the organizational processes involved in adopting new technologies. Implementation structures and processes involved in the adoption of CIM components and their integration play an important and quite autonomous role in determining the development of the organization of production and work (Hirsch-Kreinsen, Schultz-Wild 1990).

In the company under study, a number of factors were particularly relevant. These include agents and networks, resources and planning concepts.

Agents and networks

Normally the process engineering departments are the main agents for the selection and implementation of new technologies. Since these departments exist from the differentiation between conception and execution and a strong control over production processes, they tend to adopt new technologies conservatively, using Tayloristic concepts.

Because of the unusual costs and the strategic importance of the FMS in our case study, the network of internal and external agents deviated significantly from the norm. Top management was willing to take risks for technological and organizational innovation, and observed and backed the implementation process. All layers of production management, down to the foremen, participated in the planning and implementation process. Some members of this group became advocates of the concept of skilled production work. Many years of experience have shown that complex production equipment never runs the way that process engineers think it should and that, in the end shop-floor personnel have to come to terms with the inevitable problems.

The works council and the union also played an important role in the implementation process. Both parties feared large job losses, as well as deskilling and downgrading tendencies. Both expected the pilot installation to be a starting point for a rapid diffusion of FMS technology within and outside the company and therefore attached a high importance to the project. Because of public subsidies, the works council was in a strong position and negotiated a plant agreement on training, job structures and wages and the formation of a joint commission for the planning and implementation process.

A number of engineering and social science research institutes also became involved. The external experts exerted a significant influence on the technological and social aspects of the FMS. Their contribution derived not only from their competence, but also from their independent status.

All of these parties participated in the above-mentioned joint commission with an equal number of representatives from management and members of the works council. This commission was given a lot of authority on all questions concerning organizational and personnel matters. The commission, which met at regular intervals, played a significant role by clarifying questions involving labour management and problems concerned with the integration of the new production system in the company; it also acted as the central contact for the supporting technological and social research that was being carried out.

Two working groups consisting of both in-plant and external parties involved in the project were set up to investigate and report on the subjects of 'training' and 'job structure and organization'. Various departments, such as production, operations scheduling, training and manpower were represented in these working groups, which also included those responsible for the technological development of the project and the works council. Outside parties, such as a representatives of the national IG-Metall (trade union) and social scientists carrying out supporting research, were also involved in a great number of these meetings.

Planning concepts

In addition to the constellation of agents and networks described above, the planning process was of importance in the successful implementation of technological and organizational innovations. As is the case with most companies, social considerations and manpower policies did not originally play a major role in the decision-making processes concerning technological innovations; thus, for a long time there was such a preoccupation with the numerous technological questions of system design and determination of the spectrum of parts to be processed and the selection and design of machine tools, robots, etc., that questions concerning personnel policies were given little consideration. It was more or less implicitly assumed that the usual mechanism of personnel adjustment (by muddling through) which had proven successful during previous innovation processes and the application of tried and true principles of organizational development based on the differentiation of tasks and functions would ensure an unimpeded implementation process.

At a relatively late stage in the project, but not too late to have an effect (about a year before the system started to operate), planning of the organizational and manpower structures began, thus leaving enough time for discussions on alternative policies with respect to recruitment, the design of the job structure, and training policies. Technical planning was complemented by a policy of consciously taking human factors into consideration.

In the course of this process, new methods of investment appraisal were considered. Conventional methods are mainly concerned with the ratio of quantified output and factor input and thus favour structures that are strongly oriented towards the division of labour and savings in direct manpower costs. In our case additional factors which are difficult to quantify, such as the effects of alternative job structures on system availability, flexibility, etc., were also evaluated and taken into consideration in the decision making process.

Resources

Resources of time, manpower and finance that were made available were of great importance for the implementation of technological and organizational innovations. Due to the length of the project, many discussions and learning processes that would not have occurred had the participants been pressed for time, were able to take place. Also the project's significance for the company made generous in-plant personnel resources available, meaning that expert knowledge from all departments, whether their involvement with the FMS was direct or indirect, was mobilized for the project. Finally, extensive financial resources were made available; in this respect, the provision of government subsidies was no doubt important.

Government programmes

Government programmes have played a significant role in fostering wide public discussion on the subject of new technologies and work organization. The programme 'Humanization of Working Life', which was initiated by the social democratic government in 1974, played a crucial role in the discussion and critique of Tayloristic work forms and generally brought the topic of improvement of job design to public attention (see chapter 13, 23).

Of particular importance for the case study presented above is the government programme on 'Manufacturing Technologies', out of which the FMS project was subsidized. This programme was started in 1980 and extended in 1984 and 1988 with an overall financial volume of about 1.4 billion DM for twelve years. It supports technological development, technology transfer, and technology applications for the German capital goods industry. One focus is small and medium-sized companies (more than 90 per cent of the firms in this sector have fewer than 500 employees).

The main objective of the programme is to support technological innovations in research, development and practical applications. Human factors have, however, been both directly and indirectly taken into account from the start (Brödner 1987; Martin *et al.* 1987):

1 Many CIM-components, such as PPC and DNC systems, are geared towards organizational structures of large companies with a strong separation of conception and execution between the process engineering and the production departments. Often hard- and software of this type have built-in centralistic organizational concepts that are quasi passed on to the user companies (see chapter 9). The project management organization of the programme (PFT) developed several projects involving groups of producers and users to support the development of manufacturing technologies open to alternative organizational structures. Among them were projects for CNC and PPC systems.
2 Many individual projects with far-reaching technological and organizational innovations in user companies have been supported if not initiated. Among them were pilot installations of manufacturing systems with group technology and group work (production island) (see Brödner 1987) and the FMS project presented above. These projects draw a great deal of attention from politicians, employers, engineers, unions and social scientists.
3 The so-called 'Initiative Technology Transfer' is designed to transfer existing knowledge on CIM technologies to the German industry. Sixteen engineering institutes have developed model applications for demonstration purposes and comprehensive training programmes. The project management of the Manufacturing Technology Programme ensured that the initiative was not reduced to technological questions, but also encompassed information and concepts on the development of organization and personnel.

CONCLUSION

It is difficult to assess the significance of the FMS-case presented above for the further development and dissemination of group-oriented job structures in the Federal Republic. The project has had a pilot function and has served as a means of orientation for other companies with similar problems. The concept of skilled group work implemented in the FMS case is no longer a singular occurrence; similar structures have been tried out in many companies. In the automobile industry, for example, it has been possible to introduce far-reaching forms of local control on all processes within areas with flexible cells and systems.

The project has certainly given a boost to discussions on 'alternative' factory structures that have been gaining momentum over the past few years on both a national and international level. In the meantime, the discussion in the Federal Republic not only involves industrial sociologists and psychologists, but also scientists concerned with ergonomics, operations research and engineering. The fact is more or less widely accepted that traditional Tayloristic factory structures will not be able to cope with the economic and social demands made on companies and that considerable changes will have to be made in setting up the organization of the factory of the future. In this context, the FMS case that has been outlined here is of considerable value as a model for change.

12 Bringing Skills Back to the Process[1]

Rainer Schultz-Wild

INTRODUCTION

This chapter summarizes some of the findings presented in Part III of this volume from a German as well as a European perspective. It argues that the development of future factory structures is not only a question of new CIM components and systems but is also influenced by a company's organizational concepts and structural conditions. Many factors point towards reducing the division of labour and making use of skilled workers on the shop-floor. Certain conditions in the technology and implementation processes have to be met when the concept of process-related use of skills and qualifications is to be realized. The educational and training system plays an important role in determining the availability of manpower for industrial work.

FUTURE FACTORY STRUCTURES AND INTERESTS IN PROCESS-RELATED USE OF SKILLS

In the light of the imminent European unification and the increasing expansion of computer-aided advanced production technologies, the issue of future structures of work in the factory in European industry is becoming increasingly significant for politics and science. In all likelihood uniform structures of work in the factory will not develop in Europe in the near future: the differences in existing conditions and developmental perspectives between regions, countries, economic branches and market conditions are too great. Problem areas and potential solutions in the area of production technology, the availability of labour and skills, etc. also vary considerably between countries. Even if the available production technology becomes more uniform due to market conditions and/or international standardization efforts, and differing working conditions begin to converge due to national and international compensation policies, and lastly, even if certain adaptation processes take place within education and vocational training systems, it, nevertheless, remains highly unlikely that the use of technology, work organization and the division of labour within and between companies will reflect a uniform structural pattern, regardless of whether it is a small company with a narrow regional market in one of the industrially developing countries of the Common Market or that of a large scale mass production enterprise in one of Europe's industrial centres.

At the same time it is widely agreed upon among experts that there are only limited chances of realizing the concept of the so-called 'unmanned factory' within the near future and that this concept is, at best, suitable only for certain marginal areas of highly standardized manufacturing. There are also numerous voices stressing the limitations of the long prevailing Tayloristic or Fordistic model of production and work organization, which is strongly centralized and based on a strict division of labour. These limitations become particularly important within the context of the use of modern CIM technologies and the necessity for adapting to rapidly changing market conditions (see for example Brödner 1987; Piore, Sabel 1984; Warnecke 1985; Warner *et al.* 1990).

In view of these factors, particularly with regard to European industry, factory structures are being advocated which reduce rather than extend the division of labour according to functions and tasks, decentralize decision-making, and make use of skilled workers on the shop-floor (see Wobbe 1990). Such concepts are less oriented towards maximum use of computer-based automation technology including the total exclusion of human labour, but on the contrary aim for a combined use of technology and labour whereby the production knowledge and experience held by workers closely involved in the manufacturing process is systematically integrated. Strategies and structures of this type can – in accordance with traditions of organizational theory – be termed 'computer-aided Drucker' (see chapter 3).

The integration of functions and tasks on the shop-floor and the process-related use of skills and qualifications offers companies a number of advantages,[2] particularly when strategies of so-called 'hard' automation which require standardization of product and labour run into limitations and strategies of flexible automation are pursued instead:

1 Process-related use of skills can keep down planning costs and investments in hardware and software for extremely expensive automation technologies when workers instead of machines are able to take over complex tasks and functions.
2 Process-related use of skills permits savings in implementation costs and allows a comparatively rapid application of technological and organizational innovations in the manufacturing process. When extensive

basic skills already exist on the shop-floor, the introduction of new technology components will result in fewer skill deficits. This will lead to both a considerable reduction in training expenditures and a shorter running-in period since parts of the implementation work can be performed by the user company's skilled work-force.

3 In view of the prevailing high costs for hardware and software components of computer-aided flexible automation, it is of particular significance that process-related use of skilled labour can reduce the risk and duration of system failures. Skilled workers familiar with the specific manufacturing equipment can increase availability and reliability; first, they can prevent disturbances by intervening in the ongoing production process and second, breakdown times can be shortened because of their ability to perform certain repairs by themselves, thereby reducing the need to call on or wait for specialized maintenance personnel.

4 Moreover, the process-related use of skills increases the operational flexibility of manufacturing equipment for various products or product variants within a given technical range and also in terms of coordinating machine schedules and/or production capacity with the requirements of the ongoing production flow.

5 Finally, the process-related use of skilled workers in the production process can also result in lower labour costs. Savings arise in the area of technical (support) services such as process planning, programming, shop-floor scheduling and control, maintenance and quality assurance, since costs are higher under more centralistic forms of production organization. Also, fewer workers are needed than in forms of work organization with highly discrete functions using a minimum of skills, because broadly skilled workers are able to replace each other and fill vacancies caused by illness or other reasons.

All these factors point to a company interest in integrating functions and tasks on the shop-floor and in employing skilled workers in the immediate proximity of the production process. Within this strategy many variations are conceivable according to the type of product, the level of automation, the organization of production (job shop manufacturing or 'production islands'), national characteristics of sales markets, technology markets, labour markets, and educational institutions, etc. The forms of work organization and the division of labour can vary, for example between the following two basic models:

In the first case the traditional forms of job classification and hierarchical division of labour are largely dispensed with and, instead, a homogeneous group of equally qualified production workers are employed.

The group shares the different and changing work tasks within a given manufacturing area according to internal agreements and also shares the responsibility for the fulfilment of the assigned jobs (see the FMS case study in chapter 11). For example, employees such as the German skilled worker (*Facharbeiter*), with their practical and theoretical production knowledge and experience are suitable for this form of group work, after they have had some further training in the field of information systems and control technology.

The second model adheres more to the traditional form of functional and hierarchical division of labour and the job specialization resulting from it. The work-force is differentiated according to area and level of qualification; the use of personnel is based on cooperation between skilled workers and workers performing specific tasks. In this case, mixed staffing of a production team is conceivable consisting of semi-skilled and skilled workers, technicians and even engineers, based on individual job assignments, however, with close cooperation and overlapping skills (see the companies with an innovative strategy in chapter 10). This model, in contrast to the first, involves a more strongly hierarchical distribution of responsibility with a considerably restricted capacity for employees to replace one another.

International comparisons show distinct patterns within this range (see chapters 14, 19). Both models, between which numerous variation and concrete forms of design are imaginable, are suited in varying degrees to different forms of technology use and production process requirements. The first model, for example, makes greater use of the potential range for the design of work organization in modern manufacturing technology when the coupling of work tasks to the machine's operational cycle is reduced due to the automation of workpiece handling and transport. The second model possesses more advantages in those cases where gaps in automation must be bridged by routine jobs or where the use of specialists is imperative for one reason or another. However, the second model holds higher risks of a polarization of skills and the creation of permanent barriers between different employee groups.

PRECONDITIONS FOR A PROCESS-RELATED USE OF SKILLS

A common characteristic of the two models of work structuring described above is their increased 'shop-floor autonomy' by means of a process-related use of skilled labour. They reduce the significance of planning, control and supervising departments outside of the shop-floor as opposed to the more centrally oriented Tayloristic/ Fordistic factory structures. Both models correspond to a concept which is currently being discussed in

Europe as well as the United States under the heading of 'The Factory within the Factory'[3]. This concept's intention is reflected by the challenge: 'Computer Integrated Manufacturing only with Human Integrated Manufacturing (CIM only with HIM)'.[4] The realization of such forms of production structures and work organization depends, however, on a number of preconditions. In making clear what these preconditions entail, we shall briefly outline (a) market developments of CIM-components and -systems; (b) the implementation processes of new manufacturing technologies taking place within companies; and (c) the availability of different skills for production work as well as the question of compensation systems, career patterns etc.

Alternative CIM-components

In the 1980s the spectrum of products offered on the market for CIM-components and -systems expanded considerably. This holds true for design and process planning functions (CAD, CAP) as well as for production planning and control (PPC) and manufacturing functions such as machine control, tool management, workpiece handling and transport, as well as quality assurance (CAQ).[5]

In many cases the use of CIM-technologies is characterized by a kind of pioneer situation in which the technology suppliers adapt the systems' actual design to meet specific customer requirements. At the same time, however, there is also an increasing development of marketable, more standardized products requiring the system users themselves to adapt if considerable costs for tailor-made solutions are to be avoided. Moreover, in many areas in which CIM-technologies are applied one can safely say that it is the technology suppliers who control the market. This is the case for special types of products due to the high development costs of complex software solutions as well as the lack of user competency in information technology. Therefore, it is particularly important to consider which production and work organization concepts are quasi passed on to the user companies by the technology they install.[6]

CIM-components on the market can be classified by those which are more open in terms of production organization and work structuring, or those which contain strongly centralistic concepts which aim for a definite division between work executed on the shop-floor and planning, control and supervision functions.[7]

Results of our analysis point to a market dominance of CIM-components of a centralistic-deterministic nature which are exerting a strong influence on a consolidation or increase of the hierarchical-functional division of labour between conception and execution. Such solutions are usually developed and offered by large computer companies which have a strong market position and are backed by years of cooperation with the users of their systems. But it remains doubtful if these solutions meet the special needs of small and medium-sized companies (see chapter 6).

Alternative systems are usually offered by small software houses or by machine-tool producers. These systems are more open in terms of work organization and do not restrict their users to pursuing specific solutions as far as manufacturing and work organization are concerned. Therefore these systems are more compatible with forms of decentralized planning, control, and monitoring, and a process-related use of skills.

Although in the case of more centralistic-deterministic systems one cannot say that the user companies are bound to certain forms of organizing manufacturing and work, it is undeniable that the logic inherent in such solutions will influence the decisions concerning the implementation process within the company. A process-related use of skills is easier to realize or maintain when hardware and software components are designed to be easily accessible to shop-floor workers.[8]

The role of the implementation process of CIM-technologies

In spite of the constraining factors built into the CIM-components offered on the market, user companies still have considerable scope for the design of their production and work organization at their disposal. However, the question remains as to whether, and in what way, this scope for decision-making is actually utilized during the course of the implementation process.[9] Although the discussion among managers and engineers concerning the importance of 'human factors' for the successful introduction of new technologies has been given more emphasis during the past few years, on the basis of present experience, it can hardly be assumed that the development of work organization and manpower will enjoy the same careful preplanning which is given to decisions concerning technical and economic factors.

Many implementation processes are characterized by a step-by-step introduction of new technology components, whereby the existing manufacturing structures and forms of organization are retained as much as possible. For the most part, changes are limited to what is imperative from a technological point of view. This means a sort of structural conservatism. On the other hand, empirical evidence demonstrates at the same time that it is technical innovation which often initiates or is the medium of a reorganization of work processes, which may be more far-reaching than the actual innovation case. Changes of a slower and more subtle kind which lead to a gradual undermining of dominant forms of work may occur as well as real structural breaks.

The probability of more radical changes in job structures increases:

1 with the extent of the leap to automation of a given technological innovation, thereby necessitating a reorganization and extensive redistribution of work between machinery and manpower;
2 with the degree in which step by step modernization is deviated from or must be abandoned, for example due to the more system-oriented character of the new CIM-technologies being installed (e.g. flexible manufacturing systems – FMS); and
3 according to the extent in which the new technologies contain concepts of work organization contradictory to prevailing principles.

Process-related use of skills can hardly be implemented in cases where technology control systems of a centralistic-deterministic type are integrated into company structures based on a high division of labour. Centralistic control systems strengthen the planning and supervision departments outside of the manufacturing area and lead to a further reduction in autonomy on the shop-floor. The introduction of these technologies in companies where skilled shop-floor workers still enjoy a considerable scope for decision-making can result in a gradual undermining of shop-floor autonomy and can, at least in the long run, lead to a gradual erosion of skill-based work systems.

It is also conceivable, however, that due to the functional deficiency of centralistic control systems, alternative strategies will be pursued from the start of the innovation process. In this case, the preconditions in terms of organization and skills have to be systematically planned and subsequently realized during the implementation process.[10] In this context it is important which in-plant groups plan and implement technological innovations and what their specific interests and aims are. The realization of company structures based on the model of process-related use of skills is, for example, more likely to occur when shop-floor managers directly responsible for the manufacturing process have more influence than members of central departments for process planning or production control.

Availability of a skilled work-force

Apart from the structures of technology supply and the specific interests which come into play in the course of implementing CIM-technologies, the supply of personnel with different skills and qualifications is certainly significant for shaping and designing manufacturing organization and work structures. Forms of process-related skills use are easier to realize or maintain in those companies which already have a work-force with broad and relevant skills or where skill deficits can be eliminated with relatively low training expenditures.

In cases of stagnating or sinking overall employment in a company, the skill structure of the work-force is particularly decisive. This is so because an extensive exchange of personnel via the external labour market (dismissal of insufficiently trained workers, recruitment of adequately skilled) would certainly meet with resistance from worker representatives and unions and is therefore unlikely to occur.

Those companies that have centralistic, Tayloristic organizational structures and have mainly employed semi-skilled workers in narrowly defined jobs in their production departments are likely to have more problems and higher expenditures in carrying through a reorganization to permit process-related use of skills compared to companies whose production work has traditionally been largely in the hands of skilled workers. This holds true for skilled group work as well as work structures on the shop-floor based on a certain degree of division of labour. Reintegrating certain planning, control and monitoring functions with production work when computer-aided integration technology is used, results in new types of jobs and job requirements. These cannot readily be fulfilled by workers previously employed within more centralistic structures based on a high degree of division of labour and narrowly defined jobs.

Although the skill structure of the given work-force will certainly have an effect on the choice and introduction of new technologies and resulting reorganization of work processes, this factor can only be regarded as invariable from a short term perspective. Changes can be brought about rapidly by further training and retraining carried out by the companies themselves, or also – in the medium and long term – by shifts on the external labour market and within the educational and vocational training system.

The role of education and training in the development of future factory structures

Trends in technological innovation in industry are linked with varying degrees of affinity to work-force structures that facilitate a process-related use of skills. While certain deviations in the difficulty or ease of introducing a process-related use of skills exist within nations, i.e. across different industrial branches, or between companies with different manufacturing, organizational and work-force structures etc., they are likely to be greater across countries. The educational and vocational training systems, which are conditioned by specific national norms and traditions, are significant in this regard because of their central mediating function in determining the size and skill structure of the work-force available for industrial work.[11]

The expansion of educational opportunities and the opening of schools, universities and colleges for sections of the population which previously pursued other courses of education has encouraged many young people to remain within the education system for as long as possible. They may then strive for positions outside industry, particularly those outside manufacturing with its adverse working conditions (noise, dirt, shift work etc). This may cause *risks in manpower supply for production jobs*. The problem of having an adequately skilled industrial work-force cannot simply be reduced to the question of appropriate training and functioning control mechanisms in the schools and training programmes. The problem has also to do with working conditions and career chances offered in industrial jobs. Thus, the chances for recruiting workers for production jobs must be viewed, in the long run, in the light of alternative working conditions, earnings, career patterns etc. (see chapter 17).

The issue of the *content skills and qualifications* must be regarded in terms of educational and vocational training policies. There are many indications that in the course of computer integration processes the lines traditionally drawn between existing trades and occupations will become obsolete and that new qualifications and skills will become significant. The type of work involved in process-related use of skills seems advantageous to companies from many points of view and pushes towards a dissolution of traditional job classifications and demarcation lines between occupations (see chapter 11). While technical knowledge and skills related to specific products and manufacturing methods obviously remain important, data processing know-how, control technology and related technological knowledge necessary for monitoring, controlling and maintaining complex manufacturing systems are becoming increasingly important. At present no general solution exists for how such skill requirements should be grouped and combined to form new jobs and trades. Apart from the long time periods required for establishing reforms within the educational and vocational training systems, this is also related to the fact that the application of modern control and manufacturing technologies has not yet advanced beyond the pioneer stage in many factories (see chapter 9). Therefore, company experiments concentrate on the reorganization of existing work functions.

In the Federal Republic of Germany, for example, after more than ten years of planning and discussion, a bill concerning the reorganization of the industrial metal working trades within the German dual system of vocational training[12] was passed at the end of 1984. This resulted in the number of state-recognized skilled trades being cut from 42 to six, with the possibility of specializing in 17 areas (see figure 12.1). In the first year of vocational training, all trainees share a broad-based basic education covering the entire vocational field of metal-working occupations. The commonly required basic technical knowledge of various subjects was also increased and knowledge of data processing, system control and general process engineering were more strongly integrated into the training programme (compare Buschhaus *et al.* 1984).

A similar problem exists concerning the *combination of theoretical and practical aspects of knowledge and skills.* Very few of the European countries have a tradition similar to the German industrial skilled worker (*Facharbeiter*) who, in the usual 3½ years of vocational training in his trade, combines theoretical and practical skills as well as a considerable potential for adaptation and further learning on the job. This type of qualification profile certainly offers favourable conditions for the realization of work structures based on the model of work groups with a homogeneous level of skills, especially if there is a wide range of common technical knowledge shared by all workers. However, when theoretical and practical knowledge and skills for different careers are taught in different educational and training courses and in different environments (such as school and company), then more differentiated forms of work structuring are likely to occur (see chapter 17).[13]

The form in which *higher and advanced technical knowledge* is taught at institutions such as technical colleges and universities as well as the training patterns and the subjects emphasized will be significant for the development of future industrial structures. There are several reasons for this:

1 First, technicians and engineers have a considerable influence on the development process of new control, monitoring and manufacturing technology. Research traditions and forms of logic pursued will have an effect, perhaps indirectly in many cases, on the patterns of hardware and software that are developed (see chapter 5).
2 Second, the presence or absence of technicians and engineers from different professional fields, with differing orientations within company management, will be reflected in the definition of company problems and in the technical solutions developed. Higher flexibility demands, for example, can be met by different reactions, so that solutions will be sought in the area of production and control technology in one instance or in the field of work organization in another.
3 Finally, the engineers and technicians themselves constitute an important part of the company's total work-force and their supply on the

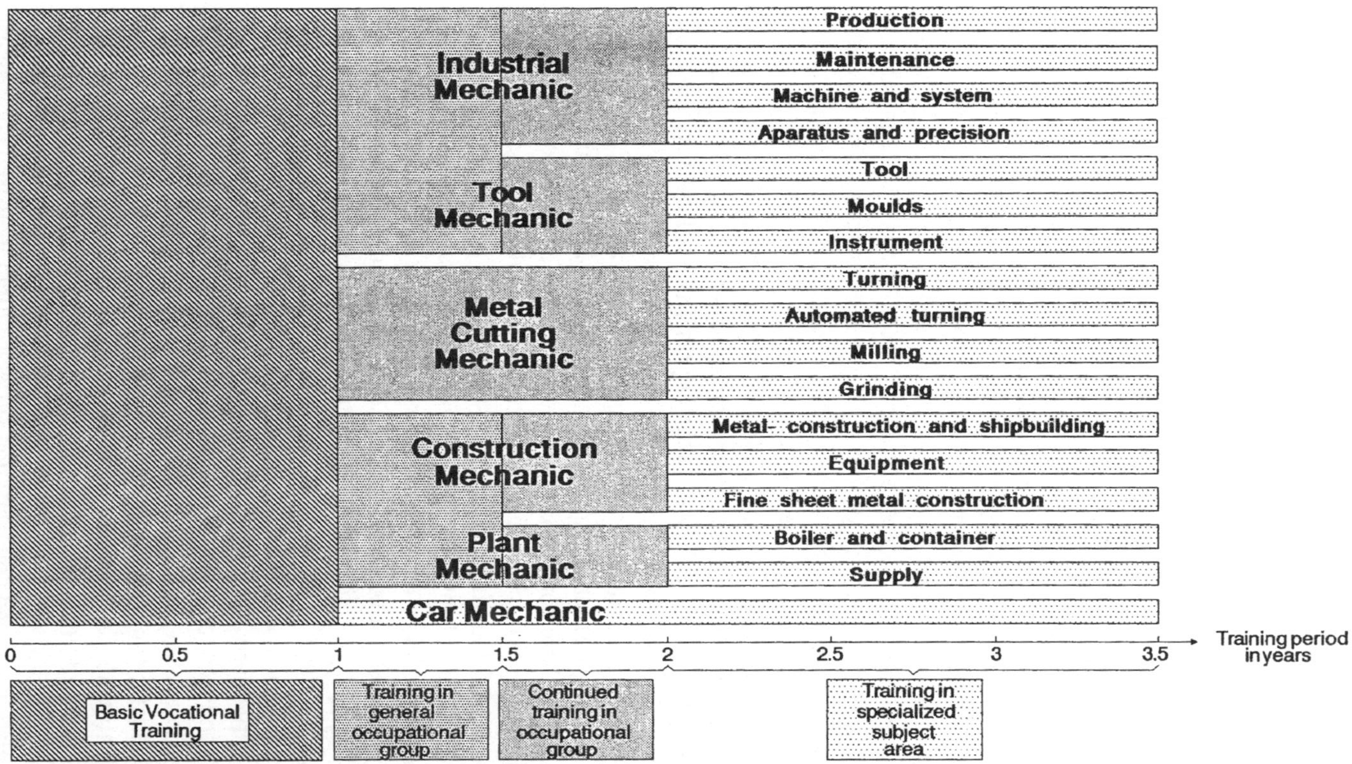

Figure 12.1 Training scheme for the industrial metal trades in the Federal Republic of Germany (as revised in 1984). ISF 1991

Source: Buschhaus *et al.* 1984; BIBB 1988

external and internal labour market effects the development of company hierarchies and job structures.

There is also the problem of communicating the relevant technological, organizational and managerial knowledge for the use of modern computer-aided technologies, especially in smaller and medium-sized companies and in less industrially developed areas of Europe. In the past, the producers of CIM-components have made some effort to offer not only hard- and software, but also the relevant 'teachware'. Also, at least in the FRG, a great number of intermediate, independent institutions for knowledge transfer and retraining have emerged in this field. But especially with the perspective of European integration there is still a broad field of action for policies in research, development, education and vocational training.

When the introduction of modern control and manufacturing technologies evolves from the pioneer phase and becomes more widespread, and particularly when forms of process-related use of skills are realized, then the *problem of adapting the skills of the existing work-force* must be dealt with in companies. When the initial introduction of new technologies only affects a few workers directly, qualification problems can be avoided and training expenditures reduced by creaming off the best suited workers for the new jobs. Such a policy of creaming off cannot, however, solve the problems that arise when larger areas of a plant are affected by innovations in technology and organization. The capacity and efficiency of in-plant and public training institutions and/or vocational retraining becomes decisive. In particular, those companies with a large number of unskilled and semi-skilled workers will be confronted with the necessity of considerable expenditures and training capacities (in-plant and/or company external) in order to achieve adequate further training suited to adult workers. The fact that further training measures can lead to remarkable achievements in skills, when certain preconditions are met, has been proved in numerous cases.[14]

Finally, it must be pointed out that forms of process-related use of skills within complex manufacturing systems necessitate *new forms of cooperation*, particularly when a hierarchical distribution of responsibility is rejected and more open and less rigid forms of job assignment are practised. The capability and willingness to cooperate are specific requirements, which, in part, contradict traditional emphases on individual learning and individual work performance. It is not sufficient simply to demand that educational and vocational training systems place a greater emphasis on collective learning processes. New forms of cooperation require the support of adequate wage and compensation systems within industrial companies. If group work is to be implemented, this also means that new career patterns must be developed or other ways and means must be found of ensuring that manufacturing work will, in the long run, continue to offer sufficiently attractive opportunities for earning and reasonable working conditions.

More differentiated observation and analysis is necessary to establish in detail how the problems mentioned here could be solved in the light of the different traditions and conditions in education and employment in the European countries so that favourable conditions will be created for the development of future factory structures.

Part IV

Restructuring of Work in Mass Production Industries

13 Work Structuring and Company Performance Policies[1]

Norbert Altmann, Peter Binkelmann, Klaus Düll

INTRODUCTION

This chapter concerns itself with forms of work organization which were introduced in the 1970s, particularly with regard to mass production and semi-skilled work, under the headings 'new forms of work organization' and 'work structuring'. They continue to play a considerable role in areas in which manual labour predominates. Moreover, they can be viewed as the forerunners of solutions in work organization during the introduction of 'new technologies'. The main purpose of this chapter, however, is to show that in their time 'new forms of work organization' not only served to solve problems of highly Taylorized work organization in the interests of workers (Humanization of Work), but entailed significant performance policy interests of the companies. This made the consequences for the workers necessarily ambivalent and potentially negative. This aspect of performance policy was never seriously dealt with in the discussions on new forms of work organization and it is one that still plays an important role today (e.g. see Parker, Slaughter 1988).

INITIAL CONSIDERATIONS AND STATEMENT OF THE PROBLEM

We set out with the thesis that enterprises adopt new forms of work organization (or 'new work forms'), which can potentially improve the work situation for the workers involved, once the utilization of labour in the production process has become problematic for the enterprise. The word enterprise is not to be understood here in its legal or organizational sense; rather, it is used to express the general interests that are pursued in the deployment of labour and the utilization of capital in the actual production process (see chapter 1). Our assumption is that problems for the reproduction of labour at the enterprise level are a microcosm of problems for the reproduction of labour in society. In society, problems get expressed in the form of politically articulated pressure to reduce stressful, overly strenuous and restrictively organized work; in the enterprise problems manifest themselves in the form of skill and performance deficits caused by the design of technical and organizational structures in the production process and insufficient provision of in-plant training measures, or as a growing lack of motivation on the part of the workers and decreasing acceptance of existing working conditions. Although problems for labour at the enterprise level have consequences for the reproduction of labour at the larger societal level, the strategies that an enterprise pursues with the introduction of new work forms are not intentionally directed towards dealing with the larger social problems for labour. The enterprise is interested in solving problems specific to the enterprise in as much as the individual and collective problems for the reproduction of labour (such as physical wear and tear, dequalification, demotivation) limit the extent to which the enterprise can fully utilize the labour at its disposal. Thus new work forms basically represent an enterprise's solutions to its problems, and although they contain a certain potential for improvement in specific work situations, they also represent new risks for the individual and social reproduction of labour.

When an enterprise introduces a new work form, it does not do so solely in order to solve problems involving the utilization of labour. Solutions to the problems of the market, production and time management are also considered. These problems arise whenever the production process, the organization of work, or demands on labour have to be adapted to market induced flexible production requirements and can also occur when attempting to gain control of the technical and time management aspects of the production process, irrespective of the dictates of the market. However, the problems involved in utilizing labour and controlling the production process are strongly interrelated. Once installed, technical and organizational structures determine how labour is used or deployed in the production process, which aspects of labour capacity are to be utilized in the production process, and which performance reserves can be mobilized in order to react flexibly to changing market conditions or to achieve the most optimal production and time management conditions.

In the light of these initial considerations, we will now turn our attention to an examination of the interests and objectives pursued by the companies in the implementation of such new work forms. We will evaluate the potential effects that these new work forms entail for the individual and collective reproduction of the workers involved and then examine the ambivalent nature of these effects.

THE MEASURES INVESTIGATED

In deciding on which concrete measures for change to examine, we proceeded with the most important area in the discussion on new work forms: namely, with new structures in sequential production processes.[2] This area subsequently became the main focus of our investigation. The new work forms that we observed fell basically into five different categories, described below:

1 The first type consists of 'traditional' group work. This form of work is mainly characterized by discretionary and cooperative work arrangements for skilled workers especially in assembling machines. We decided to include group work in our investigation in order to be able to analyze the conditions and limits the company puts on 'self-controlled group work'.
2 The second type consists of the modification of work stations in sequential production processes by restructuring the use of labour at assembly-line work stations. This form of work is characterized by the combination of certain aspects of group work with a high degree of division of labour organized according to the principles of sequential production processes. The decisive modification in this form of work organization lies in deploying labour to achieve systematic job rotation, job enlargement, job enrichment and the creation of more comprehensive training for semi-skilled workers. Restructuring measures of this kind are also based on concepts and models of the 'semi-autonomous group'.
3 Type three consists of the modification of sequential production processes in the form of jobs that are released from assembly-line-determined cycles. The basic organization of sequential production is maintained, but manual jobs are uncoupled from mechanical or automatic machining stations and transport systems and various types of buffer systems are implemented, thus allowing greater worker discretion with regard to the time taken to perform certain tasks. Combinations of this work form with number two type measures are also possible.
4 The fourth type consists of the elimination of sequential production processes through the creation of individual work stations. This work form is characterized by the consolidation of previously discrete job tasks into a single job with a corresponding increase in the scope of work and in cycle time. The establishment of parallel work stations requires the installation of special devices or transport systems in order to provide the workers with material and work pieces. The job enlargement and job enrichment (where applicable) aspects of this work form enable workers to have more room for discretion in terms of both the content and time required to perform their jobs.

5 Type five consists of technical means of eliminating restrictively organized work in sequential production processes by replacing manual jobs with mechanized and automated machining processes. These measures were only of interest for the investigation in that they demonstrated the limitations of new work forms in extreme situations (such as assembly tasks with an extremely high degree of division of labour and short cycles) and will not to be included in the following analyses.

Although the concrete forms that measures take in an enterprise are dependent upon the varying conditions of a particular company and process, the characteristics that we attribute to a particular type of work form are also basically related to company problems and strategies, i.e. to company interests. In this respect, we can make heuristic use of the various work forms to reveal significant relationships between company interests in the utilization of labour and the introduction of new work forms.

THE STRATEGIC INTEREST OF COMPANIES IN NEW WORK FORMS

Generally speaking, the results of our investigations made it clear that the enterprise's strategic interest in utilizing labour more fully and extending its spectrum of demands on performance gets concretely expressed in new performance policies. Of course, the external conditions of the enterprise (especially the market situation), structural characteristics (such as the personnel structure and investment potential, etc.), as well as the constellation of company problems (as defined by company decision-makers) determine the general thrust and which instruments an enterprise will use to pursue its interests in making more intensified demands on performance. Our study's findings and the knowledge and experience gathered from company practice in 'work structuring' revealed the following basic strategies.

On the one hand, new work forms enable an enterprise to harness individual performance in as comprehensive a manner as possible while utilizing the capacity for cooperation of its workers. The most effective way of pursuing this strategy is to introduce group work. By utilizing the worker's capacity for cooperation, this work form can allocate labour in response to ever-changing market conditions and can prevent the occurrence of friction, disturbances, and losses caused by coordination problems. At the same time, the availability of a broad range of qualified labour which works cooperatively can secure high quality results.

On the other hand, new work forms are also oriented towards a design of workplaces, in which individual performance can be utilized to the greatest extent possible, both in terms of scope and intensity. Demands on individual performance are kept independent of the work process as a whole – its technology, organization, and cooperative elements – the worker is, so to say, freed from the ties to the other parts of the production process. This strategy accommodates both the elimination of traditional forms of sequential production processes and the integration of formerly divided tasks into complete work cycles at individual work stations. This isolated use of individual performance not only enables a higher degree of flexible labour deployment, but also makes it possible to cope with technical and organizational disturbances, bottlenecks in production, and material shortages in such a way that they do not effect the overall output of a subprocess. In other words, objective coordination problems in the production process can be solved through demands on individual performance. The enhanced access to an individual's performance – for instance, in the interest of quality assurance – enables the execution of more effective error control procedures and intensifies the worker's sense of responsibility with regard to existing quality standards.

In their differing ways, both of these strategies support the interest of the enterprise in pursuing the advantages of production economy and time management found in traditional (i.e. tasks involving a high degree of division of labour) forms of assembly production processes. The enterprise tries to combine both of the work forms described above in order to achieve a cooperative orientation simultaneously with individual performance maximization. In this way, new work forms are primarily designed to absorb the problems or deficiencies of forms of sequential production processes by enabling the enterprise to make more comprehensive or more flexible demands on performance. Thus, on the one hand, labour deployment in sequential production processes can make it possible to use labour cooperatively, in the form of group work. On the other hand, the establishment of systems divorced from the rhythm of the machine cycle can make it possible to break the individual worker's ties to overall performance so that, within limits, his or her work can be used in 'isolation'. In this work form, the enterprise is mainly interested in compensating for the effects of fluctuations in individual performance on total output as well as utilizing extra individual productivity in order to improve overall output results. At the same time, self directed cooperation (such as helping, teaching, filling-in) can also be achieved.

Company strategies and the ambivalent consequences of new work forms for the workers

The introduction of new work forms can have positive effects on the reproduction of labour, as demonstrated by numerous studies and experiments (see the comprehensive literature on 'The Humanization of Work' and 'The Quality of Working Life', e.g. DFVLR/Projektträgerschaft, numerous volumes since 1981). However, as our investigations have shown, the possibilities for improving the work situation and the way labour is reproduced continue to be hindered by the strategies and conditions of demands on performance. The positive potential inherent in these new work forms has still not been fully realized and the consequences for workers are ambivalent.

From an analytic perspective, we can differentiate between the effects of particular factors on the reproduction of labour (*discrepancies*) and varying effects for different groups of workers (*disparities*). The ambivalent character of the effects of new work forms also derives, to a significant extent, from the partial or complete lack of company supporting measures when new work forms are introduced (*deficiencies*).

Discrepancies always arise when the potential for improvement in the work situation lies in just those structural characteristics of new organizational forms which make it possible for the enterprise to pursue its interests by demanding a greater range or intensity of performance on the part of its workers. This is the case, for example, when an enterprise develops a work form which allows it to utilize individual performance to a broader extent, in other words, through the maximized performance of the worker. Work forms such as individual work stations with complete assembly cycles or work stations uncoupled from the assembly line enable greater worker discretion with regard to use of time in the production process and thus improve the possibility of establishing individual variations in working rhythm and reducing stress. In almost all cases, however, the workers end up straining themselves from having to maintain performance standards or by pushing their performance to the limit in order to benefit from wage incentives.

Discrepancies also occur in work forms that utilize cooperation in performing jobs. For example, on the one hand, workers carry out tasks and deal with stress situations more cooperatively and develop more collaborative forms of behaviour. On the other hand, the risk of intensified competition between individual workers due to differences in performance ability and social pressure from the group is also present. The positive potential of group work only comes to bear if skills are homogeneous or of a high level and if the group is able to negotiate (even if informally) the performance standards of the enterprise with supervisors.

The effects of new work forms create *disparity* among workers when an enterprise tries to utilize heterogeneous skills or existing differences in performance for its own purposes. For example, a worker's position in an enterprise is determined by the kinds of demands that are made on his or her skills or performance. This becomes especially clear when new forms of work are based on a polarized division of work, which divides the workers into core and marginal groups (e.g. into working groups consisting of skilled and non-skilled workers or workers who perform simple tasks versus complex tasks, such as maintenance, setting, etc.).

Disparity also occurs in connection with company strategies that are oriented towards the isolated use of individual performance. This happens, for example, through the pre-selection of workers to new individual jobs, thereby forcing some workers into worse or marginal work situations rather than setting up a retraining programme, or by deliberately making differing demands on performance for jobs detached from the assembly line (thereby pushing the individual worker to overtax himself). Once again, this kind of situation tends to divide workers into core and marginal groups, with the danger that members of the marginal groups become redundant if the personnel is reduced. The same can also occur when the enterprise implements technical means of eliminating restrictively organized forms of work, thus creating a latent labour displacement situation which generally affects the members of marginal groups.

Deficiencies become apparent:

1 when measures concerned with the design of jobs and the work environment are insufficient;
2 when there is a lack or complete absence of the possibility of adapting skills to the requirements of the new work forms;
3 when payment policies and methods are not adapted to the new work forms; and
4 when forms of company hierarchy do not change with the introduction of new work forms.

Thus the ambivalent effects of new work forms for the reproduction of labour are also caused by deficiencies in company design measures.

The deficiencies that occur with the introduction of new work forms can be principally ascribed to the fact that an enterprise is only interested in changing the design of the workplace once worker *performance has begun to suffer* from the negative effects of physical and mental strain and the stress caused by the work environment results in productivity decreases. This implies that companies do not generally take indirect factors that endanger the reproduction of labour (such as physical strain caused by executing the work or the work environment which can result, for example, in a long-term drain on health) into consideration in their design measures unless legally required. The result is that only partial aspects of the dangers to the reproduction of labour are considered in a very short-sighted manner.

Our investigation revealed the following findings: in the machine-building and metalworking enterprises, measures concerned with the design of the workplace and the improvement of the work environment were, without exception, insufficient. High levels of strain caused by the work environment and, for the most part, physical strain, continued to exist in those work areas in which new work forms had been introduced. In motor vehicle plants, the introduction of new work forms brought about ergonomical improvements in the design of the work place, but the stress exerted by the work environment remained. The ergonomical improvements that were carried out in plants from the electrical and precision engineering/optical industries generally consisted of extended MTM procedures and were less than optimal from the point of view of union demands and knowledge available from scientific research findings.

An especially grave deficiency in company design measures when new work forms were introduced was the *lack of attention given to training* workers in the face of increasing skills and performance requirements. We were able to observe four different means of adapting skills to the requirements of new work forms which occurred in various combinations.

In the first strategy type, the effort for securing additional training is passed on to the group or worker and basically consists of 'learning by doing'. According to our findings, this was the most widely applied form of adapting skills to the requirements of new work forms. It basically consists of assigning newly recruited or unskilled workers to experienced employees. However, the core workers have neither the experience nor the ability to train others, nor are they paid accordingly. The effort required to train the inexperienced colleague leads to additional stress and also disturbs the work process (and therein lies the difference to 'on-the-job training'). In another version of this method, lower-level supervisors provide minimal guidance on a succession of work stations until a 'broad' range of training has been obtained. This form of adapting skills to the requirements of new work forms is the one mainly chosen when an enterprise seeks to utilize a diverse range of performance at a relatively low level or when workers are to be utilized flexibly by moving them to different work stations. Generally speaking, an enterprise can only implement this strategy when it is able to draw on skill reserves (usually in the form of workers certified in another industry or in crafts) of its workers. For the workers, this generally means that the enterprise makes higher demands on their performance without providing recognition in the form of in-plant requalification and suitable pay.

The second strategy consists of avoiding the necessity of providing the training required by new work forms by selecting the best-qualified workers to perform the jobs in question (creaming off). It is always pursued when higher skills or performance requirements are demanded of the workers due to the introduction of individual work stations, new forms of technology, or work stations uncoupled from the line. In such cases the enterprise selects workers with a high level of skill or experience from either the working area in question or other areas and assigns them to the new tasks. This strategy generally depends on the enterprise's use of excess skills that are not normally required in the work process (a typical situation in Tayloristic work processes that is frequently overlooked). Most of the workers who are not chosen in this selection process are either placed in peripheral jobs or their former positions are rationalized and they are moved somewhere else if they cannot be dismissed.

The third strategy concerns the differentiation of company demands on performance and the segregation of duties. In this variant, only the contents of certain jobs are enlarged and enriched and these are then coordinated with jobs of very restricted content and skill requirements. Here again the enterprise makes use of available skills in the work process in a very deliberate way. This variant is closely linked to polarized assignment of workers, in which low-skilled workers are combined in a stable core group (in group work or floating assignments at sequential production process stations).

The fourth strategy consists of the training of skills according to the specific plant and process involved. In this variant, the contents of the skills conveyed to the workers are closely linked to the requirements of the working process. More complicated forms of training (for example, by means of special training workshops, professional instructors, teaching programmes, control systems and longer training periods) are only carried out when very high quality and performance standards cannot be achieved without a more systematic training. In addition to conveying knowledge and skills connected with the process in particular, this form of training is mainly concentrated on developing speed and 'exactitude' (awareness of errors and mistakes).

These strategies of skill adaptation have the following effects on workers: workers feel they have to push themselves in order to perform adequately; they undergo mental and emotional stress and pressure ('in trying to reach the performance standard during the training period'); social conflicts erupt within groups and there is increased strain caused by the task of training 'weaker' workers or newcomers; the segregation of duties destroys any feeling of solidarity and replaces it with group egoism; marginal groups are soon faced with the increased risk of unemployment and less capable workers begin to experience a sense of 'personal inferiority'.

Another form of deficiency consists in the fact that *wage policies* obstruct the potentially positive effects that new work forms could have on the reproduction of labour and aggravate the discrepancies and disparities inherent in the work forms.

For the moment, it should be borne in mind that we found no direct connection between the various kinds of new work forms and payment policies. Group work was remunerated with both group and individual methods of payment in the form of group and individual bonuses, group piecework and hourly wages; the same also applies for the newly created individual work stations and production jobs uncoupled from the assembly line in that they were remunerated with both individual and group piecework payment.

When group-related incentive payment (in the form of group bonuses or group piecework) is used to control cooperative performance, it creates a state of competition between the workers in a group and increased pressure to maintain a high level of performance. Thus cooperative relationships which are of benefit to workers can only come into existence when they are able to influence the wage/performance ratio according to individual performance and balance out wage differences with the help of informal arrangements or wage accounting forms.

However, when an enterprise institutes 'individualized' piecework as an instrument of control over individual levels of performance, the workers have the possibility of influencing the amount they earn, but they also run the risk of overperforming in order to earn as much as possible, once again a cause of social conflict. Also control of performance by supervisors is only reduced to a very slight degree with this form of payment.

If, in the reverse case, the enterprise seeks to neutralize the relationship between performance and pay by making cooperative or individual performance independent of wages (by paying standard 'frozen' group or individual piecework wages, for example), increased job intensity, fewer pauses and extra efforts to perform in a cooperative manner are no longer compensated for by the possibility of earning higher wages. The enterprise's interest in keeping its increased demands on performance independent of bonus systems also gets expressed when workers are not moved up into higher wage grades upon the introduction of new work forms entailing higher skill and performance requirements. (In this respect, the West German categories of analytical work evaluation are especially unsuitable as a means of dealing with new performance requirements, see chapter 2.)

Finally, deficiencies in company work design measures also occur as a result of unchanged forms of company hierarchy. In most cases, the role of the lower-level supervisor is not affected when new work forms are introduced. This causes deficiencies because the positive effects such as greater worker discretion that the companies have sought to realize with the help of group work and other new forms of work in sequential assembly production, as described above, are obstructed by the intervention of lower-level supervisors who continue to carry out their traditional functions. The same applies to individual work stations and production jobs detached from the assembly line in cases where individual performance is not controlled by pay. Here the lower-level supervisors exert pressure on workers because they are responsible for the achievement of set production standards (numbers of pieces). This obstructs the possibility for workers to have greater freedom in deciding how to divide up their time in carrying out work and taking breaks according to personal preference. From the lower-level supervisor's point of view, his or her task becomes more difficult under new work forms. For example, additional attention has to be paid to jobs detached from the machine cycle to ensure that idle periods do not occur due to empty buffers. For several reasons, supervisors also do not provide the additional information that is necessary to carry out new work forms effectively. In short, they continue to act very much the way they did before the introduction of new work forms.

It should also be pointed out that companies with new work forms often attempt to keep hierarchical structures and disciplinary functions concealed through the use of new work forms by placing the emphasis on cooperative relationships between the workers. For example, group control of performance can translate into 'cleaning house', i.e. getting rid of less productive members, meaning that the task of pressuring others to perform and evaluating performance gets reallocated to the group. Another example of hidden control relationships is the company's use of the position 'group speaker'. The group speaker is supposedly representing an autonomous group, but in reality he or she acts as both the conveyer of company directives and group problems, thereby encompassing a leadership position without the recognition or the pay.

In contrast, in the case of individual work stations, in which a more direct access to individual performance and thus an increased monitoring and control of performance is possible, it is easier to legitimate explicit control relations by using so-called objective criteria (meeting the company quality standards or the allowed error frequency, etc.), thereby putting direct pressure on an individual's performance without the need for, or the intermediation of, a foreman (*Meister*).

One final point that should be made is that the ambivalencies for workers deriving from new forms of work organization limit the ability of these work forms to increase productivity. In the short term, they prevent the full utilization and development of human capital; in the long run, they bring new problems to the enterprise such as strain on skilled workers and both lack of motivation and flexibility.[3]

14 The Future of the Mass Production Worker[1]

Günter Bechtle, Klaus Düll

INTRODUCTION

The current debate on the future of work focuses in part on the future of the mass production worker. Flexible standardized manufacturing systems are calling for a changing worker profile. How is this demand being met and what consequences does this have for the mass production worker? Using information gathered in three different countries at production sites all belonging to the same multinational company, this chapter shows that the answer to these questions must consider the different institutional and historical structures that make up the context in which future forms of mass production work develop.

There is consensus in the social sciences that the historical constellation of mass production, mass consumption and widespread prosperity, including political and institutional arrangements of the welfare-state in Western Europe, has been in a state of transition since the 1970s. This is commonly seen as stemming from the new international division of labour, growing competition in world markets and from the availability of so-called new technologies (information processing and control technologies). The question with which we will be concerned here is: what will happen to Tayloristic forms of labour deployment and to the historical figure of the 'mass production worker' when their macro-economic preconditions become highly unstable? Phrased differently: if barriers to uses of technology in the production process can be principally overcome by the 'new technologies', then what forms of labour productivity increases by way of Tayloristic forms of labour deployment remain possible and necessary?[2]

We will approach this question in three steps: first, we will determine the historical-sociological location of the 'mass production worker'; his/her position will be circumscribed within a historical societal configuration whose structure is formed by the Tayloristic–Fordistic syndrome. We will then introduce some results emanating from an empirical research project, which was concerned with rationalization processes in a multinational company in the consumer electronics field. The findings deal with changing labour forms arising from the transition from highly Taylorized manual assembly work to automated assembly. On this issue, we concentrate on the national differences in solutions of company personnel policies as they are carried out in the multinational company's French, Italian and German production sites. Finally, we will touch upon the issue of generalizing about the future of the mass production worker from nationally specific solutions. In this regard, we will focus especially upon characteristics of national education and industrial relations systems.

The study begins its analysis at the macro-level, examining characteristics of a societal configuration from an historical perspective, continues at the empirical micro-level of company case studies, and returns to the macro level of historical and theoretical interpretation.

THE TAYLORISTIC–FORDISTIC SYNDROME AND THE LOCATION OF THE MASS PRODUCTION WORKER

The emergence and shaping of the mass production worker as a type was constituted through a historical, societal configuration which came to the fore in Europe after the Second World War and at whose centre lies the 'Tayloristic–Fordistic syndrome' (see chapter 3 for an explanation of this term). Characteristic for this configuration are the growing reciprocities between the macro-level of economic cycles, social milieus, political-institutional processes, and the micro-level of work and production structures within companies. Without being able to refer in detail to the inner workings of such interrelationships (see chapter 3), three systematic repercussions of the Tayloristic–Fordistic syndrome can be named:

1 *The neutralization of the contradiction between wage as a cost factor and wage as a demand factor.* Cost decreases in the piece rate or increases in labour productivity through the scale effects of mass production enable cost-neutral wage increases that – via growth in mass purchasing power – quickly effect an increase in demand; under conditions of a higher market profile, an increased volume of sales can be realistically expected, keeping the whole cycle in motion.
2 *The permanent self-mobilization of additional labour reserves.* Increasing wages and decreasing prices gradually displace traditional modes of production (the agrarian subsistence economy, artisans and small businesses). On the other side we have the pull-effect of a

relatively attractive transition to wage labour. Taken together, both tendencies provide a supply of new, quickly deployable labour for industrial production processes.

3 *The social-political compensation principle.* Indeed, increases in the work intensity of Tayloristic production do continuously endanger the reproduction of labour power. But, on the other hand, performance intensification and production increases create margins used to finance the costs of social security.

On the material base of these cycles of mass income, mass consumption, labour productivity, cost decreases, rising wages, mobilization of additional labour power reserves and the accompanying socio-political compensation for attrition induced by performance-intensity, we can determine the social location of the mass production worker with respect to his/her relevance for industrial rationalization processes and the accompanying ways that control in the company is stabilized:

1 The use of the mass production worker is characterized by a sharp hierarchical and functional division of labour, a strict organization of material and work flows, and 'time management'; the aim is the greatest possible penetration of the 'time regime' in the production process (see Benz-Overhage *et al.* 1982; Sohn-Rethel 1978).
2 An asymmetrical control by the company management underlies the mass production workers' wage-performance relation because of the way in which 'performance' is measured and recognized (and passed off as being scientifically objective). Performance defined in such a way cannot be 'negotiated'.

The highly exchangeable labour power induced by the division of labour and the control of production through time management on the one hand, and the non-negotiable determination of performance on the other, are the decisive power resources needed to stabilize company control.[3] This is the structure of labour deployment under Taylorism–Fordism and the dominant pattern in which rationalization interests get realized.

It is possible to reach the same thesis – that company control gets stabilized through the use of mass production work – through another form of argumentation. In the historical phase in which the Tayloristic–Fordistic rationalization model was able to develop and spread on the basis of the above-mentioned interrelationships between the micro- and macro-levels, the industrial rationalization process created an independent structure, whose function was the rationalization of the company organization itself. In Germany, this 'meso' structure historically found its expression in company forms such as work planning, controlling, quality control and electronic information processing. These forms are not solely oriented to the overall economic goal of the company: profitability. However, they simultaneously serve this 'meso structure,' and are bound to the Tayloristic–Fordistic rationalization model. The development of this structure goes a long way in explaining the many results of empirical social science which point to the persistency of social structures within the company and their 'conservatism of structure' (Bechtle, Lutz 1989).

The early 1970s witnessed a gradual erosion of the preconditions of Tayloristic rationalization strategies. The interrelationships between micro and macro levels, upon which the Tayloristic–Fordistic syndrome is based, increasingly ceased to function. Without going into great detail, the result of this erosion process has had a two-sided effect on mass production. On the one hand, the world market has imposed changes in production behaviour. Under conditions of decreasing market shares, the production techniques characteristic of large-scale production do not allow a further realization of economies of scale, and cost reductions are unobtainable due to company size. On the other hand, the supply of flexible automation systems has steadily increased on the technology market. The application of these systems is an attempt to solve technologically the problematic relationship between flexibility and productivity increases based on economies of scale using new control and computer technologies. The principle here is to simulate complex processes consisting of a multitude of different technical parameters (qualitative as well as quantitative product differentiation, interruptions in the production process, etc.) using computer technology. However, sweeping changes in the organization of production, especially in the way the various levels of planning and production are (inter)linked are a precondition for the profitable application of the new technologies.

ON CHANGES OF HIGHLY TAYLORIZED WORK FORMS IN FRANCE, ITALY AND THE FEDERAL REPUBLIC OF GERMANY

In the following section we will attempt to clarify the relationship between the crisis of the mass production model and its possible resolutions. For this purpose we will use results obtained from empirical research on strategies of 'flexible standardized mass production' in an international concern in the consumer electronics field.

The company is a large multinational based in France with production sites all over the world. Since the middle of the 1970s, the company has been pursuing a policy of capital concentration, first by buying Italian and German firms, and more recently (end of the 1980s) US, British and Far

Eastern firms. The goal, as with many companies in the consumer electronic industry, is to obtain the size necessary for scale production within the confines of stagnating sales markets.

Research was carried out at seven of the company's European production sites using the method of 'intensive company case-studies' (see chapter 1). Research concentrated on television assembly. Four locations with a total of 3,500 employees were in the Federal Republic; two sites with a total of 2,500 employees were in France, and one site with almost 2,000 employees was in Italy. Altogether, nearly 8,000 people were employed at the research sites.[4]

First, we will describe the initial situation as well as the most important elements of the company's rationalization strategy. In a second step, we will look at the strategy's consequences for company personnel policy in different national contexts.

The company's rationalization strategy

First, we can recapitulate the initial situation in which the company gradually developed its strategy. Competition in the world market forced investment in systems of flexible automation. These, however, necessitated an extent of investment only attainable through high-capacity utilization and the expectation of increasing sales. As a result, the company pursued a typical and often encountered crisis strategy of 'sunset' industries in this phase: movements toward world-wide concentration of capital. The idea behind the strategy is to establish a presence in new markets, so that the high investment costs of new technologies can be amortized under conditions of decreasing product life.

The company also pursued a strategy of separating distribution and production and introducing flexible standardization to make for an easier handling of variable and unforeseeable market demands. Flexible standardization usually makes it possible to selectively apply various flexible automation technologies to obtain profitability. With an increasing piece rate, investment in automation is generally most profitable where the proportion of manual labour has been high. This is especially so for assembly processes. New items offered on the technology market make it easier to transcend the traditionally strong limits to automation in assembly. However, the profitability of capital investment in assembly is increasingly risky because the additional productivity incurred by such investment – with the already high total productivity (measured in work hours) in television assembly – has reached a critical limit. Also, the comparative advantage of manual assembly is still strong. Thus, the company had to try and pursue a strategy which balanced the relative advantages of costly automation investments and the continued use of manual labour.

The company strategy can be described in the following five steps which, although they overlap with one another, also strategically build upon one another.

1 Obtaining additional market shares by purchasing European competitors active in the field of television assembly. This aimed at reaching a profitable economy of scale.
2 Separating distribution (using the original brand names) from production through the establishment of initially two, later reduced to a single production company, along with independent sales companies in the Federal Republic. This meant that parts for all brands could be produced, assembled, repaired, etc. at any company location. Production was thus protected from the demand elasticity of individual brands and could be better rationalized.
3 The introduction of a standardized printed circuit board (PCB) which, when attaching different-sized PCBs to the electronic 'insides' of televisions during assembly, can be divided and reassembled according to differing requirements. In fact, this standardization was nothing but a specific application of the world-wide practice of using a particular technological design (flexible standardization) to prevent end-product diversity from interfering with production flow. This both product-related and production organization rationalization created the preconditions for a far-reaching automation of PCB assembly.
4 The productivity derived from a successful strategy of a combined market and production economy requires either a growing sales market, or leads to the inevitable reduction of excess capacities. For the company, the contradiction between economic profitability at high piece rate production and the limited capacity of the market was insoluble. The reduction of excess capacities led to considerable lay-offs, including the complete closure of production sites.
5 Attempts were made to increase productivity further using computer technologies and organization techniques, as well as by using synergy effects to enhance the production economy: the solution was seen to lie in a 'simulated factory'. This strategy entailed the reorganization of the production flow from picture tube to final assembly 'as if we had a uniform European factory'. The slogan was: 'one production site – one product', which supplanted the original 'one production site – one brand'. Thus, there was one site for picture tube production, one for PCB production, one for PCB assembly, one for final assembly, etc.

These strategic elements and phases can be condensed into two basic theoretical arguments. The first argument refers to the crisis of mass production: the various technological and organizational forms of rationalization in assembly are elements of a comprehensive crisis strategy that seeks to flexibly standardize mass production. Achieving economies of scale through capacity utilization of increasingly fixed capital shares and a market-oriented production diversity are aimed at reaching the common denominator 'flexibilization through standardization'.

The second argument takes up a basic idea from 'systemic rationalization' (see chapter 4), but in modified form: rationalization measures of flexible standardization – with increasing automation – generate pressure to attain an integration of individual production processes by means of control technologies, and, in addition, a comprehensive systems integration of these individual processes in line with the model of a 'simulated factory'. However, this imminent pressure towards integration is difficult to cope with because limits and obstructions are of a both systematic and historical nature. They are systematic because the increase in production capacity induced by integration forces a broadening of the market in line with the laws of an economy of scale. The relevance of the market – as a system environment – is not only enhanced in temporal and material dimensions, but also, the uncertainties of market constellations react back upon the integration process, demanding new – sometimes immediate – adjustments, thus constantly calling into question systems integration along the lines of a 'simulated factory'. The limits and obstructions are historical because barriers to integration arise from production and social structures developed over the years from formerly independent businesses and their surroundings (e.g. labour markets). For one, there are disjunctions between factories with regard to the level of technology and organization (e.g. different process speeds of technological equipment). Also, forms of social integration, i.e. processes of communication and information transfer aimed at achieving greater total transparency ('trust versus competition'), remain relatively precarious under conditions of the ongoing struggle for productivity, cost and market advantages within a company. Rationalization measures are always evaluated in terms of reciprocal (dis)advantage and, if necessary, blocked ('individual company egotism' or, better: interests).

The strategy of flexibly standardized mass production, characterized on the one hand by a rationalization of the product, the process and the work organization, and the limits to an aspired inter-company systems integration on the other hand, are two lines of development central for the analysis of future demands upon assembly work.

One line arises from the tense relationship between the (logically and technologically) desired systems integration and the simultaneously emerging barriers and cleavages (e.g. inconsistencies in the organization of intra-company logistics, incompatibility of control systems, etc.). On the one hand, the systemic interdependencies between different assembly processes and upstream and downstream processes increase. This entails higher demands upon coordination and cooperation between the technical offices and the assembly area, as well as upon the production workers themselves. This is especially so when spatial and temporal buffers are largely eliminated through new logistics systems and altered flows of material. On the other hand, the above-mentioned barriers to integration raise considerable problems of controlling deviations in the production plan in terms of time, quantity, and quality. With systemic dependencies, the susceptibility of interference within complex production processes grows; deviations are carried over from one processing department to another in a snowball effect, and coordination requirements increase at an inflationary rate. The new division of labour calls for coordination through new company-wide organizational forms.

The second line of development, closely bound up with the first, arises from the structural change of work deployment in the transition from Tayloristic to automated assembly processes. Traditional assembly processes follow a sequential, mechanic and continually repeating logic necessitating a materially and temporally determined and ex ante calculable work performance. This logic and this performance structure derives from the precise determination of the labour process by the production process. Work is concretely tied to a small number of isolated characteristics of the workpiece, and entails the repetition of identical operations in identical time intervals. This stands in stark contrast to automated assembly, where the technologically autonomous production process and the 'labour process' (human intervention in the production process) are decoupled (see Bechtle 1980; Düll, Bechtle 1988). However, the above-mentioned possibility of interference between these processes under conditions of systemic interdependence, continually make renewed forms of interaction between production process and labour process (i.e. interventions) necessary. These, however, are neither calculable with regard to time, type, sequence, or duration, and are certainly not ex ante calculable. The work then necessary for a renewed interlinking of labour process and production process becomes – whenever possible – preventive trouble-shooting (Moldaschl 1989). It is no longer aimed at isolated characteristics of the object (product), but at the entire process flow or process continuity. It takes place in simultaneous coordination with other areas of deployment and rarely repeats itself in an identical form.

Both lines of development taken together – one characterized by a growing, but systematically limited systems integration, the other characterized by a growing decoupling of production process and work assignment – raise new areas of uncertainty. These new uncertainties lead to work requirements which can in principle no longer be mastered by the traditional work tasks of manual assembly work and the skills of the mass production worker. Companies must thus meet the challenge of developing new forms of skill and skill use as well as new forms of work organization that deviate from Tayloristic models. Companies' solutions in the realm of personnel policy can and, indeed, must build upon very different, nationally specific structural conditions and resources, such as organized labour markets, education and training systems, and systems of industrial relations.

Nationally specific solutions in the realm of personnel policy

Below, we will briefly describe characteristics of nationally specific solutions for coping with the new work requirements arising in automated assembly processes found at the French, Italian and German research sites of the multinational consumer electronics firm.

At the French sites, the dominant approach is the systematic further training of semi-skilled female assembly workers. This further training reaches a maximum of 1,000 hours per worker upon which a state certified document (CAP) is presented to the employee defining him/her as a 'plant electronic worker'.[5] In the French situation, this approach has a threefold meaning. First, further training serves to procure skills for production tasks on automated equipment: in French production sites, further training is considered to be an important instrument for mobilizing skill and skillable labour in the internal labour market, since skilled tradesmen usually refuse to work in production. Second, for semi-skilled workers who have undertaken systematic further training, more demanding tasks in automated assembly are seen as an advancement. Moreover, further training is used as an element of an 'alternative' social plan design,[6] the aim being to increase the chances of re-employment of semi-skilled workers in the external labour market. Third, further training in France is supported by a broad consensus in social policy. On the one hand, further training is seen as an important adjustment for the not very practice-oriented initial education and training; on the other hand, further training seems to be the only way to create the necessary preconditions for breaking up the restrictive work situation of semi-skilled workers in highly Tayloristic work processes.

Our research showed that semi-skilled female workers in French production sites – after corresponding training – were in a position to carry out more difficult tasks at automated workplaces, tasks that in German production locations usually would have been carried out by skilled tradesmen.

But this approach has certain constraints. First, the possibility for further training is numerically limited, i.e. it is controlled and applied selectively by lower management. Second, the completion of further training is no guarantee for assignment at respectively skilled jobs. Even if such an assignment is forthcoming, the employee's status within the company wage hierarchy undergoes only negligible change. Third, further training courses do not dissolve the rigid and traditional segmentation between skilled tradesmen who have been trained at vocational (*lycées d'enseignement professional*) or secondary schools and who have been assigned outside of production (in maintenance), and semi-skilled workers deployed in production.

In the Italian production site (picture tubes), the problem of process continuity and process optimization (i.e. the permanent coordination between different processing sections) is extremely important since the quality and quantity of output is dependent upon controlling these problems. In the Italian production location, the attempt is made to solve these problems by applying a particular model of group work.

In this model so-called mixed groups (of on average 80 persons) are formed, in which the functions of machine operation, machine supervision, and machine adjustment are integrated and in which job tasks are distributed autonomously within the group. Their new skill structure, consisting of specialized chemical, metallurgical, and electric-electronic knowledge, is characterized as a new 'professional matrix'. The group model's necessary training measures are carried out in cooperation with company-internal and company-external (schooling) agencies. Newly employed school graduates, for example, are trained on the job for a period of five months at 100 hours per head.[7]

The company ties the enhanced opportunities for acquiring qualifications and the effective long-term employment guarantees deriving from this form of work with company performance policy. A concrete example of the exchange in advantages for both employer and employee in the Italian case was the negotiation of a 'social pact'. Company and union representatives reached a so-called productivity agreement which contains two instruments for the enforcement of performance policy – one regulating working time and one determining wages. First, a three-shift system was introduced that extends annual working time to 291 days. The plant runs six days a week and 24 hours a day. This is compensated for by the reduction of weekly work to 36.5 hours and the introduction of a new shift model which limits individual workers' night-shift requirements (creation of a fourth shift-team); night-shift work is undertaken in a four-week rhythm on a total of six days. In this model, the maximum occurrence of night-shift work is twice a week on two consecutive nights; between night-shift and day-shift a minimum one-day rest phase is required, as a rule it is a two-day period. These rest phases are important for a majority of workers, as many still pursue private agricultural work. However, the agreement stipulates that the necessary maintenance work is carried out on work-free Sundays. Finally, a so-called productivity wage has been agreed upon, which accounts for about 20 per cent of the basic collective wage and is paid out to all workers as a supplement regardless of their wage group.

The ability to implement this agreement between the company and unions depended upon very specific conditions atypical for the other companies of the multinational firm. The picture tube production site in Italy, in comparison to other areas in which the multinational company has production facilities, receives an above-average number of orders, and has an employment level that has been stable for the past decade, and has even risen in the past few years.

The solutions that have been undertaken in German production sites of the company follow traditional patterns. Automated assembly lines are predominantly manned by skilled tradesmen (*Facharbeiter*), who often entered production in previously existing (tool)setter positions and now directly carry out production tasks as 'production mechanics' or 'operators'. Even if the recruitment of skilled tradesmen for such tasks poses difficulties due to labour market shortages, it is not – as in France – principally out of the question. In several cases companies manage with skilled tradesmen from related occupations who have gone through state-financed retraining courses.

Innovative approaches of coping with the requirements of automation work have been much less frequent in German production sites and also much less stable over time. They display features that are almost diametrically opposed to the institutionalization of advanced education and training in France and the introduction of group work in Italy.

One of the experiments was characterized by an attempt to fully utilize existing electrotechnical as well as other types of skilled personnel to supervise, control, and repair breakdowns in the automated PCB assembly. This had the double purpose of creating jobs for skilled tradesmen while at the same time fixing breakdowns quickly.

However, this experiment was doomed to failure because it had practically no back-up measures for enhancing stability. First, for reasons of performance intensification, the experiment was undertaken with purposefully tight staffing. Second, no accompanying training measures took place, so that the creation of a 'hybrid qualification' (i.e. mechanics plus electronics), which was deemed necessary, remained the exception. Third, the hierarchical levels were enhanced by institutionalizing a new level in the form of a group leader (*Vorarbeiter*) in order to avoid production back logs and cope with unexplained down-times. Across the board, this measure generated resistance by work groups and endangered their cohesion.

WHAT IS TO BECOME OF THE MASS PRODUCTION WORKER?

If we sum up the current developments of how the mass production worker is utilized in company structures across all the countries in this study, the following results emerge: first, the mass is being removed from the mass production worker; the traditional, inner homogeneity is getting lost. This goes for the entire spectrum of what comprises work: requirements, skills, workload, job security, etc. The next step of this development is the concretization of various forms of polarization of the mass production worker. Workers are found at both extreme poles of work skills: at one end workers are formed either from those who have already been pushed out of the production process, or more typically by those who occupy residual jobs which have not yet been swallowed up by automation or that repeatedly appear because they are relatively cheap. Thus, what remains is a limited number of the classical mass production worker used as a stop-gap. At the other pole, there is a small minority of workers who succeed in reaching higher positions where new hybrid skills are needed and a broad spectrum of tasks are carried out. Between these two extremes, there is a third group which succeeds in making the jump to the automated jobs. These are higher level semi-skilled positions with high performance-related burdens (especially time pressure), which were also features of the classical mass production worker. This group is formed in a more or less Darwinist selection process, in which access is based either on semi-skilled workers' superior performance records or with the help of targeted training measures.

The way these poles get distributed and weighted varies between countries. We explain this from the fact that the interrelationships between the micro and macro levels of societal production, as characterized by the Tayloristic–Fordistic syndrome, are strongly affected by historical and country-specific conditions which tend to lead back to differing industrialization processes. We demonstrate this using two aspects of these processes which could be documented from the results of our cross-country study, so-called 'representatives' of the inter-relationship between the micro and macro levels: the educational system and the structure of industrial relations.

In *France*, the method of coping with new processes by further training for semi-skilled workers derives from two sources: for one, from France's school-oriented occupational education and training which has a relatively low quantitative output, and, for another, from the fact that skilled workers with an initial occupational training do not go into direct production – at least not into mass production assembly. Both developments produce a demand for further training if new production needs in mass production factories are to be met. As early as the mid-1960s, the *Ouvrier professional de fabrication* came into being, a type of worker that is formed by a mixture of both state and company education and training. Also, a law has been in existence since the early 1970s which requires companies to invest at least 1.1 per cent of their payroll into training and further training; as a rule, investment in further training by large French companies by far exceeds this legal requirement.

Finally, in the French case we must take into account that semi-skilled workers are often union clientele, and that their further training, besides the issue of work organization and workers' participation, is a political question raised by both the union representatives of large unions in the company and so-called education and training commissions (*Comités d'entreprise*). The political influence of these representative organs, however, is low and usually does not exceed union rhetoric and radical verbal demands.[8]

In sum for France: a strong demand for further training of workers, state initiatives and union concern for the subject-matter of further training, with a simultaneously strong separation of production from so-called 'decision-making' departments, which can be counted as part of the control mechanisms of company organization.

The *German* approach's solution lies within very different structures. For really skilled tasks German companies use skilled tradesmen (*Facharbeiter*) – if they can be recruited from the labour market, which is less and less the case, especially in the mass production industry. Why aren't semi-skilled workers, utilized as in France? For one, semi-skilled workers both male and female, but especially non-German females, are the 'stepchildren' of production in Germany. The German work council member gets involved where norms exist upon which he/she can fall back on and for which he/she has legal recourse; the further training of semi-skilled female workers is not such a subject[9] (see chapter 24). And where there is no skill, there can be no claim for one: about 90 per cent of the women we interviewed in our research locations had no plans for further training and did not know of any opportunities for further training. While wage, working hours, and individual dismissals are relatively well-secured through collective agreements, such issues as work organization, performance intensification, etc. remain at the complete discretion of company management. The structural conservatism of the German model is further promoted by the traditional mode of company organization, i.e. the deterministic dictates of work planning departments and the attempt of technical offices to achieve planning perfectionism.

In sum for Germany: the lack of further training for semi-skilled workers, traditional work council functions, permanent performance intensification for highly skilled workers secured in particular by the traditional organizational structure, and maintenance of the segmentation between semi-skilled and skilled workers.

The *Italian* case revolves around explaining the establishment of the negotiated productivity and social pact. On the one hand, this involves the historical particularities of the relationship between the educational and production system in the southern Italian labour market, and a distinctive corporate structure between company, union, and state on the other.

Unlike in the French or German production sites, the Italian sites lack two groups of workers upon whose shoulders the fate of the mass production worker has traditionally rested: women (because of night shift work) and non-Italians. This means that labour problems have to be solved 'at home.'

In the Italian case, we are dealing mainly with employees whose average age is low, and who have without exception completed a two- to three-year vocational school programme. After being presented with their technical diplomas – in contrast to the French case – they are immediately deployed in direct production. When the new vocational system was introduced in southern Italy, and produced these young, well-trained workers who entered the production area, they had difficulty getting along with the dense network of middle-level supervisors who had fewer school qualifications and had attained their position by rising through the ranks. The group work model made it possible to eliminate part of this chain of command, and relegate the superfluous foremen as 'advisors'. Promotion was then more or less determined by skill acquisition, rather than by supervisor selection.

What is decisive and symptomatic in the Italian case is that structural reform was embedded in a model of 'political exchange' which became widespread in Italy during the mid-1970s, as a sort of aftermath to the workers' struggles that had taken place during the late 1960s. This model is based on exchange relations between political clientele embodied by three agents: the state, company management – represented in this case by the personnel department, which, in contrast to the German case, has a strong company-internal negotiating position – and the unions. In this triangular relationship, the company and the unions negotiated the so-called productivity and social pact (dated 1980). What was exchanged? Access to qualified work for young workers against a successful company performance policy with shift work and performance premiums. The extremely 'productivistically' oriented unions were able to legitimate this package, as it corresponded to their policy – established in the early 1970s – of egalitarian *professionalità* (occupational professionalism) which was oriented to young, vocationally trained workers in particular (see Bechtle, Heiner 1979). This pact guaranteed the maintenance and creation of employment in the region, and the government in Rome was thus willing to provide various subsidies and financial aid, a policy with returns in the form of votes.

In sum for Italy: without women and foreigners and with the help of the state, the 'social partners' (employers association and unions) were able to come to an agreement on productivity-oriented structural reform. Whether this exchange is only successful under fair weather conditions remains to be seen.

These three cases demonstrate that the mass production worker is not necessarily disappearing. Rather, the forms in which mass production work appear are getting restructured. In each country, mass production work is getting expressed through the use of a different type of worker within different types of organizational structures. This, in turn, derives from the existing and developing historical and institutional structures, such as the labour market, training and occupational system, industrial relations system, etc. found in the nationally specific context.

Although the mass production worker is getting restructured, and the forms in which he/she appear are heterogeneous, the many classical burdens and problems belonging to mass production work have not been eliminated.

15 Towards a Polarization of Skill Structures[1]

Manfred Deiß

INTRODUCTION

Research on the development of skill structures and skill requirements in industrial manufacturing, especially in light of the increasing dissemenination of new computer-based technologies and the changing nature of company rationalization processes, has yielded a number of findings, both complementary and contradictory. Some research on this issue has prompted the assessment that a reprofessionalization of industrial work structures will take place accompanied by a trend towards skill enhancement. This chapter uses the furniture industry (as a representative case of production structures in the consumer goods industry in general) to provide a detailed description of the development of skill requirements and conclusions on work structures in industrialized manufacturing. This investigation shows that, at least as far as the furniture manufacturing industry is concerned, there can be no mention of a reskilling of work, or of a far-reaching enhancement of skills.

With a turnover of 28 billion DM and 185,000 employees in 1986, the wood-processing industry (of which the furniture-making industry is the most important branch), is the most significant sector of the West German timber industry (with about 35 billion DM turnover, and 230,000 employees). Yet it only makes up 2 per cent of the total turnover of the West German manufacturing sector and is thus one of the smaller industrial branches in the economy. However, the furniture industry is similar in size to the comparatively well investigated and extensively discussed machine tool-building trade and is also characterized by a predominance of small and medium-sized firms.

In the wood-processing industry (and correspondingly in furniture manufacturing) there are a number of different occupational trades:

traditional craft trades such as carpenters, new industrial trades such as 'mechanics for wood processing', and experienced semi-skilled workers who are employed as 'skilled' workers.

It should be pointed out here, in our study, that the majority of furniture manufacturers, especially those companies geared to the output of large series, possessed manufacturing structures with a high degree of division of labour and highly mechanized manufacturing structures in selected areas. As a rule, the percentage of unskilled and semi-skilled tasks amounted to more than half of all manufacturing jobs, and could reach up to 80 per cent and more of the manufacturing personnel for mass producers of standard furniture. Only the manufacturers of exclusive brand-name articles displayed a more favourable share of skilled tasks, particularly due to their initial expansion of special manufacturing departments as a reaction to the need for achieving increased product flexibility. A larger quota of skilled work could also be found among a certain number of the smaller manufacturers of furniture of average quality. Due to their comparatively smaller workforce and percentage of mechanized manufacturing, these firms depended more heavily on the flexible utilization of skilled workers.

NEW RATIONALIZATION STRATEGIES IN THE GERMAN FURNITURE MANUFACTURING INDUSTRY

At the end of the 1970s and beginning of the 1980s, the furniture manufacturing industry was struck by a severe and prolonged economic crisis which it only emerged from in the last few years. During this slump, the number of employees decreased by about a quarter while the number of companies dropped by approximately a fifth. Saturated markets and stiffer competition for market shares confronted virtually all furniture manufacturers with the increasingly urgent need to achieve greater flexibility and reduce manufacturing costs. They reacted to these challenges with a number of sales strategies and rationalization concepts based on widely differing concepts that were either oriented to price competition and/or to product and quality competition (see chapter 6).

The companies found that conventional rationalization measures and technologies, however, were no longer adequate for coping with the manufacturing problems that had arisen. On the contrary, they only made matters worse, for they were directed toward improving individual facilities or coping with bottlenecks in production, but were incapable of bringing about flexibility and cost reduction. With the help of computer-based technologies, some companies succeeded – within limits – in simultaneously achieving flexible manufacturing and lowering costs. However, the introduction of such technological processes in manufacturing and administration gave rise to new and more extensive demands for cost effectiveness and the flexibility of production structures. Moreover, the utilization of computer-based technologies had the effect of speeding up the introduction of new and extensive control technologies and to continually integrate individual processes with one another through data technology. In this way it became possible to better exploit the potential for more productivity and flexibility inherent in these technologies.

The rationalization strategies pursued in the course of this development tended more and more toward the 'systemic' type of rationalization (see chapter 4). These new forms of rationalization are directed towards the reorganization of the entire manufacturing process and tend to encompass a company's service and supply relations to parts suppliers, as well as the interaction with furniture retailers and the manufacturers of woodworking machinery and control technologies.

This new rationalization drive manifests itself in different companies to differing degrees and rates of development, depending on their traditions and market orientations. It was the large-scale manufacturers of low-price standard furniture on the one hand, and of high-quality, exclusive brand name products on the other, which set the trends in product design and range, as well as in manufacturing methods and technologies. These companies were also the ones which implemented new technologies on a large scale in the context of comprehensive rationalization strategies. Many small and medium-sized firms that manufactured furniture of average quality and which had less market power and less capacity for carrying out targeted strategies were also put under pressure to use new technologies, at least in certain areas. It was only in this way that they were able to attempt to meet the pressing flexibility and cost constraints they found themselves confronting.

Changes in the technological and organizational structures in the furniture manufacturing industry seriously effected the development of skills in this industrial sector. The type and extent of relevant skill requirements and thus the skill structures in the companies, cannot simply be deduced from existing manufacturing facilities and the computer technologies available on the technology market, as has often been assumed in a somewhat short-sighted manner. Much more determining are:

1 the traditional personnel and labour market structures which have evolved as a result of previous manufacturing modes (mass production on the one hand, handcrafted production on the other);
2 the various rationalization measures that furniture manufacturers pursue in order to combine cost efficiency and flexibility objectives, and which are geared to various marketing strategies (price, quality, etc.); and

3 the product and marketing strategies of technology manufacturers which tend toward a 'prestructuring' of work assignments and qualification requirements.

In general, the effects of systemic rationalization strategies on work and on employees are relatively indefinite, and, partially, contradictory. The effects usually remain unnoticed or only gain attention after a certain period of time has elapsed; in many cases, they appear in other work areas or even in other companies. Due to the fact that such measures are often carried out in a tentative and isolated manner, as gradual steps in the rationalization process, their integrative potential frequently eludes recognition and their 'scope' only becomes apparent at a much later date. Therefore, many of the effects resulting from the introduction of new technologies within the context of systemic rationalization processes are not linked to the actual conversion measures, and are seldom attributed to them. The effects are also obscured by the protracted course of the implementation of new technology, as well as by ongoing processes of personnel adaptation and selection occurring concurrently, but mainly detached from the new technology implementation.

Although there is a general consensus among social scientists that the application of new technologies entails a basic freedom and variability in design options for work content, our investigations found that design options are objectively restricted to confined paths, within which particular trends in the development of skill requirements and conditions of work organization are foreseeable; moreover, despite the fact that systemic rationalization strategies are just beginning to penetrate the furniture industry and despite the many mechanisms which obscure their effects, some trends can be identified.

THE DEVELOPMENT OF SKILL REQUIREMENTS

In the course of the restructuring process in the furniture industry, which has only been briefly outlined here, the companies found themselves increasingly confronted with skill problems in their attempts to utilize new (manufacturing-) technologies and realize technological–organizational adjustment measures. This was the case for the manufacturers of mass produced furniture as well as for the exclusive manufacturers. However, their initial situations were different: the mass manufacturers had predominantly unskilled workers, while the exclusive manufacturers had mainly skilled workers. The skill deficits which furniture manufacturers (as technology users) and technology manufacturers both complained about – as did others in the manufacturing sector – were regarded, rather indiscriminately, as posing a basic obstacle to the introduction and implementation of new technologies. Our findings reveal, however, that their concern was directed toward professional skills on the shop-floor to a very limited extent.

Technology manufacturers (and especially the manufacturers of electronic control systems) reported deficiencies primarily in matters pertaining to management, organization and logistics as well as a lack of basic knowledge of information technology on the part of members of management and supervisors in user companies, a factor which they regarded as obstructing and impeding the adequate implementation and utilization of new technology. Furthermore, the technology manufacturers also criticized the lack of qualifications of service and maintenance personnel, which increased their expenditure for service considerably.

From the user companies' standpoint, the problem was deficiencies in the more traditional woodworking skills and know-how. In their opinion, such deficits restricted the traditional amount of flexibility in the manufacturing process which enabled companies to cope with frequent calls for special and rush orders and also deal with the increasing number of product designs and variants. More than anything, the firms missed the general lack of worker character traits that manifest themselves as reliability, willingness to work overtime or fill in at another work station on short notice. Such qualities are necessary for a firm to achieve a trouble-free and optimal functioning of a highly organized and 'technicized' manufacturing process as well as for the ability to smoothly deal with system breakdowns. Another problem that was revealed was a lack of leadership skills at the foreman (*Meister*) level. It was found that the foremen had difficulties relinquishing their accustomed overseeing and regulating functions and taking over new tasks of precise planning and control of manufacturing sequences and personnel management specifically directed to the requirements of computer-aided manufacturing. Deficiencies in terms of specific occupational skills, especially with regard to familiarity with machine technology and electronics, were only detected in individual cases, for skilled workers assigned to central positions.

In assessing problems of skills and qualifications, it is important to differentiate between the requirements arising during the phase of new technology implementation and the requirements of ongoing manufacturing.

During the course of implementing new manufacturing facilities, a process which usually extends over several years, high demands are made on the personnel involved as far as specific occupational knowledge and familiarity with machine and information technology is concerned. Also included are demands on the workers' capacity to organize and improvise, as well as their capacity to bear responsibility and make decisions. Therefore, only experienced and well qualified members of the staff are selected for these tasks; they generally do not come from the area of manufacturing undergoing conversion, nor are they woodworking craftsmen, but are usually members of the repair or the electrical departments. In a number of cases, their high-level skills are not objectively required for the actual operation of the new facilities. The employment of such highly skilled personnel can largely be attributed to the company's security needs (coping with complications during the running-in phase and emergency situations).

At the same time, the introduction of new technologies, taking place over a period of many years and which is geared toward achieving a systemic integration of operational sub-processes, leads to an expansion of the administrative section of the manufacturing area both in size and in enhancement of skills there. The companies require more personnel with specialized data processing skills for handling the tasks of design and control of computer-aided manufacturing systems. Usually this need can only be filled by recruiting outside of the company.

Once new technologies are running normally, user companies tend to (once again) opt for the kind of personnel utilization which restricts skill requirements to the mere operation of manufacturing equipment (the feeding and monitoring of automated facilities). This creates less need for know-how concerning materials and specific machines. Tasks such as shop-floor programming, or jobs related to the organization of operational sequences are not included in such routine operating jobs. Thus, most of the workers with special skills who were needed during the implementation phase are withdrawn and at best find themselves acting as standby personnel in the case of breakdowns or minor repairs.

The product design and marketing policies pursued by machine manufacturers lend momentum to this trend. They assume the existence of lesser skilled personnel in the user companies and seek to 'transfer' the intelligence to the manufacturing equipment. Otherwise they presuppose the availability of specialized programming personnel outside of the manufacturing area, or, to the greatest extent possible, deliver their systems with completed programs.

A similar development is beginning to emerge in the administrative areas, particularly in process planning and operations scheduling. Comprehensive or intensive demand for high-level skills only exists for few employees, while the large share of (routine) jobs are handled by relatively unskilled administrative personnel aided by computers or by employees overqualified for the jobs in question.

Under normal running conditions, specific developments have emerged regarding a shift in skill requirements. For a large share of manufacturing personnel, the installation of CNC-machines and data-processing technology has shifted the emphasis from process-related skills, to ones which concern the individual's character and willingness to perform and serve to guarantee the permanent system availability of the (integrated) equipment. These requirements are less the direct consequence of the technological innovations, than the result of new concepts in manufacturing organization which can be realized with the help of new technologies. The non-process specific worker traits are mostly expected of 'skilled workers', regardless of their training, which leads to frequent assignment to jobs for which they are overqualified, or, as is often the case with system operators (*Anlagenführer*), the additional assumption of low-skilled tasks. These so-called 'increased skill requirements' often result in, or are accompanied by, intensification of work and increased responsibility in the performance of repetitive and monotonous tasks. More challenging tasks, such as machine control and the planning of operational sequences which were formerly carried out on the shop-floor by skilled manufacturing personnel, are now transferred to the administrative areas, or technical departments.[2]

Therefore, cases in which the introduction of computer technology leads to higher skill requirements on the shop-floor tend to remain the exception (as, for example, is the case for some key positions or for CNC-controlled surface milling cutters). In the (final) assembly area, in particular, there is a marked increase in general tasks due to the extreme diversity of models and the necessity to gather pieces for individual orders. Here too, there is a tendency to transfer the machine-related skill requirements which arise in the context of new technologies to programming and repair departments outside of the manufacturing area.

The utilization of computer-based equipment and information technology systems has been accompanied by the emergence of new concepts in manufacturing organization and performance policy which have been adopted by companies in the furniture industry. These concepts induce the skill developments outlined above, or add momentum to already existing trends. This occurs in the following way: forms of manufacturing which are oriented to individual customer orders and a certain daily output are linked with comprehensive and precise planning, controlling and monitoring of the manufacturing sequence which leaves little room for deviation. On this basis, a new kind of performance policy is put into practice which combines the conventional aspects of Tayloristic performance requirements with new forms which make broad demands on performance (see chapter 13). Companies end up in the position to make more extensive and flexible use of relatively low-skilled workers, or to assign overqualified skilled woodworkers to a variety of different jobs or processes. At the same time, the latent skills held in reserve by the few trained workers, or the skilled workers in key positions, are only called upon occasionally. In the meantime, these workers may additionally be assigned to jobs which are relatively undemanding in terms of skill on a short-term, flexible basis.

All in all, this has the effect of drastically reducing discretionary tasks in the manufacturing area, and transferring the functions of manufacturing control and capacity utilization, etc., to the departments of process planning and operations scheduling. In the long run, this results in a gradual erosion of skills and qualifications on the shop-floor that were once available and necessary. This loss can lead to considerable problems in manufacturing when future technological/organizational changes have to be made to the production process. It also creates immediate problems in dealing with machine breakdowns, special orders or compensating for faulty planning.

This development also marks a gradual decline in the significance of the functions of the foremen. At present, this is taking place in the following way: companies expect lower-level supervisors who have been deprived of the discretion they formerly had, to nevertheless possess sufficient skills to deal with problems arising in the course of program-controlled manufacturing processes that have been planned outside the shop-floor area.

High demands are made on skills within technical departments, at control posts (of automated storehouses, for example) or in the administrative areas related to manufacturing. In the course of the developments outlined above, these demands either remain constant or increase somewhat when the departments take over the functions of repairing and maintaining new technical systems, or programming and control functions. In many companies, operations scheduling and manufacturing control become key departments. In such cases, lack of skilled personnel is coped with by recruiting professionals from the external labour market or by making use of the extensive service system offered by the machine and control systems manufacturers. It is evident, however, that companies in the furniture industry which are comprehensively organized in terms of information technology will reverse the increase in personnel which takes place during the implementation phase – particularly in administrative areas – at a later date. In the course of this development, greater demands are only made on a few selected employees, while the majority of trained personnel (*Sachbearbeiter*) are faced with tasks that are less demanding in content than their former responsibilities. Their jobs get more routine, at the same time that work intensity increases. (A similar 'reversal' of high skill requirements can be expected to take place in the maintenance area as well.)

As a general point, it should be remembered that more or less deliberate personnel and performance policy concepts oriented to Tayloristic principles and an extensive division of labour underlie these developments in skills and qualifications. Some of the characteristic objectives pursued include the most extensive use possible of young, semi-skilled workers at CNC-machines, rigid division of labour between machine operating and machine control, and the separation of challenging additional tasks, such as job setting, maintenance and repairs, from machine operating.

The personnel policies of the user companies conform to or complement the product and market strategic objectives of the machine and control systems manufacturers which endeavour to offer facilities that are 'user friendly' and 'easy to control'. This means that the parameters required for machining and operational sequences are either already integrated into the respective machines or have to be programmed by experts outside the shop-floor area. Accordingly, repair work can only be carried out by skilled maintenance personnel from the user company or the systems manufacturer. The objectives pursued by the product and service policies of machine manufacturers are of particular interest in this context: with the help of automatic (remote) diagnostic systems, they seek to cut back their cost-intensive service tasks, which have been extended for sales policy reasons. Thus, the tendency exists that the necessary training and ability that workers in the user companies should possess is assumed by the manufacturers of machines and control systems. In the long run, this means that lower demands are made on skills in the user companies and that the level of skills and qualifications in the repair departments also sinks.

Our findings make it clear that there is little hope of achieving integral concepts of work design that reflect alternatives in work organization and the development of skills, let alone a general reprofessionalization process of manufacturing work in the furniture industry.

In view of these developments, the occupation of the 'skilled woodworker' in its various forms (carpenter, mechanic for wood processing, experienced semi-skilled worker classed as 'skilled woodworkers', etc.) is in serious danger. Persons undergoing vocational training or working in the above-mentioned occupations no longer enjoy sufficient employment security or the chance to adapt to changing conditions. Craft and woodworking skills either play a minor role when new technology is introduced or are confined to certain niches. Placing skilled workers in jobs beneath their qualification level causes a downgrading of wages and thus a loss of income and prestige. The highly skilled jobs and new key positions in the enterprise are filled with skilled workers from metalworking or the electrical professions. The manufacturers of machines and control systems are also of the opinion that workers who have undergone the traditional training for woodworking occupations are not competent to control, maintain and repair the systems they supply.

The emerging reality is that in view of the widespread dissemination of new technologies and the rather traditional personnel policies pursued by most companies, the qualifications held by skilled woodworkers are becoming increasingly inadequate in functional terms for employment in the furniture making industry. Moreover, the vocational training of carpenters and mechanics for wood processing no longer guarantees either an employment status that is commensurate with vocational know-how and skills or medium-term employment security. This development endangers the, at present rather strong, position of skilled woodworkers, not only in the furniture making industry, but also throughout the entire woodworking industry. In recent years, a number of new initiatives have been launched in training policies with the aim of reducing the imminent risks; however, it is not yet possible to assess their scope and effectiveness.

The fact that very few furniture manufacturers carry out training measures naturally has an adverse effect on the developments outlined above. Problems arising due to employees' deficiencies in character traits, such as lack of reliability or the capacity for cooperation, are usually overcome by personnel selection (also in the context of personnel reduction). Despite the fact that almost all of the companies we examined complained that workers lacked sufficient technical know-how as far as electronics, hydraulics or machine handling was concerned, systematic or general retraining of skilled, or even more, semi-skilled workers is still virtually non-existent. Instead, most companies rely on the combination of user-friendly (simple) designs from the system manufacturers and on the short introduction to system operation provided by the manufacturer.

Only after concrete and acute skill problems arise is further training considered and then only for carefully selected workers. Generally, the training courses are carried out by the technology manufacturers, and the instruction naturally focuses on their own machine and control equipment. The same also holds true for data-processing training for administration personnel.

To a considerable extent, this situation can be explained by the fact that there are no adequate institutional facilities or financial means for the further training of skilled woodworkers and technical personnel in the furniture making branch – despite a number of initiatives on the part of respective training organizations. This deficiency makes itself felt all the more strongly due to the fact that there are no manufacturers of wood processing technology offering training measures which could be compared with the kind of further training offered by companies in the computer or the machine building branch.

SKILL CHANGES IN FIRMS SUPPLYING THE FURNITURE-MAKING INDUSTRY

In the course of the technological-organizational transformation which took place in the furniture-making industry, the need to achieve greater flexibility and cost efficiency was thrust onto the suppliers of furniture components to a great extent. This induced developments in these mostly small and highly dependent companies which markedly changed the manufacturing processes and structures as well as the skill and qualification requirements of these companies. These developments consist of:

1 a twofold process of polarization: the first is the existence of furniture manufacturers carrying out product planning and the assembly of their products while supplier firms only manufacture parts and components; the second is when the supplier market is split into two groups of companies, with suppliers of standard parts that are highly dependent on the purchasing companies forming one group, and the other consisting of firms that can be described as 'innovative problem solvers';
2 in a number of job areas, industrial manufacturing structures are replacing traditional carpentry handicraft methods;
3 on the basis of computer-generated data, a tighter technological-organizational integration of manufacturing sequences and supplier relations with the purchaser has been established; and
4 the supplier firms' capacity for achieving flexibility and increased performance, largely based on the versatile utilization of skilled personnel gets increasingly exhausted.[3]

Two factors are exerting considerable influence on skills and qualifications in supplier firms, namely the emerging trend toward industrial-type manufacturing structures, and the technological and organizational integration of the buyer company's production processes with those of the parts suppliers. Even in those companies which still have a comparatively high share of skilled woodworkers, more and more skilled handicraft tasks are being split up into individual and unchallenging operations. Step by step, traditional carpenter tasks are being devalued by the insular introduction of new technologies. Handicraft skills, oriented to producing an entire product, which many workers still possess, are increasingly jeopardized. In all supplier firm groups the trend is toward a rise in the share of short-cycle, monotonous and repetitive tasks, already relatively high in the classical kind of supplier firm manufacturing semi-finished wooden parts. This will reduce the demands for skills and therefore markedly increase the number of semi-skilled workers in the manufacturing area. Traditional handicraft skills and complicated or all-round knowledge of craftwork are losing more and more of their former significance.

The increased introduction of (partially manual) series production (when there is little product diversity), and/or just-in-time manufacturing or manufacturing to order of increasingly smaller batches generates requirements which are skill- or process-unspecific. Instead, increasing demands are made on the individual worker's reliability, adherence to schedule and personal flexibility. Thus, workers' ability to endure stress moves to the forefront ahead of 'mere' occupational skill.

Intensified segmentation of job tasks is particularly evident for supplier companies which provide standard parts manufactured in series: those which basically serve as 'extended work benches' for the buyer companies. In such companies, there are more and more manufacturing processes organized with a high degree of division of labour; handicraft tasks are divided up into separate operations, leaving only tasks that are repetitive and monotonous. Even the suppliers of high-quality furniture components have begun to display Tayloristic production structures designed for industrial style manufacturing. The result is that a large share of the handicraft work (particularly that involving surface treatment) is divided up into relatively simple and/or specialized tasks that are repetitive and have to be performed over and over again. Indeed, an increasing number of skilled woodworkers are being assigned to jobs which underutilize their occupational skills.

These developments result in deskilling, and far from slowing this process down, the increasing utilization of CNC technology lends momentum to it. In most cases, the CNC-machines consist of ordinary, stand-alone machining units for handling rather simple and one-sided tasks which require no specific machine or control system qualifications on the part of the machine operator. In order to achieve the desired degree of machining precision, however, what is most important is for workers to accumulate experience in machine use. This can only be acquired when employees are assigned to the same machine for long periods of time. More complex matters pertaining to machine technology or electronics are primarily carried out by the plant supervisor or by the, often, only all-round competent technician in the company who frequently have to consult experts from the machine manufacturers. Given the training process carried out in these companies, and in the carpenter trade, it is not possible to expect workers skilled in specific wood processing tasks to meet these sophisticated demands. Moreover, as in the industrialized furniture, manufacturing sector, the tasks that are performed in the areas up and downstream from machining operations consist of simple 'auxiliary functions' which hardly require woodworking skills, or let them be brought into play.

A broader, more challenging demand is made on skills, especially on those pertaining to machine technology, in a few, particularly innovative supplier firms whose products require the use of complex, multi-functional machine technologies. However, this does not imply that retrained woodworkers will necessarily be assigned to such jobs; in many cases these suppliers also prefer to recruit skilled workers from other industrial branches with backgrounds in machine technology or electronics.

Discovering isolated cases of increased requirements for skills and qualifications in companies supplying high-quality and innovative parts and components, cannot obscure the fact that being involved in producing the whole product – something becoming ever scarcer in the supplier firm sector – is also on the decline in these companies, despite the fact that they still employ a relatively large amount of skilled woodworkers. While it is true that all companies demand a great deal of flexibility and versatility from their workers, they usually only do so in order to make better use of capacity or to cope with urgent or short notice manufacturing orders and then only in the form of short-term assignments. The result is that well-trained carpenters, who were employed as such, find themselves carrying out simple tasks more frequently – in many cases, over long periods of time – which make little use of their handicraft skills. This is becoming all the more the case due to the fact that the suppliers' scope for determining their own manufacturing sequences is getting increasingly restricted as a result of the tighter scheduling set down by buyer companies.

These developments, which pose considerable problems for the skill structures in the supplier firms, are additionally aggravated by supplier dependence on the buyers. On the one hand, the flexibility and performance potential of the suppliers is exploited to an increasing degree, primarily because of their high dependence on their major buyers on the other hand. The problem is intensified by the fierce competition among and between supplier firms. This situation leaves the suppliers little scope to design their production processes autonomously and pursue their own interests in marketing, manufacturing and innovation.

Due to the increasing utilization of simple CNC machines, similar trends in skill loss can be expected to occur in many firms in the carpentry trade (in which series manufacturing has already been introduced in a number of companies). In view of this fact and the developments outlined above, general skill and employment problems are likely to arise for skilled workers who are trained in supplier firms, mainly in the carpentry trade. Thus the occupational orientation of the skilled woodworker and, in particular, that of the skilled carpenter, are fundamentally endangered. This is a most undesirable development in view of the fact that these workers are currently found more frequently in supplier firms than in the furniture making industry and represent a large part of the supplier firms' flexibility. Existing skills, oriented to the entire product, are fading away through insufficient and inadequate use. Another reason for the loss in skills is that specialized supplier firms and craft businesses are not in a position to provide comprehensive vocational training for carpenters or mechanics for wood processing, especially when it comes to the newest technological requirements for operating sophisticated machinery.

For the companies, especially the suppliers and certain sections of the crafts business sector, this means that a shortage of skills could arise, skills which are of essential importance for achieving manufacturing flexibility and developing innovative product and processing concepts. But especially for the skilled woodworkers presently employed or undergoing training in these firms, considerable problems will arise in the future. In view of the severe employment decline of the past years and the limited number of vacancies in the craft business and the supply sector (due to persisting overcapacities), skilled woodworkers' chances of finding suitable work in the furniture making industry that is commensurate with their qualifications after completion of their training or upon changing companies will continue to decline.

OUTLOOK FOR FUTURE SKILL DEVELOPMENTS

Factors such as the comparatively low complexity of products and manufacturing processes, competitive pressure and a certain degree of economic recovery in parts of the industry point towards rapid propagation of new (systemic) rationalization strategies in the furniture industry and thus the implementation of new forms of organization and control technologies in areas of manufacturing and administration. This implies that the effects on skills discussed here, and which are now just beginning to manifest themselves in the companies, will make themselves felt to a far greater degree in the future. Expected developments are outlined below:

1 Fewer employees will be required for purely auxiliary tasks (such as transport, etc.) as well as for handicraft work.
2 For the entire work-force, the non-specific character traits possessed by workers will become more significant at the cost of vocational skills or knowledge of processes. At the same time, know-how related to wood as a material will become less significant in comparison to more abstract skills and knowledge pertaining to machine and information technology.
3 High demands on skills and qualifications will be made on a small group of highly qualified technicians (particularly in the manufacturing and repair sectors) and on computer specialists (especially in administration).
4 A change in functions and positions will take place affecting the role of the lower level supervisor. The change will not be managed through

training or retraining, but rather by replacing personnel and transferring the control of the production system to technicians with data processing knowledge, particularly in the process planning and scheduling department.

5 A new division of labour is emerging for maintenance and repair tasks (and tasks pertaining to systems planning) whereby a great deal of the necessary machine and control technology skills and qualifications are being assumed by the service departments of the manufacturers of the machines and control systems.
6 An increase in Tayloristic principles for dividing up work tasks and organizing work will also be introduced into the small firm and craft sector (particularly the suppliers of furniture components). This will be necessary in order to master the specialized machining processes demanded by the furniture manufacturing companies and to adapt to their production sequences.
7 Unless the present situation can be improved through the implementation of comprehensive training policies, the occupation and the position of the skilled woodworker will certainly lose significance in the long run due to the increasing importance of both technology-oriented skilled workers from the metalworking trades and semi-skilled machine operators.
8 There will be an increasing polarization of qualifications between administration and manufacturing areas, as well as within the administration area itself.

The technological design of production facilities and data-processing systems as marketed by manufacturers and as (usually) demanded by user companies (see chapter 6) and the concepts of the furniture manufacturers regarding personnel and performance policies lend considerable momentum to the developments in skills and qualifications outlined above. Despite the fact that new technologies theoretically provide the potential for setting up other forms of work organization, future developments will doubtlessly be characterized by an attempt to integrate the know-how currently held by skilled workers 'into the machine' and to transfer decision-making responsibilities to manufacturing-related administrative areas. This means that, apart from a few key positions in the manufacturing area, there will be a tendency to reduce or do away with the skill requirements there.

It is not possible to make forecasts on the numbers of people that will be affected. However, it seems that in areas that are numerically small, such as process planning and scheduling or production planning and control, skill requirements will increase in the long run. As far as the majority of jobs are concerned, (which will decrease in number due to cut-backs in unskilled and semi-skilled tasks), demand for skills will tend either to decrease or remain constant.

Due to the fact that the new manufacturing technologies make relatively few demands on machine operators, the need for training measures, or for creating new occupational categories, does not arise, especially in the case of the mass producers of standard furniture. Even the manufacturers of exclusive furniture, with their complex manufacturing facilities, end up displaying both enterprise-internal and -external polarization tendencies, mainly with regard to how high-level knowledge of machines and electronics gets structured (such as between shop-floor and technology departments, or between technology users and technology manufacturers). These trends stand in the way of any attempts to enhance the skills and qualifications of the shop-floor workers or technical personnel in a comprehensive or sustained manner. The manufacturing facilities and computer systems offered by technology manufacturers are largely determined by the needs of the dominant furniture manufacturers. Small and medium-sized companies and the suppliers of furniture parts are too weak and have too few resources either to push through technologies geared to their manufacturing needs onto the market, or to enhance the skills and qualifications of their personnel on their own.

All in all, it can be assumed that neither an enhancement of qualifications and skills nor a requalification of manufacturing work will take place on a broad basis within the branch of industrialized wood processing. Our findings contradict many other assessments and forecasts of a general improvement in the level of skill and qualification as both a prerequisite and a consequence of utilizing computer technology in industrial manufacturing. On the contrary, our findings, particularly with regard to the loss of position and function of the skilled woodworker, not only indicate a trend toward the employment of less skilled workers in areas hitherto unaffected by such trends, but also seem to reflect a drastic change in manufacturing work, at least in the consumer goods industry. The push to introduce new technologies within a systemic rationalization context is lending increased significance to abstract systems-technology knowledge over specialized vocational and process technology know-how. In the long run, this could lead to a generalizing of technical and organizational skills and qualifications above and beyond the scope of individual enterprises and industrial branches, bringing about, as of now, unforseeable consequences for the character and functions of skilled work in these sectors. Whatever happens, we can be sure not to expect any marked increase in the use of skilled craft workers over the long term.

Part V

Job Structures and Skill Formation

16 Vocational Training, Job Structures and the Labour Market – An International Perspective[1]

Werner Sengenberger

INTRODUCTION

This chapter emphasizes important linkages between job and labour market structures, the type of skill and skill formation of the labour force, and the industrial relations system. Extensive standardized vocational training, regulated by law or collective bargaining, is seen as an important, though not exclusive, precondition for the development and prevalence of relatively homogeneous job and labour market structures in Germany compared to Japan or other developed industrial market economies. The important impact of the vocational training system on the organization and restructuring of production is also stressed. It is argued that vocational skills facilitate a high level utilization of plant and equipment and ease the modernization of production processes.

NATURE AND EXTENT OF VOCATIONAL TRAINING

Vocational training and the associated organization of business activity along occupational or professional lines has had a long tradition in a substantial number of countries. The medieval guilds were very much structured on vocational grounds and, also, responsible for and in control of vocational qualification of their members. There are modern remnants of these guilds, such as the craft unions in some of the Anglo-Saxon countries. However, they have vastly declined in importance and now account for only a minor portion of total employment.

It is in the German-speaking countries where, at present, vocational training of workers is most widespread and pervasive. Two reasons may be responsible for this. One reason is that in the German-speaking countries (notably Germany, but also in the area of the former GDR, Austria and Switzerland) handicraft or artisan trades organized in small firms (*Handwerk*) are still in place to an unusual extent (see Part VI). They constitute a significant and well-organized pattern of business activity even in areas which in other countries are either subject to more capitalist large-scale and industrialized forms of production and service, or in which unorganized competitive market systems based on unskilled work prevail. (Examples would be meat processing, bakeries and other food production, as well as many repair services.) *Handwerk*, next to industry, is still the largest producer of vocational skills in West Germany.

The second major reason for the relatively high incidence of vocational training in Germany is its development in big firms which started in the late 19th century and spread during the 20th century. After the First World War, vocational training was regulated nationwide by legislation as well as collective agreement with regard to the content, the curriculum, duration, quality standards, and financing. Major reforms of the legal regulation were introduced in the late 1960s, the 1970s and the 1980s, under the *Berufsbildungsgesetz* (Vocational Training Act 1969) and the *Bundesausbildungsförderungsgesetz* (Federal Training Promotion Act 1976), providing for a modernized legal framework of the organization and financial support to vocational training, respectively.

Today there are hundreds of different occupations in West Germany in which state-regulated vocational training takes place. Roughly 100 of them have more than 1,000 workers being trained at the present time. There are two major institutional forms of vocational training. One, and definitely the most important form, is the apprenticeship of juvenile workers within the scope of the so-called 'dual system' of regulated, formal on-the-job training in firms or suprafirm training centres (*Überbetriebliche Lehrwerkstätten*) and part-time vocational schooling. Around 60 per cent of those in the 16–19 years age bracket take such three to three-and-a-half year basic vocational training, for which the trainee, upon completion and examination, receives a certificate as a journeyman, or *Facharbeiter*. Of the remaining 40 per cent, a major portion undergoes university education or vocational training in special vocational or professional schools.[2]

Vocational training, especially re-training and promotional training, is also taken by adult workers (with a concentration on those aged between 20 and 40). This type of training is supported financially under the Labour Promotion Act (*Arbeitsförderungsgesetz*), partly through assistance in establishing training centres and partly by supporting trainees in terms of covering tuition and fees as well as family allowances.

As a result of an imbalance between the volume of basic vocational training and the de facto demand in the various occupations, not all of the apprenticed workers are able to find employment in their occupation. Thus, a significant number of workers are employed outside their occupational field, largely in semi-skilled jobs.

VOCATIONAL TRAINING AND LABOUR MARKET STRUCTURE

In an attempt to typologize the structure of the West German labour market we have identified three different categories of sub-labour markets exhibiting three distinct types of allocational structure (Sengenberger 1987).

First, there are markets for labourers with low-level general skills which may be characterized by a high rate of mobility across firms and industry, i.e. with little attachment to a particular firm or occupation.

For workers with specific skills we have identified two types of market structure which coexist side by side, namely:

1 firm-internal labour markets based on firm-specific skills and a mobility pattern of low inter-firm and high intra-firm mobility. In other words, the attachment of the worker is to a particular firm.
2 occupational-type labour markets which rest on vocational skills. Workers are mobile across firms within their occupation. In other words, the workers' allegiance is to a particular occupation, not to a particular employer.

It is this latter type of market structure which I shall concentrate on in the following.

Standardized certified vocational qualifications on an extensive scale have been a crucial ingredient for the development of occupational-type labour markets in Germany. By international comparison, occupational labour markets stand out as a very important structural component in the labour market as a whole.

Occupational labour markets may be characterized by their specific modes of adjustment of supply and demand. In the short run the mismatch between supply and demand in a particular occupation is adjusted through inter-firm mobility of labour; in the long run market clearing is accomplished through adjustments in the volume of vocational training.

Beyond the purely quantitative accommodation of excesses or shortages of labour in relation to demand there is a qualitative dimension for adjustment. This works through revision, extension and innovation of the skill content provided by vocational training (see chapter 12) and, also, by the substantial overlap of skill content across adjacent occupational categories (such as metal trades like mechanics, fitters, turners, tool-and-die-makers, etc.). This overlap in competency permits the interchangeability of these trades for job assignment and worker deployment to a substantial degree (Mertens 1973; Kaiser 1982).

However, whether the interchangeability of skilled workers is actually practised does not only depend on the breadth and division of skills. Rather, it depends on whether there are any institutional limitations in the deployment of different occupations. In West Germany there is, with a few exceptions (mainly outside the blue-collar sector), hardly any regulation which would significantly restrict the exchange of workers in different occupations within the firm. There is almost a complete absence of the Anglo-Saxon type of craft organization or (competitive) craft unionism using occupational jurisdictions or demarcation lines as means of job control. Consequently, there is little significant occupational segmentation or compartmentalization in the German labour market in spite of pervasive occupational structuring.

Job design, job assignment and manning levels, as well as the supply of trained labour, which in the Anglo-Saxon countries are firmly in control of craft unions, are predominantly subject to managerial discretion in Germany (for a comparison with the United States see Köhler, Sengenberger 1983; Sengenberger, Köhler 1987; Sengenberger 1987). There have been attempts by unions and works councils during the past two decades to win greater influence on these personnel policy decisions. Yet, they have not advanced very far. They have clearly trailed behind practices in many other industrialized countries.[3] The connection between vocational training and occupational type labour markets is by no means an automatic one. A number of prerequisites have to be met to assure the functional affinity between the two. There has to be an approximate structural congruency between the division of skills (in the training market) and the division of labour or jobs in the firms participating in the occupational labour market. Each employer must design his jobs or job demands in such a way that they correspond to the range of skills offered by a particular occupation. Conversely, the worker filling a particular job in a particular firm, must bring along the requisite skills for the job. In other words, there has to be structural compatibility between the firm-internal job structure and the external labour market structure.

Any individual deviation from the structural coordination of firms and markets causes friction in the occupational market. Ideally, there are no transaction costs whatsoever, either for the worker or the employer. The employer has no information cost (because the trade is fully standardized and certified and, thus, serves as a complete signal as to the workers' abilities) and no training costs (since the worker is fully productive on the job upon his hiring). Moreover, there is – as a result of standardized training – no or little differential in individual productivity among workers. Nor has the worker to bear any transaction cost in changing jobs. There are no seniority-related benefits or otherwise firm-dependent rewards that would create disincentives for the worker to quit the job to seek employment elsewhere. In short, any rule or regulation that would attach the worker to a particular employer, or bind the employer to a particular worker, is likely to produce internalization of labour and, therefore, interfere with the proper functioning of occupational-type markets.

So far I have portrayed an 'ideal-type' version of occupational markets. In reality, while there are – especially in the handicrafts sector – labour markets which approximate the pure model of an occupational market, many labour markets exhibit elements both from occupational and firm-internal allocational structures. (In the final section of this chapter some of the forces that tend to produce incentives for the internalization of training, allocation, and pricing of labour into the firm are discussed.) For now it may suffice to maintain that by international standards the occupational-type allocation pattern is fairly strongly entrenched in the German labour market as a whole.

Occupational labour markets dominate in the small craft-business sector (*Handwerk*). Beyond this, occupational structuring principles also influence the fabric of firm and enterprise internal labour markets. In West Germany, internal markets are primarily characterized by a stable, permanent core-labour force well attached to the firm and protected from dismissal (albeit usually to a lesser degree than in big Japanese firms) and highly flexible internal allocation of labour. This is made feasible, as said, by the skill-based versatility of workers. Hence, the German internal labour markets are different from internal labour markets in other countries (notably in the US) which are primarily organized and shaped through extensive and rigid rules of internal worker allocation but less restriction on exit.

VOCATIONAL SKILLS AND THE PRODUCTION PROCESS

Formal vocational training and the worker mobility which the training renders feasible, carry with them substantial sources of worker productivity, adaptability, and innovative capacity in the productive process.

The worker competency acquired through vocational training, rests on the fact that, as a rule, the worker has at his or her disposal more skills than are actually required for a particular job at any point in time. Indeed, it is the purpose of the vocational training schemes to produce a wealth and affluence of knowledge and ability that go much beyond the skill requirements of particular jobs or even job areas. This is in contrast to informal on-the-job training which, as far as skill acquisition goes, is clearly constrained by a particular work-place.

The built-in redundancy of the vocational qualification system enables the worker to master a wide variety of occupation-related tasks within and across firms and, also, to solve problems, and – within limits – transpose the problem solving experience from one situation to another. In addition, vocational training may equip the worker to acquire more easily and readily skills demanded in the future by new products and processes, or, to put it differently and more generally, to 'learn how to learn'. The learning capacity is, however, also dependent on the human resource management in the firm. It requires management to design work organization in such a way as to produce further skill and accumulate qualification.

The principles of training by which the worker acquires such versatility and diversity, or 'polyvalence' as it is also called, is based on successive exposure of the trainee to different jobs and problem-solving tasks in an ascending order of difficulty and complexity. The guided exposure to a variety of tasks takes place during apprenticeship training as well as in the subsequent sequence of jobs held by the worker within the same occupational category either within the same firm or in different firms.

The skill versatility generated by the initial basic vocational training and its extensions through practice in different firms (usually during the first half of the worker's employment life) may be viewed to some extent as a functional equivalent to the systematic practice of job sequences and worker rotation in large Japanese firms with extensive intra-firm or intra-plant mobility chains resulting from it. In both systems, there is the idea that the training process is not merely a direct function of the technical requirements of jobs; rather it is considered as having some autonomous function in the sense of producing a particular type of worker – the self-reliant, 'polyvalent' worker with an orientation towards the factory (or the trade) as a whole rather than his present job only.

Comprehensive skill may raise worker productivity in many different ways (see Part III). To mention just a few, it allows the broadening or enriching of jobs in terms of work tasks and the stepping up of worker motivation and job satisfaction. It allows the production worker to do minor repair and maintenance jobs or otherwise support the continuity of the production flow. In this way the machine down-time may be reduced, a phenomenon which is the key to a lot of productivity enhancement.

Vocational skills tend to facilitate the introduction of technical change and structural reconversion in the economy. Not only do they help the worker to adapt to new job demands associated with change. They also widen the scope for the alternative deployment of each worker and, thereby, remove fears of job and earnings losses which often produce worker resistance to change. Here again, non-competitive industrial unionism is of great importance with regard to the transfer of workers to new employment. German unions have largely taken a positive attitude towards rationalization of production and the modernization of the economy. In comparison, under the British system of industrial relations with its emphasis on the demarcation of jobs, the union organization itself has come to depend on the maintenance of job boundaries. This, of course, will make the modification of job content and the redesigning of jobs as well as the exploitation of the full productivity potential of new technology a much more difficult task to achieve.

It goes without saying that, within German industry as a whole, the use of formally trained workers supplying pertinent skills is concentrated in particular sectors. It prevails in those sectors of capital goods manufacturing which are based on unit and small batch production (see Part III). There are other industries where production is based largely on semi-skilled and unskilled workers. But even there you will find substantial numbers of workers who hold a vocational skill certificate which, however, is not relevant to the particular job they now hold. Thus, typically, many workers with vocational training in a *Handwerk* occupation tend to take up semi-skilled jobs in industry because they are not able to find jobs suiting their occupational training or because they are seeking higher earnings opportunities or other chances of improvement. Although they do not bring along pertinent skills to the new job, employers still clearly prefer them to workers without training since they are expected to exhibit some of the desired worker traits associated with formal training.

Furthermore, industries with process production technology such as chemicals, steel, pulp and paper, which formerly had on-the-job-training-systems, have created new occupations (see chapter 19) to gain flexibility and interchangeability in operations, as for example, between direct production and maintenance. This tendency is to be considered as a departure from former tendencies towards generating specialised, semi-skilled maintenance workers (Asendorf 1979).

Systematic worker versatility acquired through multiple and extended exposure to different work stations, in my view, goes a long way towards explaining the relatively high rates of productivity gains and the international competitiveness of both Japanese and German industry; a much discussed theme in recent years notably in Anglo-Saxon countries. It is important to note, however, that the productivity associated with extensive learning experience is not merely the property of the individual worker. It rests on a collective facet of common learning experience which may be called the 'technoculture' (*d'Iribarne*), a system in which both workers and managers are socialized through basic training or job experience. In Japan, the higher echelons in the firm hierarchy often start out their career in low-level jobs and rotate gradually up the hierarchy; in Germany a large proportion of managers undertake apprenticeship before they get higher degrees and assume managerial positions. Thus, they have been exposed to practical training and experience. In Britain, in contrast, production managers have fewer engineering skills and more emphasis is placed on academic achievement (Sorge, Warner 1978). Having gone through basic vocational training allows managers to have a better base for communicating with production workers and lowers the social distance between the two groups.

VOCATIONAL TRAINING AND LABOUR RELATIONS

The system of labour relations in a particular country is very much conditioned by the strategic sources by which the labour movement in general, and in particular the trade unions as the organizational centre of the labour movement, derive organizational strength and bargaining power.

Ultimately, for the unions to gain bargaining power in the labour market, the opportunity for the employer to achieve unlimited or cost-free access to and substitution of workers must be curbed in some way. This can be done in two principal ways: either by restricting the free exchange of workers through institutional arrangements, such as limitation of access of particular workers to particular firms or jobs, or by restricting the employer's freedom of allocating labour within the firm (e.g. by seniority rules). Or, alternatively, it can be done by promoting the workers' capability for mobility across firms. In this case the employer faces a limited supply of labour as well, for he risks the loss of workers or faces difficulties in recruiting them. This possibility for inflicting damage upon the employer may be used for building a bargaining position for labour. Basically, the difference in the strategic source of worker strength relates to the mode of interference with competition in the labour market: In the first instance, i.e. restricting inter-firm or intra-firm mobility, worker power is based on curtailing inter-worker competition. In the second case, i.e. boosting worker mobility, worker strength results from the enforcement or reinforcement of inter-employer competition for labour. The efficacy of the latter strategy, however, is very much contingent on the general scarcity of labour in the market or, at least, on the scarcity of particular skills that cannot be replaced without substantial costs to the employer.

In most industrial relations systems de facto union power stems from a combination of the two principal strategic sources. In West Germany, the source of collective worker strength through mobility is of relatively great importance and is, of course, related to the vocational skills incorporated in a large section of the labour force. To be sure, there are also institutional (legal) limitations or costs to worker substitution by the employer (e.g. protection from unfair dismissal or redundancy payments). But overall, the constraints to the interchangeability of workers are, by comparison, less based on formal rules and regulations than on real differences of occupational competency. Thus, for example, despite the employers' discretion with regard to the volume and type of vocational training provided, there is a limitation on the competition between skilled and non-skilled workers resulting from the significantly different attractiveness of the two categories of labour to employers.

That vocational skills are being standardized to a high degree allows for easy mobility of workers across firms. At the same time, due to a fair amount of skill-overlap of adjacent occupations and in the absence of craft unionism and the associated job control mechanism, there is a high degree of interchangeability of workers within firms which may be used by unions to pursue particular policy objectives such as employment stability and internal adjustment procedures. In other words, there exists a kind of exchange relationship between employers and workers under which the employment security of the core work-force is traded against the workers' cooperation in internal flexibility.

The interchangeability of workers with different occupational skills is not only compatible with, but supportive of, the existence of industrial unionism in West Germany and the bargaining structure which provides for large-scale regional and industrial bargaining units (see Streeck 1984). Within the industrial unions skilled workers play a major role. Union officials, and also works councillors, as the statutory representatives of the workforces on the plant and enterprise level, are largely recruited from the ranks of workers holding vocational certificates (either apprenticeship or higher vocational degrees). And they regard, in practice, the skilled core labour force as their primary clientele (see chapter 24).

Seen from an overall viewpoint of worker representation then, the strength and weaknesses of the German-type worker organization may be highlighted as follows: the strength of collective worker organization is very much linked to the versatility and inter-firm mobility of skilled workers based on comprehensive formal vocational training. Workers who do not possess formal vocational skills face an inferior competitive position in the labour market. This inferiority is reinforced by unequal representation which works in favour of the skilled groups. There may be, and has indeed been, a vicious circle operating, e.g. with regard to job insecurity. If initially employers prefer skilled workers to unskilled ones, using the non-skilled workers as a reserve for periods of high but fluctuating demand, then this inequality may be reinforced by positive feedback effects of unequal representation. As we well know, employment instability lowers the propensity of a worker to join a union and, thereby, limits his chance to voice his problems and pursue his interests. As a result, the gap in employment opportunities between skilled and non-skilled groups may be widened.

The dominance of skilled workers in collective worker organizations in Germany is, furthermore, reflected in wage differentials according to skill. For instance, the wage gap between skilled and unskilled employees is substantially larger in the German automobile industry than, for example, in the US automobile industry. In the latter, the United Automobile Workers (UAW) as the main worker organization has been much more a union of semi-skilled workers (see Köhler, Sengenberger 1983: 381; Sengenberger, Köhler 1987).

To illustrate the idiosyncrasy of the German pattern further, some international comparisons may be helpful. In Britain and in United States, the power base of craft unionism, where it does exist, rests on the inter-firm mobility of workers of the same trade or craft as in Germany, but it hinges at the same time on the limitation of access to particular jobs by other trades through jurisdiction or demarcational rules. There is, in other words, a more limited 'mobility space' opening up particular channels for worker movement and foreclosing others. The vulnerability of this type of worker control can be discerned easily. A decline in the demand for the particular type of jobs or skills – perhaps by deliberate employer rationalization policies – threatens or removes the type of jobs to which the union holds a monopoly and, thereby, eliminates its power base.

In Japan, to take another pattern variant, under the *Honko* (life-long) employment system in the big firms there is a considerable limitation of worker exchange across firms, based on human capital investments made by the employer (for details see Tokunaga 1984; Ernst 1980, 1988). The union bargaining power rests precisely on the high replacement costs the employer would incur if he exchanged labour, at least the regular workers. Permanency of employment to the *Honko* worker and employment security even in periods of slump are, on the other hand, a prerequisite for the high degree of intra-firm mobility of labour and the resulting efficiency and innovative potential.

The weakness in this system, next to the fact that only a limited number of workers gain access to big firms, lies in the dependence of the worker's bargaining power on the vitality of particular firms.

Let me summarize the last point of this section: in the Anglo-Saxon countries, although the volume and standardization of apprenticeship training is much lower than in Germany, job demarcation and job boundaries along occupational lines are used as the primary device for job control, even outside the area of skilled work. There is, in other words, much more regulation of the job than in Germany. Jobs in Great Britain are defined in terms of a set of tasks or as a job territory and viewed as a property right by the job holder whereas in Germany, jobs tend to be defined more as a range and level of technical and organizational competency. So, the difference is really between a rather exclusive job territory and a non-exclusive territory of skill or competence. And it is in this respect, where I consider the Japanese system of work-place relations, despite the absence of vocational training, to be much closer to the German than to the Anglo-Saxon pattern.

LIMITS TO VOCATIONAL TRAINING AND OCCUPATIONAL LABOUR MARKETS

As mentioned above, occupational-type labour markets coexist in West Germany with firm-internal labour markets. The relative importance of each varies by industry. The mix is related to factors such as product market structure, type of product, technology and firm size. For example, where the demand for the product is stable there are greater opportunities for firms to standardize production and use semi-skilled workers and attached work-forces. Where, on the contrary, product market stability is low, either in terms of demand fluctuations or rapid changes in product type, highly attached permanent work-forces tend to become more risky and costly to the firm and, also, the employer tends to prefer more versatile skilled workers who are better equipped to adjust to the changing demands resulting from the volatility of the market.

During the 1950s and 1960s mass-production of consumer goods was greatly extended. Production processes were standardized, the division of labour in the firms elabourated, skilled jobs were broken up into semi-skilled jobs (see chapter 3). Most studies on skill structures in that period revealed a tendency towards polarization: the previously important middle-level qualifications, notably the skilled production and office workers, were replaced by a minority of highly skilled workers (technicians, engineers) and large numbers of semi-skilled workers trained on-the-job.

Starting in the early 1970s, production in German industry has been restructured in favour of high-technology, intelligent products and product diversification; also, firms have to be much more flexible to adjust quickly to consumer demands ('tailor-made' products).

The main competitive criteran, notably with respect to exported goods, are high quality, fast and on-time delivery, and services related to the product. Given these demands, many firms believe that they are better off with skilled workers who are more capable of providing the required quality of work and changes in the production programme.

Above all, the interest of firms to develop internal labour markets depends on the business cycle or, more specifically, on the scarcity of the external labour supply. In periods of full employment and tight labour markets, as occurred in West Germany during the 1960s and early 1970s, firms make greater efforts to internalize decisions on training, allocation, and pricing of labour. Through firm-specific benefits they attach their work-forces more closely. Firm-internal training schemes are extended, and the resulting skills, on average, are more job-specific rather than occupation-specific. Moreover, firms lower their hiring standards and take in more unskilled workers. All of these measures, in some way or other, tend to work in favour of firm-internal labour markets and what has been termed 'firm-centred labour market segmentation' and, at the same time, weaken the base for occupational-type market structures.

The internalization of the pricing function of labour services during the full-employment period challenged the established lines of worker representation between the unions and the works councils. Despite the reluctance of German trade unions (operating largely outside the firms) to accept any decentralization in collective bargaining, second-round negotiations developed at plant level and additional wage elements were provided on top of the industry-wide wage rates (Teschner 1977). Moreover, during this period company agreements were concluded on recruitment, dismissals, compensation for transfers or downgrading and training.

A tendency for relaxing efforts to conduct formal vocational training and, consequently, to diminish the base for occupational market structures, may also be observed in slack labour markets. As unemployment rises and skilled labour becomes available on the external market, firms may be led to reduce the volume of vocational training and, instead, satisfy their labour needs by selection procedures.

In summary, while occupational labour markets are more widespread and more firmly entrenched in the labour market structure in Germany (and other German-speaking countries) than elsewhere, economic and institutional forces have operated to limit their incidence and viability. This explains why there has been incomplete success in public labour market policies designed to extend and consolidate occupational markets.

17 Education and Job Hierarchies – Contrasting Evidence from France And Germany[1]

Burkart Lutz

INTRODUCTION

This chapter presents results from a comparative investigation of German and French manufacturing companies in which the relationship between the educational system and resulting organizational structures within companies is investigated. French companies reveal much more pronounced hierarchical structures than their German counterparts in terms of number of levels in the hierarchy, ratios between shop-floor and office personnel and wage differentials. The chapter demonstrates the close relationship between job hierarchies and the educational and training systems in each country and draws some political conclusions. The information contained here is based on a study conducted in the early to mid-1970s. However, the methodological design, comparative relationships and basic findings have served as a springboard for a number of other studies and continue to be of relevance today.

Ever since those concerned with educational policy have become aware of its relationship to the so-called employment system, discussion on this theme has been dominated by the question of the need for graduates with specific educational degrees. Experts have based, and still base, their arguments and demands concerning educational policy on the necessities of the employment system. This is true both of the pioneers of educational expansion in the 1960s and of all those who now warn about an overproduction of university graduates. For example, in the 1960s it was argued that if additional highly qualified personnel could not be made available, further economic growth would be arrested due to an insuperable bottleneck. Today, on the other hand, it is argued that a majority of university graduates is doomed to unemployment, since there is no place for them in the employment structure.

In contrast, this chapter contends that:

1 argumentation based on need has a very weak foundation and is unrealistic if it assumes that, given a specific technology, a specific product can be produced profitably only if the personnel employed possesses specific training qualifications;
2 complicated – and elastic – interrelationships exist between the educational system and employment structure; and
3 due to these interrelationships, changes in the educational system sooner or later also influence the development of the employment structure.

This chapter will first present some results from a comparative investigation on German and French industrial companies. It will then demonstrate the close relationship between the national employment structure and the respective educational and training systems in each country. Finally it will draw some conclusions concerning the current debate on vocational training policies in the Federal Republic of Germany.

WORK ORGANIZATION AND SKILL STRUCTURES IN FRENCH AND GERMAN COMPANIES

At the instigation of a French governmental agency, an investigation was carried out in 1971–2 in seven German and seven French companies, the aim of which was to study the organization of these firms and to ascertain their wage and personnel structures. The companies investigated (in each country two medium-sized machine-tool factories, two relatively large paper factories, one medium-sized and one large plant in the iron and steel industry, and one small tannery) were selected very carefully according to the principle of 'matched pairs' and were largely comparable with regard to products, production techniques, size, regional location, etc.

Comparison of the data from the German and French companies revealed four principal differences, which will be described briefly below.

First, company hierarchy in France always includes more management positions and more management levels than in Germany. In the production and maintenance divisions in Germany there are twice as many workers per manager as in France. This reflects not only the fact that in Germany there are inevitably fewer personnel at the lowest level of management than in France, but also that the number of management levels is always greater in France than in Germany.

These differences are especially pronounced in German companies in which skilled workers dominate. Thus, for example, the production division of a German machine-tool factory employs 450 workers, but only 20 supervisors and managers at three management levels. The comparable company in France, which builds an identical machine, requires 40 supervisors and managers at eight management levels to control and supervise some 400 workers. To be sure, a majority of the German employees are skilled workers, whereas in France a majority of the employees are only semi-skilled workers or have been retrained as (semi-)skilled workers (*Anlernfacharbeiter*). These differences are quite similar, or even more striking, in the commercial management and in the technical offices of the companies.

Secondly, technical departments and offices in France (in so far as there is any technical office at all in a specific company in addition to its production division) always have a significantly larger staff, are organized in a more hierarchical manner, and are more distinctly separated from the manufacturing division than in Germany.

In the companies of the machine-tool and steel industries in Germany, 5 per cent of the employees are occupied in the technical offices and departments: in France, however, the corresponding number is more than 10 per cent (as of 1971). In the machine-tool industry these employees are, above all, concerned with development, design, and operations scheduling; in the steel industry, with technical management, production planning, and quality assurance. Departments which in Germany are responsible to the production management are organized as independent departments in France and report only to the technical directors.

In the case of one of the pair of companies compared in the machine-tool industry, both are independent enterprises with approximately 600 employees (leaving aside the sales departments, which each has organized in a different manner). The technical office of the company in Germany comprises two senior executives (head of the design department and head of the operations scheduling department) and 19 technical employees. The technical office of the French company includes 50 employees, one-third of whom exercise executive functions at three of four managerial levels. Development, design and quality are independent divisions, comparable to the manufacturing division in organization and hierarchy.

Although this was not actually the task of the present study, it was ascertained that in at least a few of the companies compared, production staff and even production managers were controlled and manipulated by the technical department to a far greater extent in France than in Germany. French supervisors – supervisor, assistant supervisor, foreman, assistant foreman – possess, on average, clearly inferior technical qualifications when compared to their German counterparts. In some of the French companies examined, we were explicitly told the technical skills are less important for a supervisor than a responsible character and leadership qualities; sometimes therefore supervisors with training in completely different fields are hired. Also as a general rule, German production and maintenance divisions employ significantly more engineers than their French counterparts; in France engineers are employed principally in technical offices or in higher managerial positions.

Thirdly, the commercial management of all French companies studied employs significantly more personnel than its German counterpart and is organized far more bureaucratically and hierarchically.

In some of the companies compared, the administration area in the French companies had up to three times as many employees as those in Germany (the German companies, however, employ fewer personnel than the French). This was found to be the case especially where the plants compared were subsidiaries of a company with several branches, in spite of the fact that (or because?) the German companies generally allow their subsidiaries far more freedom of decision than the central management of the French companies. The latter has strict regulations governing many of the operations, or controls the decision-making process itself, whereas the German branch factories are free to make comparable decisions on their own.

Correspondingly, administration in Germany has essentially only two hierarchical levels, namely those of clerical employees and department heads. Only in very large companies is there a third level (central department, division, etc.). The administrations of French companies, in contrast, generally have five or six hierarchical levels. Functions which are fulfilled in a German company by one or two clerical employees working in contact with a departmental head are organized in a French company in independent departments with a departmental head, an assistant departmental head, and several office managers or group managers.

Here, too, the different qualification structure is obviously of great importance. Whereas the clerical employees in German companies usually have gone through systematic business training, a majority of the employees in the administration area of French companies have had only lower secondary or secondary school education, have been trained on the job, and have had to work their way up step by step.

Finally, white-collar workers and managers are not only more numerous in France, they also earn more, in relation to blue-collar workers, than in Germany.

Theoretically, it would seem logical that managers and employees in technical and administrative offices in France should earn less, in comparison with blue-collar workers, and in particular skilled workers, than in Germany. If this were the case, it would mean that the respective countries make sparing use of the most costly personnel, employing less expensive staff instead.

Actually, however, the exact opposite occurs. Although technical and administrative offices in France employ about one and a half times more personnel than in German companies, the average salary of the French white-collar employee is significantly higher (in relation to the gross monthly wages of a blue-collar worker) than in Germany.

Moreover, while French supervisors and managers, in relation to the total staff, are almost twice as numerous (12 per cent) as in Germany (7 per cent), the income gap between managers and workers is far greater in France than in Germany. It only simplifies the case slightly to say that the income gap from one management level to the next is almost the same in both countries, and that the income hierarchy reflects the number of management levels, which, as we have already seen, is smaller in Germany than in France.

Thus management and supervisory functions in French companies require 20 per cent of total wages and salaries, as compared to only 12 per cent in German companies. In Germany, workers receive 75 per cent of all wage and salary payments: in France they do not even receive 60 per cent.

EDUCATION, ACQUISITION OF SKILLS AND COMPANIES' PERSONNEL STRUCTURES

In contrast to the situation in French companies, the 'German model' of company organization and personnel structure is – comparing technically and economically similar companies – characterized in particular, as we have demonstrated, by a weaker hierarchy, a less rigid vertical and functional division of work, and thus greater autonomy even in subordinate jobs. In the 'French model', on the other hand, the performance of subordinate tasks is far more subject to hierarchical supervision and functional regulation and control; due to stricter vertical (hierarchical) and horizontal (functional) divisions of work, autonomy in the performance of tasks is obviously much more restricted than in Germany.

This contrast between two types of structures in enterprises corresponds to an equally significant difference between the educational and training systems in the two countries, which results in a marked difference in the educational and skill structure of the working population. It goes without saying that the present situation in the educational system, which has resulted from the educational reform and educational expansion in both countries in the last decade, cannot be taken into account. Newer developments in the educational system had not yet shown any significant influence either in Germany or in France on the skill structure of personnel in the period in which the data quoted above were collected and for which the most recent statistics are available. This study reflects, rather, the situation in the educational system in previous decades.

French and German educational systems and personnel qualifications

If one compares the general education and vocational training of the working population of both countries, three marked differences can be observed.

Many more people have attended and graduated from institutions of higher general education in France than in the Federal Republic of Germany. The opportunities for further education within the systems of general public education were expanded much earlier and more rapidly in France than in Germany. Whereas in Germany not even 20 per cent of the working population (as of the mid-1970s) have a certificate in further education, in France the figure stands at some 40 per cent. Furthermore, only 9 per cent of the working population in Germany, as compared to 13 per cent in France, have a secondary school or university diploma including diplomas from higher technical schools and technical colleges which strictly speaking are not institutions of general education: meanwhile, 10 per cent in Germany, as compared to 27 per cent in France, have lower secondary school certificates.

This difference has not been altered by the rapid expansion of education in the last twenty years. No less than 55 per cent of young people entering the labour force in France in the years 1971–2 had completed schooling which corresponded at least to the level of the German lower secondary school certificate, or *Mittlere Reife*; in Germany, however, the number was only about 40 per cent. Moreover, 22 per cent in France, as compared to only 12 per cent in Germany, had completed secondary school education which qualified them to attend a university or college.

Many more people in Germany than in France have completed vocational training, opportunities ranging from apprenticeship or its French equivalent to university studies. Only 31 per cent of the working population in Germany have not completed some form of vocational training (a majority of these are women who otherwise comprise slightly more than one-third of the working population). In France, on the other hand, a corresponding 71 per cent have not completed systematic vocational training (with the percentage of men in this group scarcely lower than their percentage in the total working population).

The average standard of vocational training in the German working population still remains much higher if we take into account the numerous workers who had completed vocational training when they began to work – generally in handicrafts or the retail trade – but who were later forced to shift to unskilled or semi-skilled jobs. Even if we estimate this group at about 20 per cent of the working population, the percentage of workers who have completed vocational training in their field is almost twice as high in Germany as in France.

Education and training of the German working population are far more homogeneous than in France. In Germany the dominant form of obtaining vocational qualifications is vocational training within the dual system: in 1971, 62 per cent of all employed persons (born in 1918 or later) in the Federal Republic of Germany had completed an apprenticeship. And even from among those employed persons with higher vocational qualifications comprising a degree from a university, technical college or comparable French institution (in Germany almost 14 per cent, and in France less than 9 per cent), about half of those in Germany had completed an apprenticeship at some point in their training.

In France, in contrast, only 20 per cent of the working population had completed apprenticeship-type training as of the mid-1970s. The combination of such practical vocational training with the prior acquisition of a higher degree from a school of general education or with a subsequent degree from a technical school or university remains completely atypical. Much more frequent, however, is vocational training at an intermediate level in public or private schools offering a large variety of subjects, specialized fields, and levels, for which there are no real equivalents in the German system.

It should further be emphasized that in France a significant portion of the working population – more than 10 per cent – do possess an intermediate or higher degree in general education (including two years of basic university studies without the following specialized degree), although they have not had any systematic vocational training.

It should also be stressed that in France vocational qualification courses for adults, usually of six to ten months' duration, are of considerable importance, quantitatively and qualitatively. In principle, these courses belong to the sphere of further training and thus often do not lead to the same certificates and degrees as initial vocational training. For many young working people, however, they represent the only opportunity to receive any vocational training at all.

Labour supply qualification structure and companies' personnel policy

It will be immediately obvious that due to the differences outlined above in the educational system and skill structure of the working population, an employer who wishes to initiate a specific production process will be confronted with entirely different problems in France and in Germany.

The German company, partially due to its own training activities, will as a rule be able to rely on a large supply of manpower with completed basic vocational training in the form of apprenticeship as skilled worker or as specialized employee; in other words, on training which ensures a relatively high level of vocational skills which are immediately usable or can be easily activated. In France, on the other hand, correspondingly skilled manpower is structurally scarce. In addition, since most training takes place in public schools, the supply cannot be influenced by a specific company.

Thus, in French companies a greater part of the skills required can and must be taught after an employee has been taken on. This is, of course, true only under the conditions found in average companies which cannot afford to lure a large part of the scarce manpower with completed vocational training by means of especially attractive job offers. This means that a French company must attach far greater importance to the question of how to ensure that the necessary skills are acquired in the daily routine of the company, on the job, than in the case of its German competitor. The latter can largely rely on the fact that the basic vocational training of its employees will have provided them with the qualifications required by highly specific jobs, without special in-company training having to be organized (see chapter 16).

Under these conditions, the level of general education and the school degrees of its employees are of little importance to a German company. On the other hand, it is in the interests of a French employer with essentially the same business problems to use the general (social or cognitive) skills which his employees have acquired in the public school system as a basis for the acquisition of qualifications required specifically in the company.

This has the following implications for the most important groups of personnel:

1 Semi-skilled workers in Germany have often had at least preparatory vocational training in that they have completed an apprenticeship in a different field (usually handicrafts), whereas in France they have almost never had systematic vocational training or instruction of any kind.
2 Skilled workers in German companies have, as a rule, a more homogeneous and, on average, higher qualification than in France simply because they have completed apprenticeships as industrial skill workers. In France, equivalent training for skilled workers covers a narrower sphere of vocational tasks than training for their German counterparts. In addition, only a certain proportion (in many companies a minority) of the *ouvriés professionnels* (skilled workers) have had training which corresponds to an apprenticeship; all other workers classified as skilled have had at the most one year's re-training within the framework of programmes for the training of adults.
3 Nearly all German clerical and technical employees have completed apprenticeships, and technical employees in particular have also often

completed further training. In French companies, in contrast, vocational training of the employees is very heterogeneous. Often the level of general education plays an important, if not decisive, role in classification and promotion. This is especially the case with regard to clerical employees, whose vocational qualification is based primarily on on-the-job training and experience (which may be more or less oriented to the needs of the specific company). The technical employees include former skilled workers, graduates of the technical branches of the public school system, and employees who have only an intermediate or higher degree in general education, and who have acquired their vocational training exclusively through company experience.

4 Supervisors and middle-management personnel in Germany almost always have the same basic qualifications as the workers they supervise. In France, in contrast, character, rather than proven vocational skills, seems to be decisive in the employment or promotion of supervisors. Middle-management personnel, who are in any case more numerous in France than in Germany, often have a higher level of formal education than in German companies.

5 German senior executives are the most heterogeneous with respect to their educational background and training whereas, in France, they constitute the most homogeneous group. In France a senior executive is generally expected to have a secondary school diploma and, often, a university degree: among Germans at this level, on the other hand, almost all kinds of educational and professional backgrounds can be observed.

Qualification structure and company organization

At first glance, the model of company organization, personnel structure and income structure which we have labelled the 'French model', when compared to its German counterpart, seems to represent an almost pathological deviation from the principles of industrial rationality. However, a closer inspection reveals that the models in each of the two countries are in fact quite rational in the light of company interests and of the prevailing conditions on their labour markets and in their national education systems. In both cases, companies act as rationally as possible under the circumstances with which they are faced. The situation in France, which seems extremely irrational from a German point of view, is thus not the result of poor management, but rather the expression of different national conditions and, above all, differently structured educational systems. This point will now be demonstrated in more detail in the case of the manufacturing division, commercial management, and the technical offices.

An average German industrial plant with a production process typical of much of the machine-tool industry (typical, too, of many construction sites or of the continuously expanding maintenance departments of highly mechanized and automated industries) can count on an abundant supply of fully trained skilled workers who have acquired varying degrees of professional experience (either in the plant itself or in other companies with similar production structures). Consequently, they are qualified to solve most of the technical and organizational problems arising in the production process on their own initiative without this detracting appreciably from their standard performance.

Thus the company has several alternatives.

1 It can rely on a flat and relatively sparse leadership structure, in which the lower limits in the number of supervisors and middle-management personnel will generally be determined by technical and organizational necessities. For example, in every large division one foreman (*Meister*) responsible for important decisions at short notice must always be present.
2 The company can select supervisors from among the skilled workers already employed who are able to assess production problems and to exert authority over skilled subordinates by means of their extensive experience and high-level competence (generally supplemented by specific further training leading to formal qualifications).
3 It can restrict its role in operations scheduling, planning, and design to establishing basic principles and general guidelines. Technicians in these departments will nearly always be able to rely on theoretical knowledge and practical experience acquired formerly as skilled workers.

The situation in comparable French companies is completely different. Here, due to the insufficient qualifications – or more precisely, qualifications assumed to be insufficient – of a majority of the workers, the company finds itself obliged to take different courses of action.

1 It must see that the production process and execution of tasks are continuously analyzed and controlled by special technical departments employing a large staff, generally with high-level formal qualifications. In this way, the company can ensure, by means of fixed, detailed rules of procedure, that a large number of specific problems, which in Germany would almost automatically fall to the skilled worker and *Meister*, are solved at an earlier stage.
2 The company needs to ensure that a large number of technical managers with the necessary specialized knowledge is always present in the

factory to support the technical office, bearing in mind that only limited advance planning and regulation are possible in view of the diversity of tasks to be accomplished. Many problems which can be solved by qualified skilled workers and *Meister* in a German company require the active cooperation of such managers in France.

3 It has to establish a large, hierarchically subdivided group of supervisors occupying a position between the planning and operations scheduling departments and technical managers, on the one hand, and the production workers on the other. These supervisors have only very limited technical skills, but are needed in order to enforce strict enactment of the decisions taken by the engineers and technicians and precise observance of the rules and instructions which the latter have formulated. Such discipline can only be ensured by a well-staffed group of supervisors who control the work performed almost continually.

These strongly contrasting structures of hierarchical and functional division of work in German and French companies influence, in turn, the opportunities for training open to the workers. In Germany, the relatively high degree of autonomy enjoyed by a skilled worker, or a group of such workers with a *Meister*, results in a situation in which continuous acquisition of additional skills through experience and constant confrontation with new problems is not only made possible, but is also regarded as perfectly normal and, indeed, almost inevitable. Functions and working methods of *Meister* and engineers, designers, and work planners are organized in such a way that this process is supported and encouraged rather than hindered. In French companies, on the other hand, lack of confidence in the competence of workers and their immediate superiors not only reduces learning opportunities, but even appears to result in a negative selection, since the most qualified younger skilled staff apparently prefer to work in small concerns where they can expect to enjoy greater independence. It is only a slight exaggeration to say that the typical approach by French companies to the skill deficit of their workers threatens to perpetuate this deficit or even to increase it.

All of these observations are also true of the administrative and technical offices of companies in both countries. In this respect, however, those in France face a special dilemma.

If newly employed staff have not gained the requisite competence in their previous training, the company must offer them the opportunity and the incentive to acquire skills relevant to the job concerned. However, it must also organize the work itself in such a way that no performance deficits result from the qualification gaps of its employees, before these gaps can be eliminated through assimilation to and experience in company standards and routines. Both factors lead inevitably to bureaucratic structures with their twin aspects of explicit regulation and control and hierarchical differentiation. As a result of regulation and control, even employees with very limited skills can immediately be occupied in productive work by means of assignment to narrowly pre-structured and strictly supervised tasks. Hierarchical differentiation, on the other hand, provides for the necessary control and makes it possible to arouse interest in qualifications of specific use to the company by creating a system of rewards (such as promotion opportunities) in the area of employment concerned.

This practice of creating incentives in a hierarchically stratified system of jobs must, however, take into account the fact that French companies have for many years been able – and forced – to rely on a relatively large supply of job seekers with differentiated degrees in higher general education (including short-term university courses of little vocational value). These companies are thus under considerable social pressure to employ such young people: indeed, they are all the more willing to do so, since it is obviously in their own interests to exploit the differential value of these degrees and the processes of social and intellectual selection inherent in them. This, in turn, results in pressure for further vertical stratification of personnel, leading to different 'careers' for employees with different degrees, each career necessarily implying also a different level of opportunity for advancement.

In this way, however, the interest of French companies in reducing the risks and costs of qualification for office employees by relying explicitly on documented school performance comes into conflict with their interest in containing these 'unproductive' parts of the company within reasonable limits.

This contradiction may result, especially in medium-sized and small companies, in personnel and organizational structures such as those described above, in which one out of every three male clerks occupies a middle-management position.

Thus, the level and structure of the supply of skills resulting from the development of the education and training system contributes to the creation of specific forms of company organization, and of horizontal and functional division of work. In other words, the supply of skills has a strong impact on the qualifications structures of companies. Furthermore, different qualification structures require different income structures in order to function.

In so far as entering the service of a company involves the gradual acquisition of appropriate vocational skills, as is in the case in France, a corresponding dynamic income structure will be essential to encourage the learning process, either in conjunction with hierarchical advancement or independent of it. The more bureaucratic regulation and control appear necessary to ensure the operation of the company, the more the company will have to reinforce the vertical division of work and the implicit hierarchy of power and influence with a corresponding income hierarchy. The more a company is interested in winning employees with advanced degrees in general education, the more it will have to rely on salary differentiations graded according to specific careers.

IMPLICATIONS FOR EDUCATIONAL POLICY

The relationship between education, on the one hand, and the employer, on the other, has hitherto been regarded primarily as a simple relationship between customer and supplier. For the 'customer', the employer, the major concern was that sufficiently qualified manpower was available at the right time, whereas the 'supplier', the educational system with its various branches, was above all concerned with estimating and meeting the demand for its product.

The results of the comparison between personnel structures, educational systems and qualification structures of the working population in Germany and France described above can, however, in complete contradiction to this assumption, only be explained if the following hypotheses are accepted:

1 that employers are capable of achieving profitable operation and production with personnel qualified in completely different ways; and
2 that the educational system has a significant influence on the development of employment structures.

These two factors are closely interrelated. Since employers do not have a fixed need (or a 'limitational' need as it was called in the discussions on educational policy in the 1960s) for a specific amount of labour with specific training, the educational system can develop and expand independently within certain limits (to be explained in more detail below) without being completely 'uncoupled'. Since, however, employers cannot employ manpower haphazardly, they must at some point react to changes in the educational and qualification structure of the manpower available to them, by making changes in work organization, whether horizontally or vertically.

Thus, it was possible for the educational systems in Germany and France, which were probably quite similarly structured in the early nineteenth century, to develop in completely different ways. And, at first glance, it would appear that the employment structures had developed in corresponding fashion.

It would, however, be wrong to conclude, in contradiction to generally accepted ideas, that educational policy dominates developments in the employment system, and to assume that education systems can make independent decisions to which employers must simply adapt after a certain period, accepting the structures that the educational system has created.

Several recent studies (see Asendorf *et al.* 1975) have demonstrated quite clearly not only that changes in the educational system influence developments in employment structures, but that these developments in turn influence educational policy and the expansion of the educational system. Such influence is exerted through the connection between education and career opportunities as the most important determinant of educational behaviour.

This connection was long ignored, although it really was quite obvious. The massive and rapid expansion of secondary schools, such as has taken place in the Federal Republic of Germany and all large industrial countries, can certainly not be explained by the fact that the employment system needs secondary school graduates. Today we realize that the decisive impulse for this expansion came from an increasing demand for places in such schools. This demand was in turn clearly motivated by the recognition of parents and children that career opportunities are determined decisively by educational qualifications.

Thus education and employment interact in a very closely intertwined relationship. Educational policy means regulation and intervention within this relationship, and all educational demands which ignore this fact are illusive or demagogic.

The ingredients of this relationship include the organization of employment structures in companies with a view of optimal utilization of the labour supply, the influence of education and training on positions and career opportunities conditioned by such organization, and the determinants of educational behaviour induced by the latter. To varying degrees, this relationship can be influenced by what may be termed a 'meritocratic' principle, to borrow a term coined by French sociologists to describe conditions in their own country. What is meant here is a very strict conditioning (and in extreme cases determination) of the access to jobs and career opportunities by means of educational qualifications. These observations hold several very far-reaching implications.

First, the differences which exist today between educational systems and employment structures in Germany and France can be explained very plausibly by that fact that French meritocratic structures are more strongly developed. Parents and young people in France have little alternative but to think in terms of a long period of school education spent passing examinations to as advanced a level as possible. Companies, on the other hand, are forced to accept the hierarchical gradation in the education of their personnel and to adapt to this situation by creating special 'careers'.

In Germany, however, the existence of two competing qualification systems has, at least up to now, been characteristic. Set against the system of general education, consisting of the three levels of general schools and the universities, has been the so-called vocational system, consisting of apprenticeship, technical schools and today's technical colleges. This competition between the two systems probably explains why the hierarchical stratification in both is less pronounced than in France. It appears, however, very questionable whether this situation can last, as meritocratic structures apparently have an inherent tendency to expand their influence.

The educational system in France, with its strong hierarchical differentiations based on formal qualifications and selection, represents by far the most important, if not the only, system teaching professional competence (on the basis of very strict selection). Companies, therefore, have no choice but to adapt their organizational and employment structures so as to employ graduates of this system according to the special levels and profiles of the skills which they have confirmed in examination successes and the claims which they possesses on that basis.

Since only a certificate from a higher or at the least intermediate-level educational institution offers a guarantee against lifelong condemnation to subordinate, poorly paid work, the existing demand for places in institutions of higher education cannot be checked, but at the most only channelled. In the end, work of this kind – mainly but not exclusively manual – will be required and expected only of those who, as a result of unfavourable social or intellectual preconditions or character development, have been sifted out by the selection process of the educational system and labelled as inferior.

Correspondingly, expectations in the qualification potential of all those who have left the system without an intermediate or higher degree must be reduced. Consequently, companies are, it would seem, inevitably forced to accept an increase in the division of labour, limitation in the autonomy of subordinate work, and therefore, an even greater deterioration in the qualification opportunities and career perspectives offered in low-status jobs.

For these reasons, structures and trends in France should not be regarded superciliously as if they had no direct relevance for the Federal Republic. On the contrary, specific developments taking place for some time in Germany threaten to force the education and employment systems in a very similar direction if given free reign (see Köhler, Grüner 1990 and chapter 9 of this volume).

The German public has begun to realize that opportunities for those with an academic background are far more interesting than positions generally open to those who have completed apprenticeship as industrial skilled workers. The obviously meritocratic structures which have long existed in the German public service began to influence educational behaviour and career selection to an increasing extent in the 1960s. This was all the more the case as the working conditions offered by the public service to employees with higher qualifications simultaneously became more attractive. The previous independence and competitive strength of vocational training as opposed to the school and university system of general education is decreasing. The transformation of what were formerly higher technical institutions into technical colleges was only one step in this direction, though an important one. The establishment of special courses for secondary school graduates, a pet idea of some Ministers of Education and managing directors of large industrial enterprises, is another similar trend.

Economic organizations now maintain that industry does not wish this trend towards education to continue any further in view of its tendency to prejudice the career and advancement opportunities of workers with practical vocational training. Yet, in fact, this process of gradual infiltration of university graduates can be observed in all sectors. It is only a matter of time before access to top management positions will be the prerogative of degree-holders only.[2]

This process will definitely not be checked even by the rapidly increasing difficulties which university graduates encounter in their search for adequate occupations. On the contrary, these difficulties will give rise to displacement competition, whereby the stronger social position of university graduates will generally prevail over the 'mere' vocational qualification of competitors with practical vocational training. To be sure, not all university graduates will have access to relatively high-level positions, as has been the case to date. But this does not alter the fact that an increasing number of jobs will be accessible only to degree-holders.

Following this general line of thought, it is quite conceivable that present conditions in France reflect future conditions in Germany. Far from outdated, the relationships between French educational and training systems, employment structure, and company organization (with all their negative economic and social effects) are, when seen in historical perspective, most likely the precursor of conditions in the Federal Republic.

It is probably not yet too late to prevent such a development, especially since the particular constellation of the influx onto the labour market of young people born during the baby boom may create additional room for action in both educational and company personnel policy.

The opportunity, however, requires fast and decisive action, oriented to future needs. At present in Germany, the vocational system of education and training comprises part-time and full-time vocational schools, technical schools, high schools and colleges, and in-firm training. The major objective should be to strengthen these both in themselves and with respect to the system of general education so that they represent a real alternative for a future long-term situation in which fewer young people and a larger supply of training places are expected. The choice of this alternative should not carry the implication that the trainee is poorly informed, of limited talent, or lacking in ambition.

Present educational and vocational training policies, however, scarcely correspond to this goal. For one thing, authorities are exploiting the shortage of training places to delay improvements or even to revoke reforms which have already been passed. Rather than eliminating the obvious faults and weakness of the dual system, it is as if the system had already been written off for the period following the influx of young people born during the baby boom.

Secondly, instead of exerting efforts to expand vocational schools, the major increases in educational expenditure are being allocated to the universities. Moreover, despite the potential for intensified cooperation between vocational schools and in-firm training, which would be in the interests of both, the authorities have allowed jurisdictional squabbles which have actually prevented coordination of vocational school curricula and the regulations governing training. Vocational secondary schools are being increasingly misused as a reservoir for the selection casualties of secondary schools and those not accepted at the universities, in such a way as to perpetuate their inferior status in the eyes of those affected.

Finally, rather than defending and confirming the principle of a uniform standard of basic vocational training (in the form of full apprenticeship) as the most significant strength of the German qualification system, those responsible are succumbing to the temptation of hierarchical differentiation of admission requirements and final degrees. This means that the minor advantages of improved channeling of the flow of pupils and higher selectivity are being secured at the price of increased meritocracy.

It is certainly not an exaggeration to say that the present situation in educational policy is dramatic and that the decisions which must be made between two alternative development trends will have dramatic effects for many years to come: the alternatives are either the reduction of vocational preparation and initial training to the point where they become mere appendages of a broadly meritocratic system essentially oriented to the principles and values of general education, or the stabilization of practical vocational training (in institutions run by a variety of authorities) on a fresh long-term basis.

18 Mechanical Engineer versus Computer Scientist – Different Roads to CIM in France and Germany[1]

Burkart Lutz, Pierre Veltz

INTRODUCTION

There are distinct differences between Germany and France in the way industrial production technology has developed. This chapter attributes these differences to structures and processes which are represented by the terms 'social reproduction of technical personnel' and 'social structuring of technical competence and knowledge'. In France, it is argued, a strong relationship exists between the erosion of traditional skilled work, the overwhelming school-based education of technical personnel and a strategy of (forced) computerization. In contrast, the more machine-oriented development path of German manufacturing technology is explained by the relative stability of the skilled worker tradition and its connection to the training of engineers.

DIVERGENT STRATEGIES OF FLEXIBLE AUTOMATION IN FRANCE AND WEST GERMANY

Up to now, most international comparisons of factory organization, division of labour, and skill and manpower structures have been based on the assumption that the implementation of manufacturing technologies is universal in character, not nation-specific, and does not systematically vary by economy. There are a number of very good reasons for assuming that this also applies to the computer-aided technologies of flexible automation that are increasingly determining the development of industrial production.

However, a quite different picture emerges if a closer look is taken at the most important subjects of the comparative investigations carried out in the 1970s in the large industrial nations of France and the Federal Republic of Germany, namely the main aspects of technology development, structures of supply and demand on technology markets, the tenor of public discussion, the concrete situation of technological innovation within the enterprises and the public programmes in support of these innovation measures.

In France, agreement was fairly unanimous that the nation needed a strategy of technological development and industrial innovation based on the utilization of computers and computer science; for this reason, mastery of computer technology was considered decisive for the future of industry and the professional chances of the coming generation.

In West Germany, on the other hand, computer-integrated manufacturing was not so much regarded as a break from the previous course of technical and industrial development as a continuation, expansion and intensification of existing trends and practices in industrial automation and the rationalization of enterprise processes. Indeed, computer technology was primarily regarded as a useful tool that could be used for classical, application-related forms of engineering science, such as in mechanical engineering, which undisputedly held a key position in the development of future industry and technical personnel.

At least at present, these two different strategies do not display a similar degree of success. For example, in large parts of German industry in general, and the metalworking branch in particular, the development, production and application of systems and components of computer-aided automation is gaining significance and becoming an accepted part of daily life in the enterprise. The French strategy of technical innovation and industrial modernization, however, is characterized by innumerable failures, abortive CIM applications and discontinued processes of product development. Moreover, while most of the German manufacturers of machine tools have gone through an unexpected period of prosperity due to the gradual integration of micro-electronics into their products, France has had to accept the widespread downfall of its machine tool industry, despite massive state support.

These findings raise two groups of questions:

1 To what extent do these strategies depend on the national structures of industrial work and factory organization as observed in the comparative investigations of the two nations? And can a plausible connection be established between the differing significance both countries give to skilled workers, on the one hand, and computer science and scientists on the other?
2 Are these strategies of industrial innovation alternatives which industrial nations can choose? Or is only one of them destined to be successful? Could this explain current differences in the success of both strategies? Or does the French computer-centred strategy of developing manufacturing technologies (which is also being actively pursued by other

countries) open greater opportunities in the long run, despite the short-term high initial costs due to overly ambitious and therefore failed innovations? Finally, which aspects of nationally specific structures of personnel, skills and organization contribute towards finding an answer to these questions?

We will deal with these questions in three steps. The first (dealt with in the next section), argues that the striking differences in organization and skill structures in French and German companies are closely related to the extremely divergent way in which models of the 'social reproduction of technical personnel' have developed in the two countries since the end of the Second World War. In the Federal Republic, for example, practical occupational experience still plays a key role in the social reproduction of technical personnel, while in France it is increasingly dominated by a school-centred model in which the educational system has a central importance in the guidance, training and selection of future technical personnel.

The second question (discussed on pp. 281–5), which extends far beyond the traditional scope of industrial sociology and labour economics, seeks to clarify the relationship that exists between models of the social reproduction of technical personnel and what could be called the 'social structuring of technical knowledge'. The task here is to examine the premise that the increased importance of schooling for technical personnel calls into question what, in the history of industrialization, used to be the given professionalism of technical knowledge and competence.

Finally, the last step (discussed in the last section of the chapter) examines whether the strongly computer-centred strategies of technical innovation and industrial modernization as implemented in France can be interpreted as an attempt to react against the negative effects of the de-professionalization of technical competence through the expenditure of a major technological and educational effort.

DEVELOPMENTS IN THE SOCIAL REPRODUCTION OF TECHNICAL PERSONNEL: SIMILARITIES AND DIFFERENCES

Ever since the emergence of industry and industrial technology, all modern societies have been faced with the task of having to make sure that the competence required for the application and development of technical knowledge is always at its disposal. The distinct class structures of all societies that have undergone the industrialization process made the division of mental work (as carried out by the rulers and the educated) and manual labour (as carried out by the common people) seem part of the natural order of things. As professional status was by far the most widespread social expression of competence, the task of developing competence was realized historically by creating and maintaining mechanisms and structures to produce – and once the initial foundation was there – to reproduce generation after generation of technical personnel (meaning those employed for the execution of mental labour who are competent [and certified] in specific technical subject matter).

All these mechanisms and structures can be summed up as 'the social reproduction of technical personnel'. The forms of these mechanisms and structures have changed considerably in the historical process of the development of industrial societies. Through this process, clear nationally specific characteristics have emerged. This divergent as well as convergent variance can be described with two ideal typical models without too much risk of oversimplification.

The first model is characterized by the occupation-centred path of generating technical personnel. Where this ideal type dominates technical personnel are given practical tasks even at the beginning of their occupational career. Thus, they acquire much of their necessary technical competence by learning and experimenting in everyday work situations. Their learning process is generally accompanied by a more or less self-directed study of relevant technical literature and/or guidance and training from experienced technical personnel.

Characteristic of the other type of social reproduction of technical personnel is a path that can be described as being centred on schooling. In this model, technical competence is mainly acquired via classroom training. Thus, workers who gain admission to the ranks of technical personnel under this model have attained a particular level of education.

It must be emphasized, even if a further examination of the issue is not possible here, that these different forms of acquiring competence are related to considerable differences in the social backgrounds of technical personnel, in the type and degree of selectivity exercised by employers and the specific value of the worker for the employers (whose decisions are also closely connected to societal prerequisites for the realization of one model or the other).

Historically, technical personnel as an autonomous social group came into being in all industrialized nations under the clear dominance of the ideal type of social reproduction of technical personnel corresponding to the occupational centred path.

At first, technical and management functions were so closely connected that traditional forms of gradual advancement to the level of foreman (*Meister*) on the basis of practical experience were taken for granted. In large enterprises, even members of the owner's family who were destined to take over technical management functions were generally prepared for this task by working their way from the bottom up; in other words, learning from doing and assembling practical experience was even considered indispensable for technical personnel who, as an unquestioned attribute of their class position, had attended schools of higher education.

During the 19th and early 20th centuries, this form of access to the realm of technical personnel predominated to an equal extent in both Germany and France. With the development and stabilization of large industrial enterprises in both countries, a 'craft culture' arose, in which traditional craft skills and the new familiarity with industrial techniques were merged into 'professions' which, in turn, formed the basis for industrial work organization, professional labour markets and the trade unions (which were gaining in both members and power).

The categories of technical personnel still relevant today also developed in a very similar manner in both countries, partly the outcome of processes involving differing paths, but mainly differing magnitudes of upward mobility for skilled industrial workers, (to the level of *Meister*, technician and engineer). The *Meister*, for example, incorporated much of the technical competence of an enterprise well into the twentieth century. The technician differed from the *Meister*-worker due to his technical competence, as demonstrated by his occupational title, but he also differed from the engineer due to the lower scientific level of both his skill and tasks. Finally, the engineer was either designated as such by his employer or had graduated from schools of higher education such as a German establishment of technical education or a French *Ecole des Arts et Métiers*; whatever the case, he worked his way up into the higher levels of technical competence (and generally, company hierarchy) on the basis of a long period working his way through *Meister* or technician tasks.

However, from the very start, several differences existed between Germany and France that could help to explain their differences in development since the Second World War. Already during the French Revolution, for example, France had created a system of elite technical colleges which were less concerned with preparing the students to perform the functions of technical personnel as understood in the context of this chapter than with educating them to assume leading positions in the state. The prestige of these schools and the privileges that graduation from one of them entailed impeded all attempts to expand and improve the practical aspect of the social reproduction of technical personnel through education; by the turn of the century, the *Ecoles des Arts et Métiers*, which were created to fulfil this very purpose, had been integrated into the social system of higher education (and thus became inferior rivals of 'the' technical colleges). At the same time, craft culture and tradition in France was concentrated on specific industrial regions or centres and never became as widespread or taken as seriously as it was in pre-war Germany and as it still is.

The model of social reproduction of technical personnel through promotion of workers dominated in both countries until well into the 20th century. The swift and intense process of industrialization that swept Europe after the Second World War (in West Germany a little sooner than in France) was accompanied by a rapidly increasing need for technical personnel which was mainly met by promoting workers in the enterprise: Process and product innovations were no longer left to the chance ideas of unpredictable inventors; rather they were systematically developed in departments especially set up for these purposes. Moreover, the expansion of mass-produced standardized products made it necessary to undertake considerable efforts in both design and planning. The growth of mass production together with the mass mobilization of unskilled labour in industrial production laid the foundations for the triumphant advance of 'Tayloristic' forms of work and company organization. The functional division of labour that characterized such forms of work initiated the expansion of production-related technical offices and service departments.

Differences in employment trends soon became apparent between the two countries. In France, limits on the mass recruitment of technical personnel from the skilled worker pool were quickly set due to the absence of a craft culture, the small numbers and regional concentration of skilled worker training opportunities, and the lack of colleges primarily designed for ambitious young skilled workers such as the schools for technicians and engineers in West Germany. For these reasons, in order to satisfy their need for technical personnel, French industry had to fall back on well-educated employees who had either acquired their technical competence outside of the industry (such as in the army, navy, or the state railway) or through internal mobility of highly educated employees who worked their way up in the company through a system of ongoing training and promotion.

In comparison, in the decades after the Second World War, West German industry had very large quantities of technical personnel equipped with training and professional experience as skilled workers (and usually with further vocational training) at their disposal. Due to the large-scale social and professional advancement that accompanied this situation, this provided for a very effective mechanism for social satisfaction. The number of technicians and engineers employed in West German industries rose between 1950 and 1970 from 160,000 to approximately 570,000, and was primarily a result of promotion of former skilled workers through the ranks. A considerable number of workers, in fact almost a third, who were skilled industrial workers during the 1950s, had risen to the level of technician by 1970.

In order to overcome its qualititative and quantitative lack of technical personnel, France, since the 1950s, has charged its public education system with the task of training technical workers. Today, the social reproduction of technical personnel in France is mainly carried out through specialized education of young people which includes a little practical experience that does not have much significance. The almost exclusive use of school-centred education for training future technical personnel is organized along strictly hierarchal lines based on the level of general education achieved (and required) before the start of technical education, the length of time spent on education altogether and the prestige attached to the respective degree or diploma.

The stratification within the category of technical personnel (which developed under the dominance of the system of social reproduction based on practical industrial experience in accordance with the logic of a difficult professional advancement) is, in France, completely harnessed to the system internal hierarchy of the education system. This system, already in the early 19th century, was based on meritocratic principles, a characteristic which became all the more pronounced after the Second World War under pressure from the rapidly growing demand for higher education. Although, naturally in the German tradition, a certain hierarchy underlies the different categories of technical personnel reflecting the general structure of social inequality, they have not completely lost their original meaning – i.e. the differing mixtures of 'theory' and 'practice' without any systematic difference in occupational value. In France, on the other hand, hierarchical categories of technical personnel only reflect different ways of dealing with technical problems to a secondary extent. Mainly, they document the differing degrees of success attained in the acquisition of education and knowledge, the mastering of examination requirements and conformance to the values that support this system.

Although self-evident, it must none the less be emphasized that the dominance of school-centred models in the reproduction of technical personnel has far-reaching effects at the level of the skilled worker. In German-speaking countries, where occupation-centred models still play a central role in the generation of technical personnel, the chances for promotion make apprenticeships more attractive and help make the decision not to attend an institute of higher education more acceptable to both parents and children. In France, however, entering training that leads to the level of the skilled worker generally means giving up the chance of acquiring more highly valued qualifications at a later date. Under these circumstances, it is not surprising that most young people do not take such a path voluntarily, but only if they are forced to, frequently because they have failed to gain better marks in comprehensive schools. Thus, estimations of their chances to increase their educational level or strive for promotion are correspondingly (and for the most part accurately) low.

THE SOCIAL STRUCTURE OF TECHNICAL KNOWLEDGE AND TECHNICAL COMPETENCE AND THE KEY ROLE OF PROFESSIONALISM AND PRACTICAL EXPERIENCE

Technical knowledge (or, in other words, knowledge of the characteristics of materials, the laws of cause and effect, principles of technical design and of process control) is in its very nature tied to application. This is the essential difference between technical knowledge and scientific knowledge, which, at least in principle, is the product of 'pure' findings based on scientific criteria of 'truth' and 'substantiation'. Of course, technical knowledge is closely connected to the findings of science out of which it derives its laws. However, the quality of technical knowledge is not merely determined, perhaps not even primarily determined, by how 'true' (in an intrinsically scientific sense) the theory is or how 'substantiated' (i.e. in strict accordance with the rules of scientific proof) the derivation of concrete principles based on the theory is. The quality of technical knowledge is far more determined by its applicability, its instrumental value for rational action and its ability to contribute towards the solution of practical problems.

The fact that technical knowledge is oriented to application means that its cognitive substance cannot only be reduced to the application of scientific principles. Technical knowledge always involves knowledge of the field in which an application is to take place, especially when the field has not been completely defined in scientific terms. And generally speaking, the essence of what can be called technical competence lies in the knowledge of a field, in which there is familiarity with the concrete material substance specific to that field and the ability to act in an appropriate way even in cases of uncertainty.

An industrial society cannot simply assume that the knowledge and competence it needs in a concrete constellation of applications will be at its disposal in order for selected solutions to correspond to 'state of the art' technology. The ways and means that a society (or its industry) adopts to deal with the task of reproducing technical competence can be described as the 'social structuring' of technical knowledge and competence. The historical-empirical (and conceivable) forms can be placed along a continuum; at one extreme, the availability of technical competence is insured only through incorporation in the person of an individual worker, at the other, only through collective organization, that is, independent of the knowledge of any one particular person.

The greater the significance that is attached to the principle of personal incorporation of technical knowledge, the more a principle of structuring and ordering applies that can be described as 'professional'; the social structuring of technical knowledge becomes identical to the internal structure and organization of technical personnel in professions, and the way in which the social reproduction of technical personnel gets carried out becomes more important. On the other hand, the more that collective organization takes over the provision of technical knowledge, the more different structural moments and ordering principles of technical knowledge and competence come to the fore. These are almost completely independent of the social structure of technical personnel and its models of social reproduction. At this end of the continuum, since companies or enterprises are the primary means of organizing the application of technology in industrial society, rather than individually embodied know-how, such structural moments and ordering principles are determined by the enterprise and bear its imprint, as expressed by technical 'company culture' and 'style', or year-long prepared and revised internal handbooks, archives, data banks and collections of rules.

The ordering principle of professionalism divides technical personnel and their specific fields of competence into professions with clearly demarcated areas of responsibility. The more this principle is applied, the more clearly demarcated these areas become, with worker identity being primarily, if not exclusively, derived from the profession involved. This structuring of technical personnel according to profession also determines how technical knowledge gets structured. In other words, the areas of responsibility of technical occupations define (or a decision is pushed out of them on how to define) the way in which technical knowledge is clustered, i.e. which of the many diverse areas of technical knowledge belong together and which do not. The rules of the art which determine the applications of technology and technical competence are formulated, conveyed and, where necessary, enforced at the level of the professions. The norms which establish what is 'do-able' at the current state of the art, what can be risked and which problems are to be seen as priorities are set within the framework of the professions, or at least in close connection with them.

Furthermore, when the social structuring of technical knowledge is strongly characterized by the principle of professionalism, the 'cutting up' of technical knowledge and technical capability into professions forms the framework within which the acquisition of technical knowledge as an individual process of training and qualification takes place; the professions lay down the goals that the training process is supposed to conform to, they define the career expectations that can be achieved on the basis of certain qualifications and which often form the main motivation for acquiring technical knowledge, even in difficult and unsure circumstances.

In view of this situation, it is evident that fundamental changes in the prevailing models of the social reproduction of technical personnel cannot avoid having grave consequences for the social structuring of technical knowledge and competence.

Historically, technical professions chiefly developed under the dominance of occupation-centred models. The way they are divided up reflects the focus of the practical application of technology that developed over decades of industrial development, and which correspondingly displays great affinity to the division of industry into branches. The creation and demarcation of professions was primarily determined by the principle of gathering together as many tasks and problems that could arise as possible with the most important means of solving them into what then could be considered the competence of a profession. For these reasons, it has always been the convention (and still is in cases where occupation-centred models predominate) to keep the structure of professions of technical personnel identical to, or at least compatible with, the structure of industrial trades. (At the same time, this is one of the central factors which enables skilled workers to advance into the realms of technical personnel in greater numbers and is also a prerequisite for the communication of new experience and knowledge between the two groups of workers.)

The more school-centred models predominate in an industrial nation, the more two factors emerge which endanger the practical and material basis of technical knowledge structured by profession, thus calling into question the provision of what long seemed to be available as a matter of course, namely, technical competence. This is a process that does not take place immediately, but can take up to a generation – the length of time that teachers and trainers who have themselves followed the occupation-centred path continue to teach.

One factor derives from the particular logic of classroom training. It lies in the nature of the school-based acquisition of knowledge that it is generally achieved by going through a set course of study and material which draws its effectiveness from being systematic and its legitimacy mainly from its inner cohesion: in other words, by the fact that it is solidly anchored in the established sciences. The total system of tasks, problems and rules that make up problem solving, define an occupation's scope and determine its content, are, seen from a scientific perspective, completely heterogeneous and unstructured and thus are not taken seriously.

In this connection, the ability to deal with uncertainty starts to play a key role. Dealing with uncertainty involves coming to terms with only partially defined (and at the current state of science, only partially definable) facts and unpredictable events, which have to be coped with immediately and adequately, generally under time pressure. While dealing with uncertainty is almost a daily occurrence in industrial life, where it doubtlessly provides excellent possibilities of learning to cope with such situations, it can only be presented in carefully given doses in schools, in the form of fully prepared simulations developed by the teacher and course of study. In this way, it cannot possibly take on the form of a real-life, ongoing situation that involves stress, but also the opportunity to rise to challenges.

Since school-centred models of acquiring competence prevail in the social reproduction of technical personnel, considerable pressure arises to reorganize technical knowledge in accordance with the logic of scholastic education standards, which tend to be based on a scientific and systematic approach. This pressure can only be counteracted through the use of considerable social and economic resources, as in the case of the German university clinics, where medical training is complemented with practical experience.

This pressure to increase the scientific aspect of technical competence conveyed in school gets intensified to the extent that the ability to think abstractly and grasp abstractions has a central significance as a means of academic selection and criteria for academic success. And, as is the case in practically all 'modern' societies, school education takes over a key role in the allocation of social chances, thus making the educational system the potentially most important aspect of social control and selection.

From the perspective of the application of technical knowledge in society, the pressure for changes in the way knowledge and competence is acquired engenders a two-sided risk if it is not neutralized by strong counter-forces. The first can be described as the risk of de-materialization of technical competence and the second as the de-professionalization of technical knowledge.

The de-materialization of technical competence can be understood as the absence, or the reduced capability of technical personnel (or, to put it more precisely, the upcoming cohorts of technical personnel), to combine the application of scientific principles and laws with pragmatic rules in their dealings with the nature of material in their work and in adequately handling uncertainty. In this way, an important dimension which previously determined the difference between technical and scientific knowledge is lost. This dimension consists of the concrete and action-oriented relationship to any given part of the material world of technical products and processes; it is a dimension that, due to its complexity, cannot be fully understood in scientific and analytical terms and which forms the central criterion of technical competence, thanks to the supra-individual (that is in occupations) mastery of accumulated knowledge through experience.

This is closely connected with a process that can be best described as the de-professionalization of technical knowledge. The more that technical competence, or, to put it more concretely, the people who incorporate technical competence, is withdrawn from the concrete reality of a certain field of activity and is confined to mastering the general scientific foundations of technology instead, the more the professional structure of technical knowledge (which was centred on unique material conditions of a specific field of activity) will lose its character and its legitimacy. The functional foundation of the professions also disappears with the loss of competence and knowledge tied to the professions, not to mention the institutions and organizations based on professional status, such as the professional labour market, career paths that are specific to certain professions, behaviour standards and principles of action linked to professions, and finally the relationship between technical training carried out in schools and practical work experience in the chosen profession.

CLOSING THOUGHTS: COMPUTER SCIENCE AS A NEW AND UNIVERSAL FORM OF DE-MATERIALIZATION?

To the extent that it is possible to compensate deficiencies in personal training with the collective organization of technical knowledge and competence, it is possible for the de-materialization of technical competence and the de-professionalization of technical knowledge to take place in an industrial society for a long period without causing serious consequences. For example, the formation of highly structured 'internal' models of mobility in large companies, based on the acquisition of skills, enables technicians and engineers who were exclusively trained in schools to learn a good deal of the knowledge and competence once brought by the technical personnel with industrial-based practical training who were promoted. It is also possible, although at the cost of much effort (Taylor was the first to tread this path), to formulate tacit knowledge, which is a major prerequisite for dealing with unexpected situations adequately, into formalized rules. In this way bureaucratically fixed and controlled routines more or less replace what, in the framework of professional competence, is achieved through the everyday dealings with materials.

This results, of course, in technical knowledge and competence becoming more and more company-specific (to borrow a term from the theory of segmented labour markets). And indeed, only large companies which manage to steadily innovate their products and/or processes or, thanks to their superior market position, are able to pass on the very high costs involved in generating and keeping in reserve immediately utilizable technical knowledge are in the position to have technical competence at their disposal under this system; competence which is freely available when technical knowledge is structured by profession into special skilled labour markets and professionalized communication.

For this reason, an economy can only implement such alternatives to occupation-centred models of attaining technical competence, without entailing massive disadvantages and loss of welfare, under very particular conditions over the long term. These conditions mainly consist of the dominance (and efficiency) of large stable enterprises, technology development which is based on well-known 'basic innovations' in which applications are only extended and improved in small steps, and the existence of sales markets for technical goods which neither force the companies to introduce product innovation at too fast a rate nor to implement improvements in the ratio of quality to price too quickly.

Historically, these conditions existed in an almost ideal constellation in Europe in the post-war decades and in America since the 1920s (referred to as the 'Tayloristic syndrome' in chapter 3). This fully explains why many industrial nations did not seriously resist the decline of their craft culture and the rise of school-centred models for training technical personnel, and in some cases even speeded up the process.

Of course, since the mid-1970s with the erosion of the 'Tayloristic syndrome', totally new conditions and requirements have appeared on the scene. For example, structural changes in the economy have made it necessary for companies to adopt more flexible forms of organization, and technical development has been influenced by the introduction of new technologies and/or turbulent conditions on the sales markets; moreover, new forms of competition have arisen on these very sales markets, and consumer behaviour has become more and more influenced by rapid changes in fashion. This has put economies that have adopted the school-centred model of the social reproduction of technical personnel and the social structuring of technical knowledge in danger of taking a back seat in development.

There are various strategies that a nation, its political and administrative system and its industry can employ in order to react to a situation in which it will become increasingly difficult to keep the required technical knowledge at its disposal to an adequate extent and level of quality (whether for economic, social or military reasons, etc.). Of course, almost all conceivable strategies entail serious disadvantages and weaknesses in that they either take a long time to become effective or necessitate mammoth public investment or would doubtlessly awaken massive resistance on the part of powerful lobbies.

In view of these circumstances, it would be somewhat surprising if a strategy that claims to have a solution to the problem of deficiencies in competence were not met with great interest. This is even more the case when it involves the (expensive) installation of technical systems and does not entail conflict-laden and potentially risky changes to the structure of work and training. This strategy consists of the massive implementation of computers, programming and computer-aided information, control and communication systems; in other words, it is based on a totally 'new technology' that can indeed lay claim to function with considerable efficiency.

For example, technically, design offices are now able, using finite element methods, to calculate the characteristics of complex structures and functional relationships with the help of high-performance computers. This task would have required practical applications or labourious tests in earlier times. In some cases, the application of computer-aided control systems has also enabled a degree of production automation that could not have been achieved in an economical manner with the usual control systems.

However, two potential aspects of these 'new forms of technology' that are not technical in nature, but social in the broadest sense of the word, are of significance for the subject of this paper: the potential or ability of these 'new forms of technology' to generate a new form of de-materialized practice and a new form of professionalism.

For example, the foundations of computer science and the execution of practical tasks with the help of computers and computer-aided systems can be integrated into the school-centred models of training and further training with little problem. The necessary purchases are not exorbitant and students and trainees seem to enjoy carrying out practical work of all kinds on computers more than can be claimed for the other techniques and technical systems that educators have tried to incorporate into school courses.

In this way, a new professionalism that is centred on computer capability has developed surprisingly quickly (independent of the highly heterogeneous ways that people entered the field). This new professionalism incorporates a competence in individual persons that is always ready for application, with increasingly clear standards for describing and evaluating this competence, a strong professional profile, and ever improving specialized labour markets. This is all the more remarkable because originally, computer competence could only be acquired in a form that was overwhelmingly company-specific (primarily at IBM).

Thus, while on the one hand, the customary content of technical knowledge is undergoing a growing tendency toward de-materialization and de-professionalization due to the increase in school training of technical personnel, a diametric process seems to be taking place in the field of computer development and application. A very specific and structured form of work has apparently developed in connection with working at or with a computer that possesses considerable potential for socialization and qualification; it can hold its own in the structures and processes of scholastic selection and education hierarchies. It attracts particularly strong student performers and is proving to be a field in which a new form of professional competence can be acquired, with a broad range of applications and high prestige.

In this way, computer science forms the basis of a new form of competence that can be adapted to individual professions or used in a universal sense for all forms of technology. Moreover, it can be generated in the education system at a relatively low price and be made available in great quantities on the labour market, especially in view of the fact that previous forms of technical competence structured according to profession have lost ground due to the destruction of traditional forms of skilled industrial labour and the rise of school-centred models of reproducing technical personnel.

The critical question is then whether the French computer-based strategy of industrial modernization has any chance of success; whether this new form of computer-based technical competence can indeed incorporate the traditional content of technical knowledge (much more restricted to concrete applications) and thereby replace traditional forms of technical competence. The question is not, however, whether computer science will be adopted as a new, professionally structured field of study to complement traditional disciplines of engineering science (which are less threatened by the lack of practical experience in the school-centred reproduction of their personnel). Far more significant in this connection is whether this new form of technical competence can lay claim to the status of a 'meta-technology'. Although it does not have any systematic relationship to specific applications, this technical competence has a totality of universal solutions which could piece by piece appropriate the tasks that in the history of industrialization were found in exclusive areas of responsibility of specific fields and were oriented to limited applications and techniques.

Only if this claim stands on a solid foundation can an economy hope to master the new challenges accompanying the downfall of the 'Tayloristic syndrome' by defining all technical problems that arise or become more acute as tasks that fall under the competence of computer scientists and whose solution is a matter for computer applications and computer-aided systems. Indeed, it is only if this is the case that it would be legitimate for an industrial nation to entrust its industrial future to a single strategy in the face of current (or threatening) deficits in conventional kinds of professionally structured technical competence while renouncing the exploration and implementation of other strategies which, admittedly, would be more conflict-laden, labourious and time-consuming.

Doubt about the advisability of such a course is surely appropriate, for since their development in the late 1940s and during the 1950s, computer science and computer applications have always been connected with far-reaching promises and expectations (and fears) which have proved to be unrealistic in retrospect. Very little seems to indicate a fundamental change in this situation.

19 New Production Structures à l'Italiano? Similarities and Differences in the West German and Italian Steel Industries[1]

Ingrid Drexel

INTRODUCTION

This chapter deals with structural changes which production work in the German and Italian steel industries has undergone due to rationalization and societal change. Though the impulses for change as well as starting conditions in both industries were quite similar, the resulting changes were very different: in the German steel industry new problems were defined mainly as skill problems and solved as such; in the Italian steel industry it was mainly wage structures, internal mobility patterns and work organization that were first criticized and then changed. In contrast to these apparent concrete differences, on a more abstract level, the same mechanisms oriented towards a broader utilization of individual as well as collective labour resources can be identified: mobilization and development of skill reserves, motivation and flexibility, etc. Based on these findings, some more general conclusions and questions are presented regarding new structures of production work, their theoretical conceptualization, their ambiguous consequences for workers' interests and the significance of political factors for particular outcomes.

THE SUBJECT OF THIS CHAPTER

Several recent studies have noted a trend toward a new type of labour utilization especially in the production areas of large-scale enterprises. The contention is that the traditional Taylorist, restrictive use of labour, characterized by narrowly defined tasks, the separation of decision-making functions from actual production and low requirements in qualification is being increasingly substituted by one that draws more on the workers' performance potential.

Various concepts have been brought forward to describe this development which – in reality as well as in theory – is becoming increasingly noticeable, and to place it in a broader perspective. Some of the earlier concepts in this connection were: 'rationalization based on worker's skills' (Asendorf-Krings 1979, 1979a); 'new strategy of raising profitability' (Drexel 1980); 'new performance policies in enterprises' (Altmann *et al.* 1982); 'labour policy strategies for the structuring of work and the mobilization of labour' (Dohse *et al.* 1984a), etc.

The discussion today, however, is dominated by the Kern-Schumann approach (Kern, Schumann 1984a, 1985) which interprets the observed trends as new production concepts in the enterprise and predicts, particularly in prospering industrial sectors, an auspicious future. Other authors, however, dispute the validity of an interpretation of the new developments with such far-reaching implications (for example Düll 1985a). In short, West German industrial sociology is entering a debate on the extent and future of new forms of labour utilization and toward this end faces the task of developing theories to interpret the new phenomena occurring in industrial production.

This study is meant to contribute some thoughts to this discussion. It is a comparative analysis of changes which have occurred in the West German and Italian steel industries since the 1950s and 1960s – a time when their situations were largely similar.[2]

There are several respects in which an analysis of the changes in production work in the steel industry in a two-country comparison is relevant in the context of the current debates. First, the processes of (at least partial) automation in the steel industry, and the large-scale rationalization processes tied to them, occurred as far back as the 1960s. Thus, it can be seen as an early case of the implementation of non-Taylorist rationalization strategies. Secondly, a comparison of the structural changes in West German and Italian steel production can provide a new perspective on questions dealing with 'new production concepts'. By highlighting nationally specific and distinctive forms of structural change in one industrial branch, it is possible to overcome the limitations of an outlook fixated on the well-known conditions in one's own society. Both – the comparison over time and between nations – can contribute to a clearer picture of the general characteristics and connections contained in emerging trends in development and can raise new questions.

The following theses give an overview of the detailed results presented below:

1 The core of new forms of labour utilization is not primarily the product of 'management science' or 'foresightful rationalization concepts' (Kern, Schumann 1984a, 1985). Mainly, they are the product of the erosion of traditional forms of labour utilization which were part of the

rationalization process of the past; an erosion induced by multi-dimensional rationalization processes and by societal changes. The logic behind the erosion process seems to make its impact independently of comprehensive 'concepts' although its influence may be felt within and across such 'concepts'.

2 The structural changes examined originated in different areas – skills and qualifying processes in the West German, work organization and wage policy in the Italian steel industry – and then proceeded to continuously force the other areas to adapt.
3 At a somewhat more abstract level, it appears that the solutions developed in both steel industries lead back to a shared structural core. In other words, with certain exceptions, many aspects of change examined in both industries display far-reaching structural parallels. They can be analyzed as constituents of a new structural type of production work.
4 For a part of the work-force, clear improvements in the conditions for the 'reproduction of labour' (its development, maintenance, and re-generation) appear to be more or less necessarily tied to the new type of production work. However, this potential for improvement in the reproduction of labour created by the erosion of the traditional type of production work is never utilized fully. This indicates the irrevocable contradiction in enterprise interests with respect to the accumulation of capital and development of labour which tends systematically to limit the new developments in production work.
5 Political factors also play a role in determining the extent to which the potential of the new type of production work is used in particular cases and where restrictions are applied: the definition of the specific focus of the problem, the actors who have the initiative in the change process and their general and comprehensive interests and strategies are of great importance.

The following sections will begin with a description of changes in West German and Italian steel production. In order to present a total picture of the structural change in production work, and at the same time, to analyze the differences in the two cases, a condensed presentation which concentrates on structural relationships will be given. After the case studies, a comparative analysis of the structural differences and similarities of the changes follows. Finally, some theoretical and political inferences from the findings are laid out, which are presented more as questions than conclusions. This attempt to derive some general inferences is done to make the results of the analysis referable to and useful for the discussion on 'new production concepts' outlined above, even if an explicit response based on the results is not possible here.

THE EROSION OF TRADITIONAL TYPES OF PRODUCTION WORK IN THE STEEL INDUSTRY

The structural changes in qualification, work organization and wage policy, which have occurred in the West German and Italian steel industries over the last 10 to 15 years, are the result of a process that had already begun at least 10 years earlier and which is not yet finished today. It is a process in which traditional production work, and the qualifying processes that went along with it, were becoming increasingly deficient. As a result, various attempts at solutions were made, shaped by the erosion process of the old system and new problems generated by this process, but also by existing possibilities and restrictions.

The following section discusses the increasingly deficient character of traditional types of production work. The starting point is the erosion of the classical system of qualification and training, and its relationship to the most important corresponding changes in work organization, work assignment and wage policy.

The traditional system of qualification and training and its most important functional requirements

Since the beginning of this century, production enterprises of the North American and the important European steel industries have developed and gradually solidified very similar patterns of skill structures (and the work organization and division of labour tied to them) on the one hand, and processes of qualifying (based on work assignment, on-the-job-training, and internal company mobility in vertical mobility chains) on the other (ISF 1966, 1968; Stone 1975). They were characterized by:

1. a strongly differentiated and *hierarchical 'qualification pyramid'*: on a base of totally unskilled workers, several pyramids stand next to one another which are grouped around a particular production area (such as a blast furnace or a milling area), each with a number of different job grades; and
2. a process which is based on *learning on-the-job* and chances *to move up the hierarchy*: qualifications, in part very high level ones, that are mainly acquired through years of 'learning by doing' through gradual movement up the hierarchy to increasingly challenging workplaces in the same production area (for example, fourth, third, second foundry worker and head foundry worker); it was a qualifying process in which courses were the exception rather than the rule.

These models of qualification structure and training process are closely inter-related; they represent an integrated system. In practice, these structures were highly functional for the steel enterprise, in terms of a flexible acquisition of necessary skills, as well as for other personnel policy goals such as securing the willingness to perform and loyalty to the firm (Drexel 1980, 1982).

This system had some very basic functional prerequisites in enterprise structures as well as in the larger societal setting:

1 *work tasks* which made it possible to acquire skills and experience in daily working conditions;
2 a *long-term continuous assignment of workers* to the same production area which enabled the accumulation, a regulated structuring, and a generalization of experience;
3 close *cooperation in work groups* hierarchically structured by task and skill, which allowed collegial responsibility for training of new workers and making necessary corrections in their trial runs;
4 a relatively *labour-intensive production style* which made it possible to take the time and energy for such cooperative training processes as part of the job;
5 a *hierarchically structured pyramid of job tasks*, wide at the base, stretching to a point, and a similarly hierarchical, highly graded wage pyramid, which incorporated wage incentives for frequent work-station changes and the therefore necessary acquisition of additional qualifications tied to them;
6 *rigid rules of intra-company vertical mobility* (that is the promotion to work-stations with better conditions, better pay and higher job security) based on the criteria of seniority and 'doing a good job'; these rules acted as incentives for the workers to stay in one company and to achieve a high level of performance;
7 a *labour market situation* which made it possible for the companies to recruit a large number of workers who were capable of actively and independently acquiring the necessary qualifications, and then choosing the best among them; and
8 the *worker's objectively and subjectively weak position* in relation to the company which assured their acceptance of strenuous working conditions, health dangers, and a certain job insecurity typically existing in steel production.

These functional prerequisites for the system of developing and using labour, analytically separated for purposes of clarity, were naturally extremely inter-connected in a large number of ways in actual practice. Thus they partially balanced each other out; for example, cooperative workplace conditions that created good opportunities for learning through observation, could compensate to a certain extent for possible deficits in the qualification potential of recruitable labour and vice versa.

The erosion of traditional types of production work in the West German steel industry

At about the beginning of the 1960s, the existing system, as described above, was continuously, and finally irreparably, destabilized due to the fact that many of the functional prerequisites in the company and in society began to fall apart or to disappear altogether. This was due, in part, to a multidimensional rationalization process, but also to certain societal transformations.

... As a consequence of multidimensional rationalization processes

If one takes an overview of the most important qualitative changes that have taken place in the West German steel industry over the last 20 to 30 years, it will become clear that these were part of a, not yet completed, push to raise profitability.

Several aspects of this rationalization process have undermined the functional conditions in the company necessary for the ongoing use of traditional forms of labour development and capital accumulation. The most important are outlined here.

Diverse, often large-scale and *abrupt restructurings* of production apparatus and the accompanying *labour transfers* impede the practice of long-term assignment of workers to the same production area and thus erode the preconditions for gradual accumulation and generalization of experience and knowledge. This makes the procurement of high-level skills from experience impossible. In the wake of frequent transfer of workers to different production areas, skill structures oriented to particular production areas become problematic: when workers are transferred to different areas of the plant, they cannot enter at their current grade, but must start training again 'from scratch'. This situation and the accompanying loss of income, stirs up considerable conflict from the more qualified workers and lower level supervisors and therefore, often creates problems for the company. The high degree of differentiation makes it difficult for management to keep track of the qualification potential available in the enterprise and consequently to place the right people in the right job.

Increasing mechanization and automation makes *learning only through experience inadequate*: growing complexity, integration of processes and lack of visibility of production processes makes it difficult, or even impossible, to learn by observing; at the same time, theoretical knowledge becomes necessary which principally cannot be acquired through experience.

The reduction of the number of job grades *threatens the traditional step-by-step process* of acquiring higher qualifications through upward mobility: the distances between the remaining job grades get larger, and that makes the transition from one grade to the next harder. The transition can no longer be achieved as quickly and with so much certainty exclusively through individual efforts of learning through experience with support from colleagues.

The reduction in the density of the work-force in particular job grades and the resultant *work intensification* changes the *conditions of learning in the cooperative work group*. The growing strain on all workers engaged in production tasks has the consequence that there is less time and energy available to help qualify new workers through explanation and correction of their work. Also, errors, which always occur in first attempts by inexperienced new workers, are less tolerated as capital intensive production makes breakdowns and repair more and more expensive.

. . . As a consequence of societal transformation

It was not only changes within the company, but also in the society that contributed to the erosion of traditional forms of labour development and the extension of labour utilization.

There was a considerable *decline in the supply of labour available* for the West German steel companies through the middle of the 1970s in both quantity and quality. The overall economic expansion, and thus the intensified demand for labour in more attractive industrial branches, demographic developments, and increasing participation in higher education diminished the size of the available new labour pool. In particular, the traditional recruiting reservoir for production enterprises, consisting of young workers who had been trained in craft or small enterprises in professions such as butcher or baker in which they could not find jobs after the apprenticeship period, was reduced. Therefore, the steel industry increasingly had to recruit foreign workers with no previous occupational training. These newly recruited workers had greater problems obtaining skills in the course of the work process through their own initiative, which also limited opportunities for company selection.

The social problems arising from *typical working conditions in* steel production (plant-specific, even production area-specific training; shift work; very strenuous working environment, etc.) increased rapidly and were also *less and less accepted*. Skill devaluation through transfer, wage losses, premature health problems and the consequent danger of losing one's job, stirred up resistance on the part of the production workers and reduced the attractiveness of jobs in the steel companies on the labour market and therefore their recruiting chances considerably.

The *position of workers* vis-à-vis the company, and the ability to push through their demands became *stronger*, due to the labour market situation as well as for political reasons. This limited the freedom of the company to decide how the social problems stemming from the various work and rationalization measures should be dealt with: the formerly common practice of pushing everything onto the workers' shoulders in the form of wage decreases and dismissals was becoming less and less possible.

The erosion of the traditional type of production work in the Italian steel industry

The picture yielded by a general examination of available analyses (in the form of case studies, see Michaeli *et al.*; Butera; Barisi; Villa), shows that the developments in Italy have considerable parallels to those in West Germany, with, however, some differences.

First, the Italian steel industry also experienced pressure to *raise profitability*, and companies, as in Germany, tried to achieve increased profitability through a variety of measures. However, in Italy, the construction of new highly automated enterprises, and, in part, changes in the range and structure of existing enterprises and their automation processes, stood more strongly in the forefront than the reduction of personnel.

The Italian steel industry, like the German, experienced a lasting transformation in *labour supply*: one reason was the erosion of state-run systems of vocational training (which occurred in certain traditionally industrialized regions in which large numbers of students were favouring schools offering a more general, academic education); another reason was the limited reservoir of qualified workers trained in small industrial or craft firms in the unindustrialized regions of the new industrial sites in the South. Furthermore, it was caused by the changes in character traits demonstrated by the newly recruitable workers, who were considerably higher educated compared to those in the past.

Finally, partly as a consequence of these labour market developments, and partly independent of them, there were changes in the *political power relations* between company and labour in the Italian steel industry and in what workers were willing to accept.

On top of these developments, which were more or less similar to those occurring in West Germany, came the effects of the 'job evaluation' introduced in the Italian steel industry at the beginning of the 1950s; this wage system supporting Tayloristic work organization was apparently much more sweeping and damaging than in the German steel companies that used it.

These developments resulted in serious problems: the worst problems are outlined below, gleaned from existing Italian case studies:[3]

1 Of particular importance was the increasing difficulty of transferring workers to the necessary extent either up the hierarchy or across production areas. This so-called *'rigidità'* of labour derived from skill deficits, especially an insufficient range of knowledge. Another and perhaps greater reason was workers' resistance to transfer. In a period with a growing need to transfer workers, both presented a problem.
2 The problem of *quality in work performance* increasingly made itself felt, a result of both skill deficits and motivation problems.
3 The traditional hierarchy, especially *lower- and middle-level managers*, was increasingly questioned by higher level management as well as the unions. Central points of criticism were the technical incompetence of, in particular, the foremen (*capi*), the restriction of their functions to maintaining discipline and the increasing bureaucratization of company management.
4 Working conditions and their negative effects were less and less accepted. This was expressed in high turnover rates of young workers, in growing criticism of health risks and in emphatic demands for changes in the organization of work.
5 Workers in general showed a greater *readiness to enter into conflict* and had objectively and subjectively better chances to articulate and carry out their demands. This mainly political factor also intensified the 'rigidity' of labour.

Due to lack of information, it is not possible to estimate how widespread and characteristic these problems were in the Italian steel industry. However, the number and content of experiments initiated by the companies in the second half of the 1960s to improve work organization lead to the assumption that a sharp erosion of traditional production structures took place. It is especially apparent that the emphasis and type of criticism articulated from the end of the 1960s on by worker representatives and unions, and their demands for change, were a reaction to existing, broad-based problems (even if they naturally intensified the situation at first).

THE NEED FOR NEW SOLUTIONS

The crisis of the traditional system exerted pressure on the enterprises by creating a multiplicity of problems, which tended to intensify each other. Change was inevitable. Attempts to deal with the situation by implementing isolated 'small-scale' solutions proved inefficient. The necessity for a 'large-scale' solution increasingly made itself felt. The solutions which were conceptualized and, to a greater or lesser extent, put into practice are, at least at first sight, vastly different.

Solutions in the West German steel industry

In West Germany, the principal method used to solve the problems was the *creation of a new vocational training scheme*. New systems of training leading to new qualifications gradually brought about changes in the overall structure of qualifications, in the patterns of work assignment and in the wage structures. Some details are outlined below.

For years, enterprises had experimented with 'small-scale' solutions offering courses for lower-level supervisors and top-level production workers, and introducing a certain formalization of the traditional 'learning by doing', and in some cases also job rotation. Since these experiments proved to be insufficient, a comprehensive solution was created: a new state regulated vocational training scheme was developed which, over a period of three years, led up to a specifically created craft occupation (namely that of *Hüttenfacharbeiter*, i.e. skilled steel worker) and which was standardized throughout the steel industry. The level of theoretical and practical knowledge imparted during the training process is quite high; it comprises the entire production process, i.e. iron production, steel production and milling.

The introduction of the new *Facharbeiter* qualification level had an impact on the entire hierarchy of qualification within the enterprise, at levels both below and above it: it was not only the junior staff who went through the new type of training and benefited from it, but also some of the older workers who had been with the enterprise for years, and had been trained on their jobs. However, only a certain proportion of production workers received vocational training, given the ongoing need for unskilled labour. Besides, despite union opposition, numerous steel enterprises introduced the second class qualification of *Hüttenwerker* which is reached after only two years' training. Even below this level, enterprises recruited unskilled workers to be trained on the job. They gradually gain skills and rise slowly in the hierarchy, but in their careers could only reach the level of skilled worker (*Facharbeiter*). For those who wanted to work as lower level supervisors (assistant foreman or foremen [*Meister*]), vocational training as well as additional courses became compulsory.

As a result of all of these changes, production workers within the enterprise tend to be polarized, with well-trained skilled workers on one side and unskilled workers on the other. The career prospects of these two groups differ considerably.

Solutions in the Italian steel industry

The Italian steel industry implemented a number of far-reaching *innovations in wage structure, work assignment, the organization of work and* – albeit with comparatively little emphasis – *training*. Over many years, steel enterprises tried to find solutions for their problems by experimenting with new forms in the organization of work. At a later stage, union demands for new forms of work organization, wages, and mobility were fulfiled to a certain extent, only to be subsequently modified in the interest of the employer. The adoption of a far-reaching innovation process triggered in large measure by union demands is especially valid with respect to the partly state-owned steel industry. Since most of the material that is available deals with this sector, the following section concentrates on this particular area.[4]

The following transformations can be regarded as the most important elements of the Italian solution. The extent of their implementation differs greatly across enterprises, and since the beginning of the 1980s has been destabilized.

In 1970, a common wage system for blue-collar and white-collar workers, the so-called *inquadramento unico*, went into effect. There was a reduction in the number of wage groups to a total of eight (five for blue-collar workers and three for white-collar workers), as opposed to the 24 or more discrete levels contained in the former rates. This resulted in a considerable compression of the wage hierarchy.

In 1972, a wage agreement containing *new regulations* concerning wage structures and promotion within the enterprise was concluded. Of these regulations, the following are relevant in the present context: the transition from the lowest to the second lowest level was to go into effect automatically for all workers after one year's employment. All other transitions from one wage group to a higher one were to be regulated according to whether or not the jobs formed part of a 'professional family', i.e. a hierarchically structured job scale where the additional knowledge necessary for a particular job is based on that necessary for the job immediately below it in the hierarchy (e.g. fourth, third, second, first smelter). If this was the case, the enterprise was obliged to create the opportunity for every worker to gain the skills necessary for the job immediately above his own in the hierarchy. The worker had to be in a position to do work which entailed tasks of the higher level job, and when he had proved his abilities, he had the right to advance to the higher wage group. In this manner, the wage 'careers' in this area were clear to everyone and had a secure basis, whereas up to then they had been rather uncertain, depending as they did on contingencies in the enterprise and the decisions of lower-level supervisors. The fact that the acquisition of knowledge and training took the place of 'loyalty' as a criterion for promotion gave the workers, as well as their representatives, who were in charge of supervising the fulfilment of the wage agreement, an influence over mobility within the enterprise.

In cases where a job was not to be part of such a hierarchically structured 'professional ladder', additional skills and rights to promotion could be attainable through vertical and horizontal rotation. An increase in wages could be achieved, if the worker was prepared to continue working on a job of higher value in spite of having completed a certain fixed number of rotation days; or, conversely, if he was prepared to go back to lower jobs even after having achieved a higher qualification, if the necessity arose.

In cases without the possibility of such rotation, jobs could be enriched in order to create career opportunities. In areas where even this was impossible, as well as for particularly stressful jobs, it was agreed that every worker was entitled to move to another job within the enterprise at the end of a certain period.

Training measures were to be offered in all cases either in addition to, or instead of, those described. Successful participants could, after a certain period, be entitled to advance to the next wage group. In addition, it was agreed that workers with reduced performance should no longer be downgraded. Instead, the enterprise should cover their retraining costs.

These career regulations and the common wage structure for blue- and white-collar workers resulted in a marked rise in the average wage and a drastic reduction in wage differentials.Within a short time a large number of workers were now in the higher wage groups, often above the wage group associated with their actual workplace.

The new agreements however, did not bring about an increase in efficiency and productivity to the extent expected by the employers. They also did not generate the improvement in the organization of work and subsequent increase in the workers' *professionalità* that the unions had hoped for: higher professional skills and consequently a greater use of worker discretion within the production process. The training measures agreed upon were put into practice to some extent; but when unemployment rose sharply, they were revoked very quickly in several cases. Later, even the equalization of the wage structure was counteracted by the employers through renewed differentiation strategies.

The unions reacted to these developments with two new demands: firstly, a new 'professional model', a type of job which was to incorporate the tasks of blue- and white-collar workers and allow blue-collar workers access to wage group 6, for which they had previously been ineligible; secondly, the recognition, valuation and enhancement of the workers' collective *professionalità*, i.e. their collective ability to control the production process and to cope with the problems of qualification and training by themselves. The concept of collective *professionalità* was becoming more and more central to the unions' aims; its given principles were to be made more explicit and recognized by the enterprise.

At the same time, the concept was to be enlarged by a programme for a basic transformation of work organization towards *unità operative*: teams in charge of clearly defined stages in the production process and with a collective responsibility to other teams and to management. Although they would at first include workers with different skill levels, the *unità operative* was to be regarded as homogeneous and was to, in fact, gradually achieve this homogeneity. The concept of *unità operative* made changes in work organization a principal objective of the unions' demands: the division of production functions into individual tasks, which was the basis of the traditional type of work organization and its logic of individualization, was to be replaced by a new logic which was collective (although limited to the particular team).

A new wage agreement concluded in 1978, and even more so the one concluded in 1981, achieved a breakthrough: a specification and extension in the implementation of these innovations. The *unità operative* should eventually become the general basis of work organization. Furthermore, an – albeit limited – number of professional profiles was defined which were accorded a claim to wage group 6 for blue-collar workers. The jobs rated as wage group 6 were, however, excluded from the *unità operative*. As a result, the possible job rotation, training and 'career' prospects for those who worked in the *unità operative* were significantly limited. Above all, the enterprise had the right to prescribe technical and economic goals to the *unità operative* which were not open to negotiation and/or modification after conflicts.

CHANGES IN THE WEST GERMAN AND ITALIAN STEEL INDUSTRIES – SIMILARITIES AND DIFFERENCES

To summarize: far-reaching changes have occurred in both the West German and Italian steel industries; some of them have been realized fully, while the course has merely been set for others. Their main impact has been felt in different areas (qualification and training in West Germany; work organization, work assignment and wage structures in Italy). For this reason, they seem to have very little in common.

But, as has been suggested above, this is true only at a very superficial level: at a somewhat more abstract level, if we trace the transformations to their core, the principal problem-solving mechanisms reveal surprisingly extensive similarities. We can identify the following six central problem solving strategies – elements, as it were, of the steel industry's 'new production concepts'.

Common problem solving strategies

1 The traditionally strong vertical and horizontal divisions of the production process into a variety of individual tasks have been reduced, making *broader skill profiles* possible: In Italy this was done by uniting several functions in *unità operative*, with their trend towards a homogeneous qualification of all team members, while in the Federal Republic of Germany production tasks – albeit only the more demanding ones – were united to form the 'professionalized' function of *Hüttenfacharbeiter*. This horizontal and vertical 'de-differentiation' of the employees' (formal) skill structure abolishes the traditional limitations put on the individual worker's qualification and performance potential to some extent. Broader and more comprehensive potential than that regulated by the traditional division of labour has been released, mobilized, developed and utilized; or in some cases, where informal modifications of the formal division of labour had taken place before, these were now made explicit and given official recognition. Thus a necessary, but not sufficient, condition has been created for a broader sphere of activity for the individual worker as well as for more flexibility in the utilization of all production workers.

2 An attempt has been made, through better work content, i.e. broader, less monotonous tasks and more decision-making, to *motivate workers* to feel a *greater identification* with the goals of production and to accept responsibility for larger units of production. In the Italian steel industry this has been achieved through an expansion of functions in the *unità operative*; in the West German steel industry, through a concentration of the more interesting and responsible functions in the jobs for skilled workers *(Facharbeiter)*.

3 The acquisition of broader and higher skills has been encouraged by establishing a close and *formalized connection between working* on a variety of tasks within the production process, which increases the acquisition of skills, *and 'better'*, i.e. faster moving and more secure, *careers*. In Italy, the new wage group structure, the accessibility of wage group 6 to workers, and the regulations concerning mobility within the enterprise provided by the wage agreements have encouraged workers to keep acquiring skills either by working on different jobs within the

enterprise or by attending courses. In the Federal Republic, the same effect has been achieved through the introduction of vocational training and the constitution of a professional profile with the status of *Facharbeiter* (skilled worker), whose career prospects are generally better and more certain than those of a semi-skilled worker, as well as by making the attendance of corresponding courses a condition for promotion to group leader and foreman (*Meister*) positions.

4 In both cases, a short- and medium-term *mobilization* of the workers' *latent qualification and training potential* in an upward direction has been effected: in Italy, this has been done through the mobility regulations; in the Federal Republic through adult training, i.e. the chance for some of the semi-skilled workers to receive training. This allows, in both cases, the creation of qualification reserves, i.e. potential that is significantly higher than the skills called for in the present job. Similarly, in both cases, this produces certain selection and marginalization tendencies for the weaker workers (although in Italy cases where workers fail to rise in the hierarchy are supposed to be, after a certain time, investigated by workers' representatives who are designed to counter this effect).

5 It is not only the workers' ability to work in various positions, but also their *readiness* to do so, that is encouraged through the connection between better jobs and wages and job flexibility: former 'rigidities' and obstacles to flexibility on the part of workers are reduced, in Italy, through a decrease in wage differentials (which creates an increased margin for transfers without consequences on wages), through the mobility regulations, and through the possibility for workers to be grouped in wage group 6; in the Federal Republic this is effected through the broad-based vocational training which makes multiple tasks gratifying rather than strenuous. This allows a fuller utilization of a worker's potential within the enterprise in terms of quantity as well as a higher intensity of utilization in terms of quality. Yet, in both cases, it is the enterprise which determines the extent to which the potential created in this way is actually utilized.

6 *Production work is re-evaluated* through broader work tasks, greater discretion ('responsibility') and varying job assignments as well as an improvement in wage and career opportunities, be it in the shape of mobility regulations and the opening up of wage group 6, or be it in the shape of the status and career of a *Facharbeiter*. This makes for an improvement in the acceptance of production work within the enterprise as well as in its attractiveness on the labour market.

The common denominator of all the measures described is, then, a reduction of the restriction of both the utilization and the development of the individual and collective labour force contained in the old Tayloristic system.

Structural differences

Apart from these parallels, there are two notable structural differences between the German and the Italian problem-solving strategies.

Firstly, in Italy, the improvements described tended to include *all workers*. In West Germany, on the contrary, only those workers who had received vocational training were included; the situation of the semi-skilled workers very likely even tended to deteriorate, as their career opportunities were cut short. Thus, at least in terms of career prospects, a polarization of the German production workers became probable.

Secondly, in the solutions adapted by the German steel industry, there was nothing that corresponded to the recognition of the inherently *collective character of qualification* as expressed in the concept of *professionalità* with its implication of shared control of the production process. The orientation of the formal work organization to this collective quality of work as expressed in the Italian work-form *unità operative*[5] also does not exist in the German solution. Certain cases of group work, if anything, can be regarded as steps in this direction.

SOME INFERENCES AND OPEN QUESTIONS

A new structural type of production work?

The major result of the present comparison is that in both steel industries widely differing transformation processes – promoted principally by different social forces – have led to the appearance of 'something new' which in its central mechanisms is remarkably similar to the developments contained in 'new production concepts'. How is this 'something new' to be theoretically understood and categorized?

It has been demonstrated that a shared characteristic of the analyzed transformations is a more thorough utilization of productive labour. It is important in connection with the debate referred to above that there is not merely a broader utilization of extant labour potential, but also a *broader development and a mobilization of labour potential*. In addition, it is not the individual worker's labour that is at stake, but that of workers' groups (either that of teams or the entire work-force in an enterprise).

The comparison also shows that the processes of development, mobilization and intensified utilization require and cause changes in all of the dimensions of the enterprise's structure that have been analyzed here: transformations not only occur in the qualification structure, but also in the areas of work organization, work assignment patterns and wage structures. The analysis demonstrates a close connection between the concrete forms of these four dimensions: a necessary inner consistency and coherence between specific patterns of work organization, work assignment, qualification and wages.

If this observation can be generalized, one could assume the existence of a limited number of historical structural types of work. The recent developments could be read as the beginnings of a *new historic type of production work.*

This term yields two distinct advantages compared with the term 'production concepts': first, it stresses the clear qualitative difference between two types of production work at the level of their structure and their inner logic, instead of defining them in terms of external features and quantity (such as 'more qualification', 'higher productivity', and the like). Second, this term emphasizes the objective character of the development of the production process instead of linking it to its perception and planning by the enterprise's management. The term 'new production concepts' coined by Kern and Schumann which is ambiguous in this respect, conceals just this paramount difference between objective and subjective strategies of capital.

If a historical typology of production work holds true, a question arises which is also of political significance: to what extent do the assumed imperatives of consistency and coherence determine the design of a given production process in detail? The question is not a new one: how large is the scope for shaping a given production process (*Gestaltungsspielraum*) or, in Kern and Schumann's words the 'spectrum' of the new possibilities for development?

A new unity of the conditions of production and reproduction at a higher level?

The comparison between the two steel industries suggests that there are certain constraints for *consistency* between a specific type of *production* on the one hand, and the level and form of the *reproduction* of labour on the other.

This becomes particularly obvious through a comparison of the sequences of transformation in the enterprises analysed: the West German steel industry tended to begin by transforming the production process, improving the qualification, wage and career prospects only at a later stage. In the Italian steel industry, by contrast, owing to political pressure exerted by the workers and their organizations, the improvement of the workers' opportunities for advancement (wages and career prospects) came first. It was only then that the enterprises started trying to assume control of and utilize these improvements for bettering productivity.

In both cases, however, the analysis has also shown that the consequences of the new type of production work for the workers have been *contradictory*: first, the potential for an improvement of the production workers' working and living conditions inherent in the new solutions was not utilized fully. The enterprises have striven to limit this potential in different ways: in West Germany, the rise in qualification and the extension of decision-making responsibility connected with it, was limited to a small number of production workers while the others tended to lose out. In Italy, after a period of homogenization, some (few) elite workers were given significantly better opportunities than the other workers. Trends towards the equalization of the job and wage structure were counteracted by the creation of new formalized intermediate levels, such as the two years training for specialists *(Werker* instead of the three-year training for *Hüttenfacharbeiter* in the Federal Republic) or new intermediate wage groups in Italy. The improvement of training conditions for production workers in long-term employment does not benefit all workers automatically, but requires a high level of personal commitment, activity and staying power. In view of the difficult working conditions typical for the steel industry this necessarily leads to selection and marginalization.

In addition to this, there are *new risks for the workers*. The possibility of utilizing labour more efficiently both quantitatively and qualitatively which results from better training and more flexibility, naturally leads to a reduction of the total number of workers in an enterprise. For those still employed, a more flexible utilization of their labour and higher responsibility for the production process can – apart from certain positive aspects – result in growing health problems, especially those related to stress. Where formalized qualification measures take the place of traditional means of promotion within an enterprise, this 'appropriation of training' by the enterprise takes the control of the training of junior staff out of the hands of the experienced workers. Thus, they also lose the opportunity of educating younger workers politically and integrating them into the plant's traditions and colleagual codes. In addition, an increase in the importance of the workers' personal initiative as far as training and careers are concerned tends to sharpen competition between young and old workers in favour of the former; the protection of older workers from competition from younger, stronger and better performing workers that the traditional seniority principle provided, has been lost (for other new risks, see chapter 13).

In view of contradictory tendencies of this kind, a possible connection between new forms and levels of production work on the one hand, and better reproduction conditions on the other, should not be overestimated. The one does not automatically follow from the other.

There are also theoretical considerations forbidding an overestimation of 'necessarily' positive consequences for the workers. If under certain conditions, the enterprise is interested in utilizing the workers' productivity potential as broadly as possible, and is hence also interested in its development, the opposite is true as well: In order to keep reproduction costs low and to secure power, the enterprise is also interested in preventing, or at least limiting, the development of labour. As a consequence, it will be interested only in a limited utilization of labour.

This contradiction in the enterprise's interests is irreducible. Depending on the historically existing economic and political conditions, this contradiction takes ever differing forms: different compromises between restriction and development.

New structural types of production and reproduction can hence be analysed systematically as products of specific historical compromises between these two general principles. Only in this manner can the limits of potential improvements in the reproduction of labour be viewed systematically, and only in this way can phenomena such as the implementation of such improvements and the enterprise's attempts to revoke them be explained. Finally, it is only in this way that forecasts can be made which, alongside the progressive aspects of the new production structures, systematically integrate counter-tendencies and the limitations resulting from such tendencies.

The significance of political factors

The results of this comparison allow certain inferences and raise certain questions regarding the significance of political factors, especially regarding the *'social actors'* who have the initiative in the change process and the *area* in which a discussion and solution of problems is held.

It is clear that it was of considerable significance for the way structural changes were set up that new solutions to problems were defined, developed and propagated in the Italian steel industry primarily by workers and their organizations, while this was done primarily by company management in the West German steel industry.

What significance does the area in which complex problems are primarily discussed have for further developments? In the West German steel industry the solution centred on the area of qualification and training – with the introduction of formalized vocational training, while in Italy – more or less as its functional equivalent – new rules of mobility (see pp.

298–302) were created. Does this fact lead to differing forms of a more extensive utilization of labour and to differences in the scope of the new trends? Questions such as these, which cannot be answered here, are of great significance for analyses of comparable processes.

Qualification has acted in West German discussions as the proxy for other problem areas for a long time. To speak about qualification and training problems in Germany always implies criticism of unsatisfactory work content, limited decision-making opportunities, restricted job enhancement, absence of responsibility in job tasks, low wages, lack of job security, and bad chances on the labour market. But, which problems for work and living conditions have remained in the shadows through this extensive definition of the qualification problem, and what consequences does this have? In the light of new developments, should not the identification of worker interests with better skills that has been taken for granted up to now, and the special weight given to skill as a way to evaluate change be re-examined? Isn't this the lesson of the trend toward an intensified use of the higher skills of fewer and fewer workers contained in the 'new production concepts'?

If this is so, a shift in the accentuation on the significance of qualification is necessary. Towards this purpose, a look at the solutions of the Italian unions and the role that qualifications play within them could be useful: the concept of collective *professionalità* stresses the workers' capacity for control. And next to their individual capabilities and their ability to cooperate, it stresses the capacity to cope with the organization of the production process collectively. As shown, the collective *professionalità* was concretized through the *unità operative*, which were defined as being homogeneous and were, through their own initiative, to produce a real homogeneity including all production workers. Skill requirements were defined in close connection to other demands in a way which not only hindered further differentiation through qualification, but also counteracted existing differentiation. Concrete demands for changes in company skill and wage structures were integrated into a broader political perspective, one which transcended the goal of securing the individual worker's immediate reproduction and aspired to worker solidarity and a defence against work-force fragmentation.

These points lead us to a final conclusion to be inferred from the comparison of the two steel industries regarding the analysis of a new type of structure of production work which may be developing: *Political culture, political consciousness*, and *political perspectives of the worker movement* are also significant for whether, and in which forms, new trends for the expanded utilization of the productive power of labour come to be implemented.

Part VI

Large and Small Firms

20 The Development and Structure of Small-Scale Firms[1]

Stefanie Weimer

A specific institutional framework has contributed to the fact that the small firm sector of the economy has been able to hold a comparably strong position in the Federal Republic of Germany. In fact, for several years a new expansion of this sector has been registered. The following chapter investigates the reasons for the increase in jobs in small-scale companies. It is demonstrated that although the ongoing industrial restructuring process provides new opportunities for small firms, there are certain structural developments which in the long run may contain disadvantages for employment and quality of work in this sector.

INTRODUCTION

In spite of all the forecasts voiced by those involved with theories concerning the process of economic concentration, an economically vital and quantitatively significant small-firm sector has held its own in the Federal Republic of Germany, as the following data will underscore: in 1986 in the manufacturing industry sector, 69.9 per cent of all companies, that is a total of 23,500, had between 20 and 100 employees. These companies employed a total of 1.1 million persons, which amounts to 15 per cent of all persons working in the manufacturing industries sector. The number of small-scale companies in the manufacturing sector having fewer than 20 employees totaled 51,000 in 1987, amounting to 330,000 employees. Additionally, there is the craft sector[2] which consists almost entirely of small-scale businesses and which employed nearly 3.5 million persons in 1986. These data give an impression of the significance of the small-scale company sector in the Federal Republic of Germany.

The following discussion concentrates on small firms in the manufacturing sector. However, it should be kept in mind that, as in other industrialized nations, certain parts of the service sector are mainly the domain of small firms in Germany. For instance, in the so-called 'other services' category of the industrial classification system, that is all private services that do not fall under the categories sales, banks, insurance, transport or communications, more than 90 per cent of the firms had fewer than 20 employees in 1987. Approximately 3.2 million persons were employed in these firms in that year (see Tengler, Hennicke 1987; Deutsche Bundesbank 1988).

There is no single dominant type of small firm which can be distinguished from large firms on the basis of certain characteristics – e.g. wage levels – and which would make it possible to speak of a dualistic business structure based on the size of the firm. In the Federal Republic, the spectrum of small-sized companies ranges from crafts and artisan businesses which – often with a high percentage of manual work – produce for a local market, to small industrial or crafts businesses using state-of-the-art, often capital-intensive manufacturing equipment for the production of highly specialized, innovative and quality products geared for both national and international markets. There are small-sized companies which are capable of offering their employees working conditions comparable to those of large-scale enterprises, in terms of job security, wages and social security benefits. Some even offer conditions superior to those in large-sized businesses with regards to the scope for taking independent action on the job and the level of skills maintained. However, there are also small-sized companies which have relatively low wages, unstable job conditions, and work burdens which are no longer found to such an extent in large-scale companies (see Weimer 1983).

INSTITUTIONAL FRAMEWORK OF SMALL FIRMS

What is the institutional framework which has contributed to the formation and stabilization of a strong small-scale company sector in the Federal Republic of Germany?

First of all, it should be pointed out that there are few or no exceptions made for small-scale companies as far as important regulations in social legislation or the system of wage determination are concerned. Therefore small-scale companies were never allowed to operate in a protected sector of the economy. The employer contributions to social security have to be paid by all companies and remain the same regardless of company size. There are some specific exceptions related to company size with regard to dismissal provisions, regulations on work safety and co-participation rights of the representatives of employee interests, but these pertain mostly to very small businesses with less than five, or with less than 20 employees. In the Federal Republic of Germany, wage agreements in the different sectors are concluded centrally for regional wage zones. Within these wage zones, which often cover a wide geographical area, the same contractual wage applies to all firms covered by the agreement. No allowances are made for the profitability of individual firms or for companies of different sizes. The fact that the wage differential that exists in the Federal Republic between large and small-scale companies is very small in international comparison, is a result of this centralized wage system.

As paradoxical as this may sound, the situation which imposes higher costs on small businesses has contributed towards improving their competitiveness in the long run: small-scale companies cannot sit back and enjoy a competitive advantage resulting from lower wage costs. In order to be able to compete, they have been forced to enhance their productivity to a level commensurate with that of large-sized companies – which means continuously modernizing manufacturing facilities, or seeking gaps in the market for quality products. In order for such competition strategies to be successful, small-sized companies must be able to draw on a source of skilled, adaptable and versatile workers. Qualified personnel of this kind can only be recruited by small companies from the external labour market, and they can only retain such employees by offering working conditions and a stable employment relationship which is comparable to the standards set by large-scale companies.

Another important condition is the existence of a vocational training system in the Federal Republic – of the formal, off-the-job type – which is particularly well adapted to the needs of small-company manufacturing with a high level of skill and craftsmanship. The transmission of vocational skills while work is actually going on, which is a typical characteristic of training within large firms, is in many respects dysfunctional for small firms: it is too protracted, and the gradual acquisition of skills requires the division of labour to be much more developed than is the case in many small firms. An off-the-job vocational training system provides training with a standardized content, which is not firm-specific. Skilled workers trained in this way can be utilized with a high degree of efficiency in the manufacturing area within a short period of time. Since a large proportion of industrial training takes place in craft and small-scale firms, the small firm sector always was able to gain access to skilled workers, despite the migration of the skilled labour force from the craft and small-firm sector to the large industrial firms.

A third factor responsible for the position held by the craft sector in the Federal Republic is the formation of strong collective organizations serving craft businesses. While these influential self-help and support organizations do have a central umbrella organization, they are structured so as to reach all the way down to local levels, where they encompass every single craft firm. The collective organization of the craft sector did not take place without the aid of the state which created the essential preconditions for this development by authorizing compulsory membership and the transfer of state functions to the self-governing organisations in the craft sector. The tasks carried out by these organizations comprise the entire system of vocational training, trade inspection, further training, company consulting, as well as the activities of advisory committees and expert groups etc.

On the basis of this organizational structure, further forms of collective self-help came into being. Thus individual craft and artisan business sectors can draw on the support of strong cooperative organizations, which were originally founded with the aim of securing more favourable conditions through collective buying. In the course of time these cooperatives have extended their area of involvement considerably. Today they are active in supporting and carrying out further training, company consulting and directing industry specific research institutions.

THE QUANTITATIVE DEVELOPMENT OF THE SMALL-FIRM SECTOR

An increased number of employees have been registered in the small-firm sector in the past 15 years in the Federal Republic of Germany. According to an investigation carried out by the Federal Labour Office, there was an increase of 580,000 jobs in small-sized companies with fewer than 20 employees during the years 1977–85 in all industrial branches, including crafts and services. During the same period of time, companies with 20–99 employees reported an increase of 89,000 jobs, whereas companies with more than 500 employees dismissed 225,000 employees (Cramer 1987).

The reasons for the increase in jobs in the small-scale company sector remain largely unclear. The structural change toward the service sector is one, but not the only cause, for the trend toward smaller company size since small firms are also appearing in branches of the manufacturing sector – in areas as diverse as machine-building, electrical engineering and precision mechanics, wood and plastics processing, the foundry industry and lock-smithing trade, as well as in structural steel and light metal engineering and in the food processing industry.

The small firm phenomena could be a result of the increased establishment of newly founded firms. Empirical data show a continuous increase in the foundation of new firms registered in the Federal Republic since the mid-1970s (see Szyperski, Kirschbaum 1981; Clemens *et al.* 1986). The question, however, is what the relationship is between this development and the increased number of jobs in the small-firm sector. One point is that investigations have revealed a close connection between increases in new small businesses and the high rate of unemployment. Apparently an above-average share of the newly founded firms in the craft sector were set up by unemployed people (Weizel 1986). This means that at least part of the employment growth in the small firm sector due to the establishment of new small firms could be a temporary phenomenon. Another factor indicating the unstable character of new small firms is that their increase is accompanied by a simultaneous increase in the number of insolvencies since the mortality rate of newly founded firms is high. Nevertheless, the number of new entrants has been exceeding the number of failures since the beginning of the 1980s. However, even given a positive gain in the number of new small firms, the overall job potential created by them is small; as empirical studies show (Weizel 1986), the majority of newly founded businesses were one-man firms. Only after the second or third year do newly founded firms, that have survived that long, exhibit modest effects on employment. It was in particular newly founded firms in the craft sector that raised their level of employment (including the founder) about 2.5 times (Weizel 1986).

The contribution of new firms to the process of modernization of the economy on the whole should also not be overestimated: the largest number of newly established firms was in the artisan and craft business sector, followed by the commercial business and service area. The industrial manufacturing branch accounted for a very small percentage, and the number of high-technology-based firms (HTBF) was also comparatively low. An assessment of the new entries to the Commercial Register in 1983 revealed – based on a favourable estimation – that the percentage of HTBFs came to 3 per cent, amounting to 1030 HTBFs from all of the firms newly established in that year (Anglo-German Foundation 1988).

Another explanation for the increasing number of jobs in small-scale companies is that the already existing small firms are able to expand their employment level because these businesses have been more successful in adapting to ongoing structural change: the change in the structure of demand towards individual, more highly differentiated products, the increased competition from imports of standardized mass-produced goods, and the technologically determined decrease in the importance of economies of scale. All of these factors have been favourable to small firms because they correspond to the small firms' specific advantages such as flexibility, creativity and closeness to markets and customers.

However, there are other economic indicators which would contradict the adaptation to structural change thesis: during the same period of time in which the number of persons employed in the small-business sector increased, the process of economic concentration in the overall economy did not slacken. Large-scale enterprises were able to further boost their share of investments and turnover. Moreover, the latest investigations have shown that the difference in profitability between large and small-scale businesses did not diminish between 1979 and 1985 (Irsch 1985). The increased economic significance of the small-sized companies seems to have been limited to the production of goods and services without manifesting itself as increased financial power. It seems as if a process of reorganization in the overall economic division of labour between large and small-sized companies is taking place, which is partially responsible for the increase in employment in the small business sector. This process consists of large companies contracting out parts of their manufacturing functions to small firms and thus creating more jobs there while, however, retaining their economic influence (see chapters 4, 16).

SOME STRUCTURAL DEVELOPMENT TRENDS WITHIN THE SMALL-FIRM SECTOR

Besides the quantitative expansion of the small firm sector, there are some qualitative trends which might lead to considerable structural changes and transformations in the near future. On the one hand, these trends are leading to an expansion of employment in small firms, but, on the other, constitute a threat to the conditions of that employment and are bound to have an effect on the quality of jobs in such firms.

There are a considerable number of small companies within the Federal Republic which are manufacturing for a national as well as a local market and which are competing directly with large-scale companies in their respective product market. For this reason small companies are experiencing increasing pressure. Two examples of this are presented below.

In the Federal Republic of Germany certain areas of food processing have traditionally been the domain of small craft businesses. As opposed to the practice in other industrial nations, the mass processing of meat products and breads and pastries has not been able to establish itself on a broad basis. Small-sized businesses have managed to hold their own in this sector with the help of a competitive strategy which is rooted in quality standards and a highly diversified and individual range of products. However, these traditional craft firms and small industrial companies with their largely regional markets have recently been experiencing increased competitive pressures exerted by large trading firms, discounters and shopping centres. These enterprises, operating on the national market, either have their own production facilities or cooperate with large manufacturing companies in the food processing sector. The result is that the number of companies and employees in the small company sector are dwindling, and a considerable process of concentration can be observed (see Mendius *et al.* 1987).

The second example comes from the furniture industry, a branch which is characterized by small and medium-sized companies in the Federal Republic, and which has been experiencing a severe structural crisis since the end of the 1970s. The causes for this lie in excess capacities, growing import pressures, and cost pressures exerted by retailers organized in large purchasing groups. This branch has also been profoundly affected by the change in demand for more diversified and individualized products which create problems of expanded product variety, shorter manufacturing cycles and smaller batch sizes. In the furniture industry in the Federal Republic, a drastic process of contraction is going on (between 1980 and 1986 there were 21 per cent fewer companies and 25 per cent fewer employees registered) which is accompanied by a process of concentration (see chapter 6). The number of small furniture manufacturers, who succeeded in establishing themselves in a market niche are few indeed. A large number of small furniture manufacturers unable to hold their own on the increasingly complex final product market, are attempting to survive by becoming suppliers of semi-finished products. The component supplying market in the furniture industry is growing because many large manufacturers are trying to cope with the increased demands for flexibility and the increased cost pressure by contracting out a growing number of steps in the assembly process to other companies. These subcontracting companies, which usually have fewer than 50 employees, tend to be extremely dependent on one or two customers. They often fulfil the function of an 'extended workbench' and are highly replaceable (see chapter 15). Relatively few of these small-sized component suppliers have succeeded so far in securing a stronger position based on their own specific product and manufacturing know-how. It would appear as if in this particular branch, the small-scale businesses have not been able to capitalize on the trend toward more individualized and diversified quality products.

A second developmental model which portrays the role of small firms in the Federal Republic is the relationship to one or several large-scale companies based on a division of labour, whether as components suppliers or as downstream service firms concerned with the sales, installation, repair and maintenance of the products of one or several large manufacturers. The automobile repair trade is a classic example of the latter type. An example of components suppliers is the network of suppliers to the automotive industry located in Baden-Württemberg. These firms, some of which are relatively small, are autonomous in both economic and legal status, and drawing on their own specialized product know-how, usually supply several large manufacturers.

This well-established division of labour between large-scale companies and their often small or medium-sized component suppliers is at present undergoing a fundamental change. In many companies a drastic reduction of in-house production is currently taking place. The decentralization of manufacturing by contracting out certain steps in the assembly process to components suppliers is just getting underway in the Federal Republic of Germany. At the same time the suppliers are becoming more strongly integrated into the rationalization strategies of the large companies and are thus being confronted with new and very far reaching demands. Just-in-time delivery, data links with buyers, the takeover of quality assurance and control, and increased involvement in research and development are aspects of production the suppliers must take into account. More and more large companies are pursuing a policy of single-sourcing, i.e. buying one part from only one supplier. The contracting out of various steps of the assembly process by buying complete components and aggregates instead of single parts is also on the rise.

However, it is questionable whether all subcontractors, particularly the smallest firms, are in a position to finance the necessary changes stemming from the new demands. As a result, small contracting firms could be squeezed from the market or pushed down to the third or fourth level of the supply chain (see chapter 22). One consequence of these trends can be seen among the suppliers to the automotive industry in the Federal Republic of Germany where a wave of mergers on a national and also international level is taking place.

If the small firms among the suppliers want to adapt to the new demands of their customers, they are forced to expand. Meeting the new demands also threatens to change their internal character; the introduction of flexible manufacturing systems, necessary to satisfy the increasing demands for product flexibility, strongly integrates suppliers into the planning and manufacturing organization of their customers. Thus, they have to alter their work organization and adapt it to industrial standards. For small firms this means: centralization and formalization of operational communications and decision-making processes, and increased standardization and division of labour. This, in turn, jeopardizes their fundamental economic advantages, namely their flexibility, individuality and the comprehensive use of qualified, skilled workers.

It is in those areas involving the supply of differentiated and high quality custom-made and small batch products, for example equipment parts for manufacturing facilities, or in the tool manufacturing and mould-making sector, that small firms will most likely maintain a good position as suppliers to large enterprises.

Also small firm industrial branches fulfiling downstream service functions for the products of large companies are at present confronted with a widening gap between the technological development of a given product at the manufacturer's level and the skills of the personnel in the servicing firms. This problem, which is generated by the increasing share of micro-electronic components and modern control technologies in many products, has become particularly acute in office machine repair, and is also evident in the heating and ventilation systems branch, as well as in automobile repair. In the case of office machine repair and the heating and ventilation systems branch, the result has been a loss of market shares to 'outsider industries.' Lack of expertise also makes it difficult for existing firms to expand into new fields and take advantage of potential new markets, for example, that of energy consulting for the heating and ventilation systems branch. This, of course, simply enhances market losses and the possibility of decline. In the office machine repair sector efforts are currently being made to combat the existing skills deficit by developing target-oriented further training programmes – a joint initiative involving industrial associations, further training establishments and also a number of large manufacturers of office machines.

One of the most widespread developmental models within the small-firm sector is the so-called industrial small-firm district. Do such cooperative structures play a significant role among small businesses in the Federal Republic? There are, in fact, some examples of regional cooperative organizations of flexible, specialized small firms, for example the cutting implement industry in Solingen or the textile machine industry in Baden-Württemberg. At present we do not know terribly much about the history and functional mechanisms of these small firm centres since research in this field is just commencing in the Federal Republic. Empirical research on the example of Baden-Württemberg carried out by the Science Centre Berlin (Maier 1987) shows that industrial policies adapted to the needs and requirements of small firms and which favour concrete measures rather than financial assistance have played a significant role in the development of this industrial area.

Baden-Württemberg's industrial policies have been an excellent example of this. Very early, as far back as the 19th century, it was recognized that specialized production for market niches is dependent on the international market. Thus, industrial policies in Baden-Württemberg were developed to support small firms in their export activities and assist their participation in international trade fairs. Also, the establishment of decentralized credit institutions rooted in the region helped small firms gain access to the capital market. One of the most important achievements of governmental industrial policy in Baden-Württemberg was the development and further expansion of a comprehensive infrastructure for industrial vocational training and further training. This guaranteed the supply of qualified, skilled workers, which in turn is the prerequisite for the manufacture of specialized, high-quality products. These training centres, which are organized on a supra-company level, are sponsored by the self administrated organs of commerce, the various Chambers of Commerce, professional organizations and partially by the municipalities. They are decentralized and situated at local levels, which enables them to cater directly to the specific needs of the companies in their vicinity. Local companies often participate in the conceptualization and execution of vocational further training measures. The existing infrastructures also greatly facilitate and accelerate the introduction of new technologies in the small firms of the region. Thus, for example, these training centres offered the first courses in the Federal Republic for CNC-turning and CNC-milling. In so doing they cooperated closely with the manufacturers of such CNC-machines and the potential local user companies. This way the companies were given the chance at a very early stage to become familiar with this new technology, to test its application potential for their company and to solve the related training problems.

CONCLUDING REMARKS

In conclusion one could say that there is a broad foundation of small firms that have evolved historically in the Federal Republic of Germany, and that this constitutes favourable conditions for the further development of the small firm sector. However, as I have argued, there are some types of small firms in the Federal Republic currently facing severe problems of adaptation. To what extent small businesses will be able to cope with these problems and make use of their market opportunities will essentially depend on whether they succeed in solving the problem of the introduction of new technologies and whether they continue to supply qualified skilled workers by modernizing the form and content of vocational training and further training measures. In striving to achieve these objectives, cooperative solutions are likely to play an increasingly significant role in the future among small firms in the Federal Republic of Germany.

21 Small Supplier Firms at the Crossroads – Meeting New Challenges through Cooperative Retraining Schemes[1]

Hans Gerhard Mendius, Stefanie Weimer

INTRODUCTION

With market requirements changing, inter-company cooperation is becoming more and more important as an instrument, especially for small businesses, to assert themselves on their traditional markets. Very little empirically based knowledge exists about the dissemination, actual forms and structural preconditions of cooperation between small businesses. Theoretical concepts are also in short supply. Taking the example of a special group of small businesses – small and medium-sized suppliers – and the example of a particular company function – further training – we shall examine the opportunities for implementing inter-company cooperation and the possible consequences for employees in terms of the value of the resulting qualifications on the labour market.

At present small subcontractors are under heavy external pressure to adapt to new demands from their customers. New requirements from predominantly large customers include greater flexibility of delivery, the undertaking of quality assurance tasks, integration in the customers' computer network, delivery of preassembled modules, and the contribution of their own development services. Therefore, small businesses are being forced to modernize their production technology and to restructure the entire organization of their company operations. This is bringing about profound changes, not least in the qualifications required.

Workers with the right modern qualifications are generally scarce. Small businesses in particular have difficulties in recruiting them because big companies are usually able to offer better terms of employment that look far more attractive, at least at first sight. An alternative is to provide further training for the existing work-force. But it is precisely small businesses that usually have relatively little experience when it comes to staging further training courses themselves and to making use of external offers of further training (see BIBB/IAB 1987). Against this background the question arises whether cooperative methods might not be able to help solve the problems of adaptation:

1 in the form of inter-company cooperation on further training among several subcontractors, and
2 and in the form of assistance for subcontractors from their big customers in solving their qualification problems.

NEW QUALIFICATION REQUIREMENTS IN SMALL SUBCONTRACTING FIRMS

The new demands made by customers affect very different company levels and various groups of workers. New qualification requirements arising from the introduction of formalized and documented quality assurance processes for raising product quality are a central aspect of the new demands. Depending on the products and the production process, it is necessary to have a knowledge of measurement methods, statistical evaluation methods, the operation of computer-controlled testing systems, the compilation of inspection plans and so on. A company's further training needs depend on the existing qualification structure (polarized production work dominated by semi-skilled workers or a preponderance of qualified skilled workers), and on how the new requirements are distributed among existing activities and positions. This distribution means for example, whether quality-control tasks are turned into an independent centralized function, or whether duties of quality inspection and documentation are integrated into the activities of the production department.

Like other companies, small subcontractors tend, in our experience, toward centralized solutions. They set up new departments for quality assurance and man them with specialists, who are equipped with the necessary additional qualifications in training courses. In those places where quality-assurance departments, however, already existed, there is a notable trend toward reintegrating these functions in the production department. Some companies discovered that the awareness of quality in production decreased with centralization of the quality functions because the workers no longer felt responsible for quality, seeing it as the sole duty of the special department created for the job.

Where production is based on skilled work, the integration of quality assurance is less of a problem. Care and accuracy are components of the skilled worker's training and form a basis for imparting a knowledge of quality-inspection methods. A need for new qualifications arises mainly in cases where more demanding quality inspection duties (e.g. inspecting the dimensional accuracy of workpieces with complex geometries) are to be integrated in semi-skilled activities.

Higher demands for quality can also produce a need for new qualifications in pure production tasks. Especially in production processes with highly repetitive components and a high proportion of semi-skilled work, e.g. in forges, plastic processing or the production of lathe-turned parts,[2] problems of qualification and motivation can lead to considerable quality defects with expensive follow-up inspections and repairs. In such cases, an enrichment of semi-skilled functions could prove sensible, also from a cost saving perspective. At present, acceptance of these considerations continues to meet with resistance among decision-makers in small businesses. Their implementation is made all the more difficult by the fact that there is practically no counselling available to small and medium-sized businesses on questions of work organization design.[3]

In the small businesses examined in this study, a need for training also arose from the increasing use of program-controlled equipment and integration systems, a trend that has proceeded at an above average pace even in small craft sector businesses operating as subcontractors (see Halder *et al.* 1988: 33 ff.). Training needs, when using CNC machines, depend a great deal on the way that technology use and workers are organized. With so-called decentralized use of CNC machines, the workers in the production department perform the work planning and programming duties themselves, and possibly even the short-term trouble-shooting. The production personnel require a knowledge of the way that CNC systems work, a knowledge of programming, and key qualifications such as e.g. the capacity for strategic, systematic, abstract thought (see Sonntag 1985; Hoppe, Erbe 1986). With the concept of work organization that emphasizes centralization, however, programming and trouble shooting are taken care of by the job planning and maintenance departments; activities in the production department are reduced to operating and monitoring the machines. Whether or not concepts of this type give rise to new qualification requirements for workers in production depends on the extent to which they are left to perform functions of adaptive precision control and program optimization (see Part III).

Although a decentralized concept of worker deployment is considered particularly well-suited for quality production in small businesses (see Erbe *et al.* 1989), it was apparent that in a number of supplier firms – after an initial phase with shop-floor programming – the programming function shifted from the shop-floor into a central job planning department or the design office when the number of CNC machines increased, and at the latest when CAD was introduced.

The introduction of computer-based design methods (CAD), which can be found in small companies whose components call for a considerable input of their own design work (e.g. in tool and die making for the working of metal and plastic), also leads to new qualification requirements in small subcontracting businesses.

The new technologies, especially 3D-CAD and interlinked CAD/CAM, confront these small businesses with a training effort that is unprecedented in its intensity and structure, and which produces appreciable problems for personnel policy. If the system is to be put to optimum use, the lengthy and very expensive theoretical basic courses of instruction must be supplemented with extended training periods in the enterprise. Moreover, those needing the further training are the designers and those highly qualified skilled workers who form the interface in the job planning department between design and production. This means that to carry out these training programmes the company would have to release urgently needed skilled workers and specialists who perform critical functions for a considerable period of time. The fact that there is a lack of independent further training options in this area creates a dependency on the very specific training offers from the system manufacturers. Yet, the new qualification requirements are highly company-specific in nature. As a rule, CAD/CAM solutions are developed as customized systems, duly modified for the given specific requirement – making qualification requirements equally specific. Especially when implementing such technologies, a precise knowledge of company procedures and design is necessary to be able to adapt the programs. For this reason alone, there are limits to solving the problem of qualification through recruitment of new staff from outside the company.

On the whole, the training needs of small subcontractors show parallels to comparable small businesses, but the pressure and pace under which the necessary modernization measures are taken and the resultant problems of qualification are very much greater.

THE ADAPTATION OF QUALIFICATIONS AS PRACTICED IN SMALL FIRMS

Coping with new qualification requirements revealed patterns in small firm further training attitudes that were familiar from other investigations (see e.g. von Bardeleben *et al.* 1989; Koch 1988):

1 Limitation of employee participation in the training measures to the absolutely essential and, as a rule, to a few mostly young workers with good preliminary qualifications and a willingness to take part in further training. It is not rare for small firms to rely successfully on 'interested'

workers trying to obtain new qualifications through individual further training activities in their own time and at their own expense.

2 The dominance of short-term on-the-job courses of instruction of a rather unsystematic nature. The instruction of the majority of workers affected by the new technologies is usually undertaken by those who previously attended a course. This group thus assumes a multiplier function.
3 Low utilization of external further training options which run for longer periods, are more formalized, and issue certificates.
4 The great importance of training courses offered by the manufacturers of the equipment being installed.

Some businesses have in fact organized further training courses themselves, drawing on professional, external further training expertise. In these cases the self-devised course content is generally closely tailored to the respective company's needs of adaption. In general, however, the considerable need for requalification does not get reflected in further training programs that are either quantitatively or qualitatively well-developed. On the contrary, the outlined response patterns to training needs tend to produce appreciable problems. They result in excessive demands on the workers involved during the introductory phase, which often persist, albeit latently. And because no certificates are issued, there is no sure guarantee that the newly acquired qualifications will be duly rewarded financially, especially upon moving to a different firm. On the whole, unskilled and semi-skilled workers have only a slight chance of taking part in the training measures. For small businesses, this can have negative consequences such as a higher fluctuation of employees, a high degree of absenteeism, more faulty pieces, low job flexibility of workers with excessively narrow qualifications and, as a result, less-than-optimum utilization of capital-intensive equipment.

These shortcomings exist even though – except for a few regional bottlenecks – a wide range of training is offered that is specifically geared to current technologies.[4] The fact that this expansion of the further training market has failed to leave its mark on small businesses points – among other things – to discrepancies between the content and organizational structure of existing training offers on the one hand, and the small businesses' specific requirements for further training on the other.

In certain respects the demand for further training among small businesses is:

1 *specific*, i.e. its content is often related to a production process applied by just a few companies, or to highly specialized products or materials, because small businesses often pursue pronounced market niche

strategies. Moreover, unlike a big company, a solitary small business is unable to fully utilize such company-specific further training measures to the fullest extent, and so it cannot push for such specialized training to be offered on the further training market. It is also,

2 *company-individual*, in the sense that what is company-specific is not the content of the knowledge and skills to be taught, but which relevant parts are selected from the more general field of knowledge, the combinations in which the particular knowledge is demanded, or the group of recipients for whom the specific knowledge is to be prepared.

Precisely because the resources for a long trial and implementation phase do not exist, small businesses rely on the supply of further training courses which deal with concrete company application problems. The timing of many further training programmes – full-time courses during working hours – makes participation difficult, because small businesses usually work in single shifts. Personnel resources are scarce and do not allow the release of staff for lengthy further training courses during working hours. In small businesses there are practically no stand-by workers or planned manpower reserves such as are customary in big companies. The problem is all the more acute in subcontracting companies because they are often placed under considerable time pressure by their customers, are committed to tightly scheduled delivery obligations, and are often required to perform additional services at short notice.

What would be most suitable for small businesses in terms of training to meet their new production requirements would be training options that are flexible in content and in organization:

1 as a supplement to the prevailing programmes of standardized courses, further training institutions should provide more measures that are tailor-made, company-individual or geared to the needs of a small group of businesses. Of course, this in no way excludes the use of any available 'standard modules'. Likewise,
2 further training institutions should offer more measures that are implemented on-site at the workplace, or meaningful combinations making use of both the training centres and companies. In this way the training would be more adapted to company processes and could also exploit the advantages of the workplace as a place of instruction.[5]

INTER-COMPANY COOPERATION ON FURTHER TRAINING

The development of suitable further training programmes for small businesses involves considerable costs that are not affordable for individual small businesses. One way to provide tailor-made training for small businesses all the same is for several small businesses to join forces in organizing and carrying out such measures. Towards this purpose they could make use of the experience gathered by various small businesses in organizing such measures on their own.

The simplest form of this type of cooperation on further training would be an ad hoc association of several neighbouring companies, which would then join with a further training institution in conceptualizing measures for the purpose of solving a qualification problem affecting them all. Such an ad hoc association could also include agreements concerning the mutual use of facilities and machinery in the participating companies for further training purposes.

A more heavily institutionalized variant would be a permanent co-operation – including a further training institution – aimed at a fundamental and lasting optimization of the actual further training practices in the participating companies. The job of the further training institution would be to ascertain each company's training needs, to draw up a long-term training plan, and to develop a coordinated programme of measures. The companies participating in the cooperation would make mutual use of each other's companies as places of instruction.

Cooperation on training tasks already exists in vocational training, in which different companies take on particular parts on a revolving basis. Training associations of this sort have established themselves in the past few years and are stirring growing interest.[6] The reason is the increasing tendency of particular companies to specialize on one product/one production process, while simultaneously quality requirements for basic vocational training are rising, due especially to the restructuring of training in the metalworking and electrical professions. This has made it more difficult for the individual company to conduct a complete, high-grade course of training. The question is, to what extent are the positive experiences in basic vocational training transferable to cooperations on further training.

Among the supplier firms there are isolated cases – and some miscarried attempts – of cooperative actions, too, but in areas quite different from training, e.g. the exchange of tools, joint use of capital-intensive test equipment, etc. Although the idea of an association for further training has elicited a certain interest, there are also considerable reservations. The doubts and apprehensions concern mainly:

1 a fear of competitive disadvantages resulting from a drain of know-how to partner companies, especially with regard to association concepts involving the reciprocal use of production capacities for further training purposes;

2 a fear of skilled workers being poached by cooperation partners;
3 the additional outlay of manpower resources and time arising with the organization of an association; and
4 a lack of willingness to cooperate anticipated among potential partner companies.

The difficulties of implementing cooperations on further training appear to rest on two levels.

The first lies in the object of cooperation itself, i.e. further training: the idea of further training has almost no roots in small businesses, not least because no concrete experiences exist to evaluate its potential for solving qualification problems. There is a lack of consensus, therefore, concerning the strategic importance of the cooperation issue itself. For the decision-makers in small businesses, further training is often a low-priority task; its economic utility is not always clear to them.

There is not enough pressure to change the current practice of further training, in part because problems of adaptation resulting fully or partially from problems of qualifications are not registered as such. The short-comings of the current response patterns for dealing with changes are too well concealed. Consequences such as the overtaxing of workers and the resulting deficiencies in the production process and the product are taken as given. For subcontractors, problems responding to the need to adapt qualifications, rank far behind other adaptation-related issues. The association idea is probably more successful in the basic vocational training sector because it helps to solve a major problem facing the company, namely the impending loss of company training capabilities. The fact is that having an apprentice programme remains the most important channel of skilled worker supply, especially for small businesses. This manifest company interest in training associations in the basic vocational training sector appears to outweigh even company fears of a know-how drain or the poaching of fully trained workers, which play a role here too.

Cooperation on further training hits on a by and large unstructured field. Further training is far less regulated by law than basic vocational training. The existing training regulations give more or less detailed specifications regarding the content, aims and methods of vocational training, which all contributes to simplifying the practical implementation of cooperations in this sector in as much as it provides a framework of content and organization.

Cooperation works most easily where it deals with fields of activity which are defined, which are capable of being performed outside the company, and which are far removed from the core area of production (e.g. sales, advertising). This is not the case with further training activities that are designed in close consideration of company practices, specific applications, coping with a company's practical production problems, and with mutual use of each other's companies as a place of learning.

A second level of restriction in the implementation of cooperation on further training lies in the particular group of businesses being investigated in this study, i.e. small subcontractors.

Cooperative strategies are always difficult where the partners in the cooperation are (potential) competitors. In the case of small subcontractors the supposition was that a basis for cooperative behaviour existed in the similar customer demands directed at these businesses and in their accordingly very similar training needs. The heavy regional concentration of some of the investigated subcontracting sectors suggested that channels of communication already existed between the companies.

Evidence shows, however, that in most cases, these favourable foundations are not utilized, but are overshadowed by the pressure of competition. The market for subcontracted parts tends to be tighter than markets for final products. Many of the small subcontractors have a limited – and sometimes identical – circle of customers. In a number of the small subcontractor supply branches the level of interchangeability is quite high; the applied production processes are similar, and only a few companies have specialized know-how at their disposal. Overcapacities in some areas and buyers putting greater pressure on prices and terms has triggered a kind of predatory competition. Against this background, the fear of a 'know-how drain' is particularly pronounced. Under these circumstances, a supply of workers trained in new methods, e.g. through a company's own further training efforts, becomes a competitive advantage. There is no particular evidence that communication exists among neighbouring competitors in the same region. Under these conditions it is not possible for trust – one of the fundamental prerequisites for a successful cooperation (see Müller 1987: 219) – to develop.

Admittedly, the considerable external pressure does favour cooperative strategies even among subcontractors, but such cooperation is typically between companies involved in different stages of the manufacturing process or which manufacture complementary products. For example, greatly reduced innovation cycles and the trend toward a coincident development of parts and tooling in the automobile industry are forcing the tool and die makers to cooperate earlier and more closely with the builders of the machines into which the tools and dies are to be fitted. Customer demands for JIT delivery of complete components are triggering joint ventures among medium-sized subcontractors, who are setting up joint assembly works at the doorsteps of their customers.

A more or less socio-psychological barrier to inter-company cooperation lies in the fact that the companies under examination are small businesses. The owners of small and medium-sized businesses typically place a very high value on independence and autonomy of decision. Accordingly they greatly fear having to surrender something of this entrepreneurial freedom of choice in a cooperation.

The above-mentioned obstacles and stumbling blocks to cooperation on further training between small subcontractors make one thing very clear: Cooperation between small businesses is not going to develop from the companies' own initiative – at least not in the further training sector. Coöperation on further training needs a push from outside. This makes it a marketing job for the further training institutions, but it also needs individual promoters to put it into practice.[7]

FURTHER TRAINING – AN AREA OF COOPERATION BETWEEN SMALL SUPPLIERS AND BIG BUYERS?

In the previous sections it was explained in detail that one of the key factors behind small companies' training needs are the new demands being made by their big customers. It is a known fact that in the automobile and electrical industries in particular, but in mechanical engineering, too, company further training has been well developed and updated continually for many years. These big companies are accordingly equipped with substantial resources of infrastructure and manpower. They also possess appreciable know-how as regards the development and adaptation of suitable training methods and curricula for adults. Seen in these terms, it seems obvious that customers and subcontractors could cooperate intensely on the training of workers to respond to new requirements. This is all the more true considering that at least part of the training needs in subcontracting businesses is owed to the requirements of new technologies that had to – or still have to – be mastered by the manufacturers themselves.[8]

The mere possibility of cooperation between a customer and his subcontractor is not enough, however. The cooperation must also be meaningful to both sides. For companies operating under market conditions this means that the cooperation must be profitable for all concerned. The suppliers' interest in cooperation with customers gets awakened when it proves to be the most favourable way to respond to requirements that have to be met in order to assert their position on the market.[9] Provided cooperation with customers on further training promised lasting help compared with other options, suppliers would probably be prepared to shelve their various doubts, such as whether customers are in a position to provide training that matches the supplier's specific problem, whether training measures will be used to increase supplier dependency on the customer, whether supplier specific know-how – and even employees – might be poached. Some suppliers feel that the simpler solution would be for the customers to peg back their excessive demands and so get rid of the specious qualification problem. It is not clear where the customers interests' stand.[10] They often stress publicly how interested they are in the existence and development of an efficient supply sector and hence in cooperation, and that the price of subcontracted products is by no means their only consideration in dealing with suppliers. But the impression which small suppliers obtain based on their regular negotiations on delivery terms, often looks completely different.

What then is the real basis of the customer/supplier relationship? A subcontractor is attractive provided he can offer better services at the same prices or the same services at lower prices, compared not only with the competition, but also with the customer's own production. In making his calculation, the customer is strongly influenced by such factors as position on the sales market and the degree of utilization of his own capacities. At the moment, most big companies are producing at the limits of their capacities, which is why reduction of in-house production is not only to save costs, but also to gain workers and material capacities for an expansion of production. As such, the situation for subcontractors is particularly favourable. The process can be reversed, however, if the suppliers' customers face a decline in demand. In this case, in-house production can be applied as an instrument to improve a company's own capacity utilization and to avoid expensive, often highly problematic staff reduction measures.

A further point to consider, however, is that just a limited part of the subcontracted production will lend itself to reincorporation. First, the company has to have the right production facilities at its disposal; second, and more importantly, the work-force has to possess the right production expertise, and this becomes a greater and greater problem, the longer the subcontracting has been taking place. On the other hand, experiences of large companies with subcontracting are not always positive, and often there seem to be great differences in the way company departments assess the success in reducing in-house production. What the purchasing department – whose importance increases with a decrease in in-house production (and vice versa) – considers to be a success can be evaluated quite differently by the production departments or by research and development. These departments can then be the source of important alliances with, for instance, a works councillor looking to increase in-house production.

The comparative *advantages* of small subcontractors are generally considered to be: the willingness to implement changes quickly and to satisfy special wishes (here the employee qualification structure plays an important part); a leading edge in production gained through specialization; a transparent organizational structure, with easy access to contact persons in a position to make decisions; and last but not least, their more favourable cost situation (mainly due to lower wages and social benefits).

The *disadvantages* of small subcontractors as compared with bigger suppliers and a customer's own production are often seen to lie in product quality, quality assurance and quality documentation, the use of modern design methods, and incorporation in the customer's computer network.

Big customers' attempts to enforce their quality standards unilaterally by specifying guidelines (their own quality manuals, stipulation of specific general standard requirements) can be viewed as a response to problems they were having with the quality of bought parts (not least those supplied by small businesses). It very soon became clear, however, that simply sending along the documents is usually not enough, so they provided suppliers with 'courses of standard instruction' on the quality standards and procedures of quality assurance. However, these were not adapted to the specific needs of small businesses. A number of big customers have also staged so-called supplier conferences as forums for general discussion and information.

What is not common practice, however, is to provide comprehensive and timely information about impending new requirements, in a form related to the small supplier's specific situation and in a manner he can use, accompanied by a discussion of their implementation. Many technical services and departments for the quality assurance of bought parts are already providing help with the changes needed in production (albeit the help is frequently not preventive, but responds to problems after they have arisen). The help usually stops, however, at technical and organizational counselling, and reaches almost exclusively the top levels of management.[11] The further training of junior managers and production staff would be a field where a form of cooperation attractive for everyone could take place, especially since customers are gradually realizing that a great many roots of the problems are to be found at these levels. It would be easier if there were contact persons who were responsible primarily or exclusively for cooperation (and not for control), which seems to be the trend in the departments for supplier development recently created in a number of companies.

The big companies in the sectors examined have been organizing a great diversity of training measures for their work-forces for quite some time, with increasing involvement of workers in production. It has been, and still is, necessary for the big companies, too, somehow to deal with the new demands deriving from their relationships with suppliers. But despite the existence of considerable potential, the inclusion of suppliers in big company training measures is still in its infancy, in large part because up to now the idea of cooperation on training between subcontractors and customers has hardly been discussed. But probably the main reason is the failure to adequately illuminate the possible benefits for all concerned. Cooperation on training is attractive when it helps to increase the advantages of subcontracted supplies and services for everyone concerned. The supplier's interest results from his deficiencies in qualification; the customer's interest grows when he can count on 'surplus earnings'.[12] This can be achieved in a number of ways: the first possibility is to set cost-covering or profit-making prices for the provision of training services. Hardly anything is won this way in the long term, however, because the subcontractor will ultimately add these expenses to his prices and pass them on to his customers, thus closing the circle. This solution becomes attractive (regardless of the question of how the costs of the training measures are borne) when costs are avoided as the result of cooperating on training. This is the case when quality is improved, e.g. by a decrease in the percentage of parts unfit for assembly. The savings to be achieved in this way are not limited to the value of the faulty parts themselves (which is charged to the supplier), but above all from costs incurred by the refinishing, reassembly or even production stoppages which arise. Even if, through a redistribution of the training costs, the subcontracted product itself becomes more expensive, this outlay is more than compensated for by savings in extra quality. To be at all able to identify and utilize such margins, the customer has to carry out a complete cost calculation that also covers all the indirect costs of quality defects. And to actually be able to pursue this quality- and qualifications-promoting avenue, it is also necessary to inform the sub-contractor about the productivity potential waiting to be tapped by quality improvements, and – in the event of success – to allow him to share in the benefits. The current extent of cooperation on training would tend to suggest, however, that this is not always the case. Still, a number of examples and starting-points do exist.[13]

One automobile manufacturer has 'opened' up special courses on a wide scale to subcontractors. Designed to pass on information concerning a quality (assurance) guide-line drawn up and implemented by the auto manufacturer, the courses focus more on the dissemination of knowledge about quality standards and less on concrete help in creating preconditions for sensibly implementing the quality requirements. So far the course lacks an audience-specific orientation (the same course is held for people from a great assortment of companies with a great divergence of preliminary knowledge). Moreover, the more the course involves small suppliers (especially at the second and third step down the supplier chain), the greater have become the problems of instruction, so that the possibility of a 'preparatory course' is now being considered for these prospective participants. At the moment many suppliers tend to view this training as an obligatory exercise with limited practical value. The organization of the courses takes place under the sole responsibility of the training department. From the customer's other departments such as Purchasing, Quality Assurance, Supplier Development, etc., there is no feedback as to whether and to what extent the training has actually reduced any quality problems, nor any follow-up considerations about an appropriate restructuring of the measures given any problems.

A number of other companies occasionally give subcontractors the opportunity to send staff to training courses intended primarily for their own work-forces. In these cases, it is possible to go into the specific problems of the 'external' participants to a limited degree. Aside from the various instructions concerning quality guidelines, the emphasis is on courses for CNC, CAD, pneumatics and hydraulics, SPC and introductions to the use of data processing equipment for a variety of purposes (mostly PC courses).

The ambitious plans being discussed by a major auto maker would go much further. Training measures are being considered as a worthwhile activity for subcontractors as well as for the customer. The manufacturer realizes that all of its departments affected by supplier relations need to cooperate closely if the project is to have a chance of success. In other words, the provision of training measures for subcontractors has to be integrated in an overall concept of supplier development, so that in addition to the department with direct responsibility, at least the quality assurance department for purchased parts and the actual further training department should participate. The concept has not been fully implemented yet, but in a first step the company has allowed subcontractors to attend courses from its own further training programme that seem appropriate. At the moment it is not possible to judge how well the programme is being received, nor can we tell whether the measures are being adapted to the special needs and qualifications of the participants (which would certainly be sensible), or whether full use is being made of the experience gathered from the subcontractors.

Extensive support services are being provided – also in the area of training – to a specific group of suppliers by another large company. The targeted group are suppliers who have taken over production (with the necessary equipment) in 'traditional', technically developed areas which do not demand new, specialized know-how, and which the customer has discontinued himself but still requires. In this case, however, the reciprocal ties have a quality extending beyond the 'normal', mainly market-oriented supplier/customer relationship and so cannot be considered exemplary or generally desirable.

Another company sees its first and foremost duty as lying in instructing the some key figures in the subcontracting companies in its philosophy of quality, and in getting them to accept the necessity to perform systematic quality assurance. Enormous deficits in this respect are thought to exist among the topmost executives in small supply firms. It is believed that the need for training cannot be acknowledged until quality consciousness has been aroused. Then the supplier is helped to devise his own training measures, and he is also given the opportunity to participate in the customer's measures.

These few examples illustrate that there really are starting points for cooperation on training measures between customers and small suppliers. However, to realize such a cooperation and have it meet either party's advantage, a number of conditions need to be guaranteed. This means, above all – as already suggested – illuminating the potential for savings. For the capacities of big customers to be utilized successfully over the long term by small subcontractors there has to be real cooperation on the structuring and implementation of measures. In other words, the customer must do more than simply make his technical capacities, training concepts and materials available; he must also be willing to modify, extend or simplify his measures in accordance with the needs of his new clientele. This means establishing the adaptation needs of small suppliers and mobilizing suitable expertise. Above all what this requires is that the departments involved exchange experiences with the customer and co-operate. As mentioned above, supplier development is only beginning to establish itself as an independent function. As a rule, there is no direct contact, and virtually no indirect contact, between the customers and the supplier's further training departments. Nor can it be taken for granted that the quality assurance, supplier development, and further training departments get together at the same table. Joint discussions between these departments and the supplier do not take place at all.[14]

Possible steps in this direction are: carrying out a thorough evaluation of the measures already being applied, with particular emphasis on interviewing participants from small subcontracting firms; utilizing the experience of staff from the customers' quality assurance and supplier development departments; and tapping the knowledge of those subcontractor executives who already have experience in carrying out training measures. Of course, representatives from training institutions could also be drawn into the conceptualization of proposed programmes. And last but not least, it is important to take stock of the qualification situation at the subcontractors, i.e. to draw up a record of existing qualifications and to determine the employees' willingness for retraining.

Cooperation between customers and suppliers could be made particularly attractive, if it proved possible to draw together further training associations of subcontractors and big customers, for whom a majority of the associated businesses are at least potential suppliers. This could then unite synergetic effects in both vertical and horizontal dimensions.[15] But, there is still a long way to go to reach this point. Whether and how cooperation on training between small suppliers and big customers will develop depends, on the one hand, on the efforts of those involved. More important still, however, is what will happen to the status of subcontracted services, and this, in turn, is greatly influenced by the development of demand on the customers' product markets.

CONCLUSIONS

Cooperation between companies for the training of workers in small supply firms is certainly a sensible approach with high potential for dealing with new requirements. However, left on its own, cooperation on further training will not broadly establish itself. Cooperative training programmes require continuous management, with duties which extend beyond organization and coordination to the development and continuous 'maintenance' of communication channels and the motivation of the companies involved. Completely new tasks are facing the professional associations and further training institutions.

The activities of the further training institutions and the picture they have of their role must change and become more company oriented. The further training institution must develop into a company qualification consultant, with knowledge of company problems in production and application. Besides offering further training in their own institutions, these further training institutions must become more involved in the principles and methods accompanying learning phases that take place within the company. Where specialized problems of company training go beyond an organization's know-how, it will have to make more use of external lecturers and act as an interface between demand in the companies and supply on the further training market.

Up until now, the professional associations have been less active in the further training field than in vocational training. This is the outcome of limited manpower resources and evolved historical responsibilities, but it also has to do with the prevailing belief that further training should be regulated via the market and so does not require intervention by the management and labour associations. Important functions of the professional associations could be to either initiate further training associations themselves or to give active support to their promoters. Where further training associations already exist, the professional associations must advise the further training organizations in determining the sector-specific requirements for further training and in future developments of relevant production technology.

Cooperation on further training between small businesses is, in principle, possible in all sectors. In the subcontracting sector there is also the possibility of cooperation with big customers. As we have shown, points of approach certainly exist or could be created. But here, too, the initiation of suitable initiatives could greatly profit if suitable promoters were to approach the customers. In other words, the first precondition is discussion with the customer about the idea of improving the quality of subcontracted products by providing training programmes for the suppliers' work-forces. A starting point could be the purchasing departments or the quality assurance divisions, which would then need to coordinate – and this has been the exception up to now – with departments such as supplier development and especially those responsible for further training. For the customer, it would certainly be easier to cooperate with existing and specially created further training associations, simply for the synergetic effects (each has several hundred suppliers), as opposed to forging only bilateral contacts. Here again, the professional associations could act as an intermediary. Unfortunately, there is presently a lack of manpower and material resources in the professional associations and in the further training institutions for these spheres of activity.

If we were to try and estimate the possible effects of cooperation on further training in terms of its effect on the labour market, we could say the following:

1 Without suitable precautions or incentives, there is a risk with further training activities in small companies that unskilled and semi-skilled workers will not be covered at all or only insufficiently, that instruction will be given only in qualifications that are oriented directly to applications and restricted to 'essentials', that these qualifications will turn out to be extremely company- and plant-specific, and that no company or, more importantly, inter-company certificates will be issued.[16]
2 The continuous and comprehensive adaptation and improvement of existing qualifications under prevailing working conditions and with emphasis on the company as the place of learning is absolutely essential in light of emerging requirements, and it cannot be substituted by a (naturally meaningful) 'modernization' of basic vocational training courses nor by training activities isolated from company activity.

3 Inter-company cooperations on further training can help small businesses to pursue the training of their work-forces. Through the broader mobilization of expertise, further training associations afford better chances of avoiding the problems involved in internal company further training activities. Thus the activities undertaken by such organizations can be seen as a contribution to improving the efficiency of the labour market. In this context, to assess the value of different types of measures would require not only an exact analysis of various training components, but also an investigation of their labour market value (i.e. the supply/demand ratio for these workers on the labour market).

In terms of reinforcing desirable effects, or preventing or mitigating problematic developments on the labour market or in educational policy, the question arises as to whether it is possible or necessary to seek government promotion of further training activities. In the Federal Republic of Germany, the structuring and implementation of primary vocational training as part of the dual system is a task performed with major state intervention. This takes the form of specifying standards for training courses (with the participation of labour and employers and of considerable financial support e.g. through the state vocational schools). Further training, on the other hand, is left up to the workers and companies.[17] All the key socio-political groups agree that a comprehensive system of worker further training is imperative for the national economy's competitiveness. Assuming, therefore, that the status of further training will continue to grow in relation to primary vocational training, the question arises as to whether state participation would not be meaningful here, too. The statutory standardization of further training in a form comparable with primary vocational training is evidently not a serious consideration. On the other hand, there have already been political and scientific discussions on the question of controlling further training via public funding, and even attempts at putting it into practice.[18] These attempts have demonstrated that a contribution to the 'modernization of the national economy' can be made by promoting further training measures, and that some meaningful results for labour market policy can be achieved, e.g.:

1 a commensurate participation of otherwise disadvantaged groups (unskilled and semi-skilled workers, women, foreigners, persons with limitations in the jobs they can perform);
2 measures whose content is designed better and which have an adequate time plan;
3 acceptable conditions for participants (payment, claims on time, etc.);
4 better opportunity to obtain – wherever possible – certified qualifications in demand on the labour market; and

5 incentives for making a joint-company response to the training problem instead of solving it by exchanging workers via the labour market.

The promotion of cooperation on further training between small businesses and other small businesses on the one hand, and between small businesses and their big customers on the other, could help to achieve these desirable effects. Possible promoters would be the professional associations and the further training institutions, but also any existing self-organized initiatives of small businesses arranged for other purposes.[19] It is conceivable that such groups would set up adequately equipped experimental models for inter-company cooperation on further training. The success of such measures does not depend only on the extent of the promotion, however, but is conditional above all on precise control, exact monitoring and, in particular, the full involvement of everyone in designing the measures.[20] A further factor of vital importance is a commensurate 'public relations' that publicizes the possible advantages beyond the circle of those immediately involved. If there is a realistic chance of starting up a self-support process, a fixed period of state promotion of company training measures can be well justified on the basis of educational and labour market policy considerations.

22 Small Firms in Big Subcontracting[1]

Klaus Semlinger

INTRODUCTION

During the last few years many large industrial companies have attempted to reduce in-house production in favour of outside supply from independent subcontractors (outsourcing). Cost reductions and gains in flexibility are the arguments for this increasing inter-company division of labour. To what extent, however, there are general gains in flexibility and efficiency in the economy as a whole, or whether the result is only a shift of costs onto weaker and less powerful enterprises, is still an open question. This chapter argues that there is a highly unbalanced distribution of flexibility options and necessities to adapt as well as of chances and risks within the new subcontracting networks. In particular, the majority of the small and medium-sized suppliers and their employees seem to be negatively affected in their future prospects.

Small firms are currently regarded as the new hope for achieving full employment. However, a closer look reveals that – at least in the FRG – the favourable trend in small business employment does not apply to all small firms. Many existing small firms have experienced job losses and the total increase of employment in the small firm sector is primarily based on newly founded businesses in the service sector. Additionally, even given small firm employment increases in the 1970s, large enterprises were able to increase their share of the gross national product during the same period so it can hardly be said that large companies have lost their competitive strength (Bade 1986).

In fact, favourable development opportunities for small firms – particularly in the manufacturing area – are not predicted to come from their direct competition with large enterprises, but in a complementary division of labour. Recent developments in the craft sector, which consists mainly of small-sized firms, are claimed as empirical support for these forecasts: during the past few years, mainly those craft businesses have prospered which act as suppliers to industry or are involved in the distribution, marketing, maintenance and repair of technical consumer durables manufactured by large enterprises (Marahrens 1980; RWI-Handwerksberichte 1988).

In order to 'sail in the wake' of large industrial enterprises, however, craft businesses acting as suppliers to industry are forced to adapt by industrializing their manufacturing structures (see Rühle von Lilienstern 1986: 54). The problem is that the introduction of the technical modernization and adaptation measures demanded by their large customers often leads to conversion problems which endanger the very existence of many small firms (Bullinger 1987). In the following I will go into this issue in more detail by describing the position held by small firms within the supply system of large industrial enterprises, their perspectives, and their risks of adaptation.

THE SMALL SUPPLIER FIRM – A DEFINITION

When we speak about small firms, it is often unclear what the term actually means. In many cases, it is also even unclear which level of organization is being referred to: the local plant, the (multi-plant) firm in a legal sense, or the (multi-firm) company as an economic entity. For the issues being dealt with in this chapter, it is of decisive importance to know whether the small supplier is an independent enterprise or a subsidiary of a larger company – or in the extreme case a subsidiary of the customer itself. In the following, when I refer to a small supplier firm, it is to be understood as a legally and economically independent small enterprise.

How should firm size be measured and exactly how large is a small firm? In many cases, the sales volume and/or the number of employees are used as the criteria for defining company size since these data are the most readily available and the easiest to use. The assumptions underlying the use of these data are that a firm's turnover is an indication of its economic power and that the number of employees reflects its capacity. Obviously, across different industries, using these criteria to draw the demarcation line between small and large firms is problematic. Therefore, it is necessary to add qualitative criteria and even then it is unavoidable to leave room for some ambiguity.

Given the large number of qualitative attributes that vary by firm size, we have to restrict ourselves to a selection of criteria to define small firms. Since this chapter's focus is the effects that a more or less close relationship with one or several large buyer companies has on management and employment policies in small supplier firms, those characteristics of small firms are selected which are of central importance for these policies, and which would tend to be affected by the exchange relationship with large buying companies. Thus, in this chapter small firms are companies:

1 with management and control structures oriented towards the owner-enterpreneur (or a single managing director) with an only weakly developed hierarchy and functional division of management;
2 in which the enterpreneur/directing manager is directly involved in company operations;
3 which do little planning and where operational procedures and rules of decision-making are only slightly formalized;
4 in which the exchange of information occurs in a relatively direct and personal way with a low degree of documentation and standardization;
5 which employ a comparatively large share of skilled workers in an organization of production with a low degree of division of labour and
6 with fewer than 100 employees.

To conclude the terminological preliminaries we have to say something about what we mean by subcontracting or outsourcing. Here too, there is no satisfactory, generally valid definition of the term. Outsourcing is a substitute for in-house production. In theory everything can be supplied by external sources (Männel 1981). This definition, however, encompasses a highly heterogeneous field of subcontractors for which it will be difficult to identify a common pattern of structure and development. Therefore, the concept of subcontracting/outsourcing used in this chapter is limited to the supply of/demand for components and parts which are directly integrated into the buyer's product without much further processing and it includes other goods and services only if they are used exclusively in the manufacture of a given product or in a specific production process. This definition includes work for commission orders (*Lohnfertigung*) as well as the make-to-order production of parts and components and the supply of moulds and other special tools. All services relating to finance, administration, maintenance, and other overhead services are excluded from the definition as well as the supply of basic and raw materials, operating resources and energy, and, finally, all non-product-specific construction services and equipment.

Thus, small supplier firms examined in this chapter are (formally) independent companies with the characteristics outlined above and which are members of an integrated inter-firm manufacturing group (*Produktionsverbund*) of at least one large buying enterprise. Furthermore, this chapter will concentrate on firms that directly supply to large-scale production.

OUTSOURCING AS A SPECIFIC FORM OF DIVISION OF LABOUR

Thanks to learning curve effects and other advantages of specialization, division of labour is associated with productivity gains. This not only holds true for the division of labour on the individual or plant level, but also when it is organized at the inter-plant level, in which case additional advantages, such as those resulting from optimal localization of tasks (*Standortoptimierung*) can be realized. Based on this, the American automotive industry – the dominant industrial model well into the 1950s – had already established a highly differentiated system of specialized manufacturing sites back in the 1920s, whose status were that of subsidiary to the mother company (Feuerbaum 1956: 36). The concept of outsourcing, i.e. the division of labour over and above the individual enterprise level, goes one step further, in that work is not only divided up, but also allocated outside the company. Thus, the restriction or even the reduction of in-house production in favour of outsourcing cannot be explained by the productivity effects achieved by the division of labour alone.

In the same way, the reason why enterprises decide to purchase parts from independent suppliers cannot be explained by additional economies of scale that can be achieved by standardizing components which than can be used for different products. The advantages of mass production for larger markets, i.e. decline of marginal unit cost, a more rapid capital turnover and improved control over capacity utilization could all be achieved by a corresponding in-house differentiation as well. Furthermore, the in-house approach has the advantage of assuring more reliability of supply and more direct, hierarchic-bureaucratic means of control. In fact, vertical integration and diversification have always been, and still are, a preferred way to secure the advantages outlined above.

Meanwhile changes have occurred in these preferences: conglomerates no longer aim for 'stochastic economies of scale', i.e. the mutual compensation of contrasting sales fluctuations, but seek instead to generate synergistic effects and the consolidation of 'strategic competence' (Grabher 1988: 76). Many large-scale producers, in particular, have started with a reduction of in-house production. Obviously, vertical integration and inter-company division of labour are strategic options (Sengenberger, Loveman 1988: 36), and the choice between in-house manufacturing and outsourcing is a management decision which has to take a multitude of factors into account. Before discussing the causes for the increasing importance of outsourcing in greater detail (see pp. 349–51), I will briefly lay out the basic rationale behind outsourcing.

Scanning the criteria cited in managerial literature for deciding the make-or-buy question, and reviewing relevant reports from industrial practice, four main aspects stand out, namely:

1 *Availability*: In terms of size, employee skills, machinery, product and manufacturing knowledge – is the company in a position to cope with the task, or would it be necessary first to build up new and/or additional capacities or competence?
2 *Feasibility*: Does the establishment or re-structuring of the old manufacturing system meet with resource bottlenecks, legal obstacles (e.g. patent or environmental law) or with other problems of implementation (e.g. conflicting departmental interests and opposing works councils)?
3 *Profitability*: Given the potential for in-house production and considering the investment costs which may be necessary to build up an adequate capacity – would outsourcing be more economical or at least of comparative advantage so that a re-allocation of company resources will lead to efficieny gains?
4 *Suitability*: How do outsourcing and in-house production compare in terms of uncertainty (supply bottlenecks, know-how drain, fluctuations in demand) and other more intangible factors (such as company image, goodwill market policies, organizational climate etc.)?

These criteria sometimes overlap, and in some cases they may even contradict each other. In the end, they are weighed according to whether the decision to be made concerns old or new goods and services, or the question of increasing outsourcing or expanding in-house production. The criteria do make it clear, however, that apart from cost advantages, outsourcing offers considerable gains in flexibility. By opting for outsourcing instead of in-house production, fixed costs are turned into variable costs, thereby reducing the risks of capacity utilization and amortization especially in cases of uncertain, irregular, low and/or temporary demand.

THE POSITION OF SMALL FIRMS IN SUPPLIER SELECTION

Companies that wish to profit from outsourcing must select their suppliers carefully and subject their make-or-buy decisions to constant scrutiny. A multitude of criteria and procedures can be used to assess suppliers. The main criteria of assessment which can be singled out in practice are price, or more generally: the value obtained for the money expended, reliable delivery and the quality of the delivered goods (see Arnold 1982: 80). There is no doubt that small firms can cope with these demands as well as large firms. In addition, the terms of transaction can be dictated to small firms more easily because their strategic market position tends to be unfavourable compared to that of large purchasing companies (due to their small market share, with the large enterprise customer accounting for a sizeable share of their turnover) and because they have less room to manoeuvre (due to their limited financial power and their smaller capacity for planning and marketing research). On the other hand, there are at least two important reasons why small firms are less attractive for large companies looking for suppliers.

First, there are the considerable R&D-efforts which are necessary to achieve a decisive lead in terms of product and/or process know-how. Small firms are usually unable to undertake such efforts when they call for high investments in equipment or require systematic, highly differentiated or long-term research (Brose 1984; see Acs, Audretsch 1986). Second, small firms hardly have the capacity to satisfy the quantity requirements of large-scale producing enterprises in the case of more complex parts and components, not to mention the requirements, when an even larger sales market is to be served to take full advantage of all possible scale economies. Providing a supplier with technical assistance, however, to improve its manufacturing competence would detract from at least some of the supposed advantages of specialization which are given as a reason for outsourcing, while dividing up the subcontracted goods between several small suppliers would jeopardize learning curve effects and economies of scale and additionally, would result in higher transaction costs.

Therefore we have to think of direct supply (in the above-defined sense) from small firms as being mainly restricted to the niches of the inter-company manufacturing network of large companies: small firms will be primarily engaged as subcontractors to supply small parts, custom series, special, yet less complex, manufacturing and development services, and to cover peak demands for or to bridge short-term supply gaps of such products. Therefore, although large enterprises have a considerable number of suppliers, among them many small firms, in terms of the value of goods or services supplied, the main share is allotted to only a few large supplier firms.[2]

Thus, it is to be assumed that it is a mixture of rational calculation and pure habit that accounts for small firms playing a role as direct suppliers to large enterprises at all – especially to those who are engaged in large-scale production. In many cases, business relations have only managed to survive because the relationship evolved over a period of many years and has been cemented by personal contacts. Nevertheless, there are two reasons commonly cited for using small firms as subcontractors: their supposed low overheads and their high flexibility.

The lower overheads of small firms result in part from their lower administrative costs, but a considerable part is also due to their lower wage and non-wage labour costs. Lower (non-) wage labour costs, however, do not derive from higher productivity, but are the result of small firms' less comprehensive social services and fringe benefits and the, on average, inferior health and social security benefits given to small firms' employees (see Giebel 1985). Empirical investigations have shown that large enterprises organized as a multi-(small)-plant company can achieve a higher degree of technical efficiency than small single plant enterprises, yet are often unable to convert this higher efficiency into greater profitability due to their higher wage and non-wage labour costs (Aiginger, Tichy 1982: 70). Accordingly, when large companies place orders with small supplier firms on cost saving reasons, they are not only doing so – either explicitly or implicitly – to benefit from the organizational or technological efficiency of these firms, but to profit from the fact that the employees in these companies only modestly participate in the yields generated by their firm's efficiency.

Likewise small firms' flexibility cannot only be seen as a productive feature of their specific organizational structure: there is a big difference between flexibility in the form of 'active versatility' (*Beweglichkeit*) on the one hand, and 'passive pliability' (*Beugsamkeit*) on the other (see Semlinger 1988). The first type of flexibility is based on a high capacity for reaction and adaptation, resulting from efficient information processing, quick decision-making, and highly skilled employees capable of handling a wide range of tasks. With this type of flexibility, it is possible to quickly recognize and exploit small and temporary market niches and to respond at short notice to rush orders. In comparison, the second type of flexibility is characterized by the ability and readiness to submit to outside pressure, and to accept long-term risks and/or to cutback existing company goals. Considering the distribution of market power between large enterprises and their small supplier firms and the types of functions fulfiled by the small supplier, it is not difficult to imagine that a good deal of their attractiveness results from their greater readiness (passively) to adapt to the demands of their large-enterprise customers, particularly with regard to fluctuating capacity utilization and price concessions.

This is not to maintain that all small supplier firms are inherently in a precarious situation of dependency. For example, it is possible to cite a number of small, specialized firms which have succeeded, at least in the short term, in becoming indispensable to their customers. However, they are, in fact, the exception to the rule, as interviews with industrial associations and major German automobile manufacturers have confirmed. Companies of this kind might be held up as the heroes of modern 'small firm lore', but they can hardly serve as a general model. Likewise it can also be assumed that the 'sweat shops', which are situated at the other end of the spectrum, cannot be expected to play much of a role as parts suppliers to enterprises engaged in large-scale production, since the competitive advantage of these firms with their outdated manufacturing facilities, smaller proportion of skilled employees and comparatively large share of temporary workers is mainly restricted to their highly flexible capacity which can be almost arbitrarily expanded or diminished.

The majority of the small firms involved in direct parts supply and product-specific toolmaking are located somewhere between these extremes. Their personnel and machinery standards qualify them for orders which are more demanding in terms of quality and adherence to scheduled delivery dates. The functions they fulfil, however, could be relatively easily replaced by in-house manufacturing or by other suppliers, explaining why purchasing companies exert considerable control over prices. When customers request short-term delivery or changes in product specifications, such firms achieve the flexibility demanded of them on the basis of their organizational structures (efficient communication and fast decision making), as well as their skilled work-force. Thus, their position within the inter-company manufacturing group is not only based on advantages in terms of overhead, but also on company attributes which – with reference to the literature (see pp. 343–4) – we hold as the typical characteristics of small firms. The following section will deal with the changing demands and requirements in the subcontracting markets that are likely to have a considerable effect on these attributes.

MODERN PURCHASING POLICIES AND THE PERSPECTIVES FOR SMALL SUPPLIER FIRMS

Outsourcing as a strategic marketing and rationalization device

The manufacturers of industrial mass products find themselves – given the present distribution of purchasing power on the consumer markets – confronted with both market saturation trends and considerable price pressure exerted by new (international) competitors. They react to these challenges with product differentiation and further product development (increasing product diversity and shorter product life cycles) in order to stimulate and satisfy new markets and evade price competition. At the same time, they step up their rationalization efforts. Now that advanced automation technology in many cases has reduced direct processing time to the near minimum attainable – at least close to an economic optimum – enterprises are turning their attention to the concept of 'systemic rationalization' (Altmann *et al.* 1986). Systemic rationalization seeks to reduce processing stages by means of more manufacturing-oriented product planning and increased utilization of microelectronic technology, while striving to cut back administrative coordination work, idle time in manufacturing, as well as buffer stocks and thus the corresponding capital costs, with the help of modern logistic concepts and increased utilization of electronic data processing.

Thus, while market forces and marketing strategies compel enterprises to achieve greater product differentiation and higher product quality, there is also a call for a better fitting integration of the individual manufacturing steps due to the acceleration of throughput times and the reduction of buffer stocks induced by the new rationalization strategies. In other words, increasing complexity in operational processes has to be matched with greater demands on reliability. This does not leave the inter-company division of labour untouched. On the contrary, reducing in-house manufacturing and making increased use of outsourcing are important devices to cope with these contradicting demands. This means that greater demands will be made of suppliers; the criteria of supplier selection increases in number and pretension: flexible supply does not only have to be carried out within shorter time frames, but also has to be concerned with greater variations in ordered volumes and design of parts and components. The deviations now tolerated in terms of delivery dates, amounts delivered and product quality gets narrower. Finally, the accompanying services now expected include more R&D contributions, better financing arrangements (e.g. short-term credit), intensified quality monitoring and documentation, increased stockpiling and improved communication facilities.

Modern purchasing policies are heading toward an intensification of inter-company cooperation referred to as 'integration of suppliers' (Gelder 1986: 35) and 'dissolving the boundaries of the enterprise' (Becker, Weber 1986: 37). In contrast to the usual processes of vertical integration, however, there is no economic and legal merger of the firms involved. With the help of new information and communication technologies, large enterprises are using the outsourcing strategy to establish a supply system that combines the reliability of vertical integration with the advantages of a market exchange relation (i.e. market pressure on price and performance) thereby minimizing the purchasing company's need to make risky investments and the danger of having to commit itself to a particular course of action. Thus, powerful purchasing companies are able to shift – at least partially – the burden of coping with the increasingly complex and dynamic market and production requirements onto the shoulders of their downstream suppliers. In doing so, they succeed in maintaining or creating greater room for their own flexibility while passing flexibility pressures and the associated risks to their suppliers.

At present, the extent to which the cost savings and the new range of flexibility that large purchasing companies achieve by expanding outsourcing at the cost of their suppliers is an open question. Undoubtedly, there are real productivity-related advantages of an inter-firm division of labour, and the active flexibility potential of small firms does exist. In view of this, it might be that the ongoing restructuring of intercompany manufacturing groups will, in fact, result in real flexibility and productivity gains or in decreasing flexibility and production costs, respectively.

Due to the fact, however, that these inter-company manufacturing groups do function under the governance of market mechanisms – contrary to intra-company hierarchical structures – the purchasing company can instrumentalize market competition to put pressure on the autonomous supplier without any consideration of the effects on the supplier's working conditions, social structure, profitability or market position. Accordingly, whether the ongoing restructuring will be a zero sum game in which large companies take their profit at the cost of their (small) suppliers, or whether a real surplus is made depends on the market position of the parties involved. Only in a balanced market situation or when both parties are of similar strength will the higher flexibility afforded by an inter-company division of labour coordinated by market mechanisms be entirely based on a voluntary utilization of each party's potential for 'active flexibility'. In case of unbalanced power, when the weaker party does not have the option of entertaining other business relations (access to other markets barred, high investment costs, etc.) it often will be forced to adapt by mobilizing its capacity for 'passive flexibility'.

The very goal of strategic buying, however, is geared toward shaping the relevant supplier market into a buyer's market in order to control the exchange relation. After all, it is the competition among suppliers on the open market controlled by the demand side (the purchasing company) that makes outsourcing so attractive in comparison to vertical integration. Inter-company systemic rationalization requires strategic buying – at least to a certain extent; in any case it is a device that maximizes the flexibility and profitability gains obtainable by systemic rationalization. In view of these facts, the much-acclaimed 'partnership' in the subcontracting relationship often ends up being rather one-sided, even if few will go so far as a purchasing manager of a large German enterprise, who stated that 'we must aggressively and resolutely pursue the goal of transferring the suppliers' profits to us' (Hartling 1986: 42).

The need for suppliers to adapt

The interests pursued by purchasing companies described above are by no means new, and endeavours to this end have always existed in one form or another. However, the increasing internationalization of supply markets and modern information and communication technologies have provided purchasing companies with new means of realizing these objectives. Therefore, it comes as no surprise that in the last few years there have been increasing complaints by suppliers about large enterprises' attempts to dictate prices, and quite often, business and trade associations have to mediate between suppliers and large-scale purchasers on disputes over conditions. The main points of conflict are:

1 demands to gain insight into the technical and economic conditions at the supplier firm;
2 reliability and completeness of order specifications;
3 use and observance of commercial rights such as patents, licences, and the protection of company know-how;
4 price setting and price adjustment clauses; and
5 liability and warranty issues.

This last point again touches on the distribution of risk; since this is a central aspect of intensified purchaser–supplier relations, we should take a closer look at it. Concentrating on those changes which 'naturally' emerge from present trends in outsourcing, a transferral of sales and manufacturing risks to the suppliers' shoulders should be mentioned first: with the shift from contracts specifying quantity to contracts only setting a quota of a more or less unspecified demand volume (some of which run for several years), the purchasing companies can restrict their purchasing obligations to their actual sales prospects. While short-term quantity-based contracts did not provide the supplier firms with the certainty of receiving a follow-up contract – which could be regarded as a normal market risk which could be counteracted through appropriate sales efforts – the new type of quota contract ties up the supplier's manufacturing capacity, without the security of knowing whether it will be utilized as planned.

Additionally, increased control and monitoring risks emerge which result from the new logistic concepts pursued by large purchasing companies which pressure the supplier companies into shorter supply intervals, smaller batches of supplied parts and zero defect delivery. Particular problems arise (a) when purchasers demand parts to be delivered within periods shorter than their actual throughput times at the supply firm; (b) when batch sizes are called for which are smaller than the batch sizes optimal for manufacturing; and (c) when the purchasing company requests delivery to be synchronous with its own manufacturing processes while dispensing with the inspection of incoming goods.

Finally, the reduction of the purchaser's own in-house production is resulting in more than just the transferral of direct production tasks. By demanding that their suppliers fulfil an increasing amount of research and development, purchasing companies also place part of the onus for constant product improvement onto the shoulders of their suppliers. Of course, this can potentially open up new market prospects for supplier firms which possess the relevant know-how. However, it also means an increase in risk: the need to contribute towards product development by employing new technologies (such as microelectronics), new materials (such as plastics) and more complicated manufacturing methods (as in surface treatment), within increasingly shorter periods not only demands considerable R&D-investment, but also means the adoption of the accompanying R&D risks.

In order to deal even partially with all of these additional risks, ensure more precise control of their own manufacturing processes, and achieve the closer communication required by the purchasing companies, supply firms will have to:

1 increase the degree of automation in their own production;
2 utilize state of-the-art planning and (quality) control methods; and
3 implement computer-based communication and information processing systems (see GEPRO 1985).

These requirements are so closely inter-related that they cannot be fulfiled in isolation from each other. Moreover, all of them are linked with considerable investments in equipment and facilities as well as personnel skills, and in some cases also necessitate considerable internal company restructuring. Therefore, the willingness to adapt is not sufficient; it must be accompanied by a corresponding capacity to adapt. This cannot be expected from all suppliers to the same extent. Thus, the intensification of inter-company division of labour entails very different prospects for different kinds of supplier firms.

Effects of adaptation on small supplier firms

In view of the financial and organizational efforts required for achieving a tighter inter-company logistic network of production, it is not surprising that until now these efforts have been mainly concerned with supplied parts which are of particular significance both in terms of value and/or volume, and have focused on suppliers with whom the purchasing companies maintain a continuous and significant business relationship. Even in the future, it can be expected that the more ambitious integration concepts which dominate the current discussion (such as 'just-in-time' delivery and on-line integration of company data systems) will remain limited to comparatively few suppliers. Yet even then, the other suppliers will not be unaffected by these measures. In fact, it is quite conceivable that integrated suppliers will take over supply tasks formerly fulfiled by other, less integrated supplier firms, as is already reflected in the observable practice of cultivating only one supplier for a particular part or part family (single sourcing), and/or the attempt to move increasingly toward the purchase of pre-assembled components.

All in all, large supplier firms are most likely to be the ones to profit from this development, not only because it is easier for them to make the necessary investments, but also because single sourcing for large series manufacturing leads to an increased minimum of production and supply capacity. Additionally, there is an indirect reason for the trend toward using large supplier firms: the new integration technology minimizes the supply risk involved in outsourcing even with 'critical suppliers', i.e. suppliers whose economic existence does not depend on the particular supply relationship because new communication systems make this risk more calculable. Therefore, the extent to which control and influence need to be secured through economic dependence of the supplier is reduced. This meets with the rationale of increased outsourcing (see pp. 349–51), according to which a high degree of suppliers' economic autonomy is regarded as being desirable. In fact, a number of well-known, large companies have stated that they would like their purchasing volume to amount to no more than 20–30 per cent of a given supplier's turnover, in order to avoid assuming direct or indirect responsibility for the supplier's capacity utilization.

Thus, the outlook is an increasing process of concentration in the direct supply to large purchasing enterprises which will leave no room for small supplier firms of the traditional kind. At the same time it seems highly questionable whether small firms will be able to retreat to the second rank or to positions further downstream without undergoing the processes of adaptation outlined above: if all 'first rank' suppliers have to comply with the demands their purchasing companies make on them in terms of more flexibility and greater reliability, they will certainly want either to increase in-house production of tasks formerly subcontracted to downstream suppliers to make their position more secure or, as far as possible, to pursue an outsourcing strategy analogous to that of their large customers. Thus, while it is unquestionably demanding to adapt, it is not necessarily an advantage for weaker supplier firms not to be directly included in the current integration strategies of the large purchasing enterprises: firms standing at the sidelines may find themselves out of the supply business altogether in a very short period of time.

However, as soon as small supplier firms undertake efforts to adapt, they are quickly confronted with financial obstacles, and they must also solve considerable organizational and structural problems. Whereas medium-sized firms can deal with these challenges within the framework of a steady organizational development, small firms will be forced to undertake profound shifts in their internal structures which might result in a completely changed organizational character. What ends up being affected are precisely those aspects of their organization which – apart from low wage costs and overheads and their specific potential for 'passive flexibility' – are regarded as the strengths of small firms and which establish their position as suppliers (see pp. 346–9), namely, their less formalized information, decision-making and planning systems; the more personal nature of their internal communication and social relations; a comparatively low degree of division of labour and the considerable emphasis which is placed on skilled workers with professional competence and vocational experience.

The implementation of modern planning, control and (quality) monitoring methods means a stronger formalization and integration of the respective processes. This will threaten to restrict the level of autonomy, discretion and responsibility of the individual worker, and is likely generally to limit the potential for making spontaneous and improvised changes in production as well, i.e. it might negatively affect central attributes of small firms' 'active flexibility'.

Additionally, a more formalized and automated production process as well as the expectation of faster reactions to purchasers' requests for supply deliveries or order modifications will affect the internal communication structures. The danger is an increasing anonymity in company-internal communication between management and workers as well as in the exchange of information on the operational level. The personal character of intra-company coordination will be superseded by procedures which are determined by superficially objective constraints and requirements due to external demands. This, in turn, will also affect industrial relations within small firms (see Mendius *et al.* 1987: 180): a shift towards more formalized or externally controlled production processes, and the resultant restrictions to autonomy and informal arrangements will most likely meet with resistance from the workers. Although this may not necessarily lead to higher exit or recruitment problems, it could result in a 'compensatory' loss of in-house cooperation, and thus, undermine another essential element of small firms' 'active flexibility'.

Finally, all of the above-mentioned factors are likely to affect the skills profile of the personnel employed in small firms. The utilization of state-of-the-art manufacturing and control technologies requires additional skills and qualifications which on the one hand extend beyond the traditional framework of skilled worker training requirements and which on the other hand could have a more company-specific character. This not only raises the problem of how these additional skills can be imparted within small firms, but also the question of transferability of such skills to other firms. Currently it is skilled workers with qualifications which are acquired according to uniform guidelines and which are therefore relatively easy to assess and utilize in different company contexts, that form an essential precondition for the capacity of small firms to adjust rapidly to changing personnel requirements via the external labour market. At the same time the non-firm-specific character of skills helps prevent the loss of a particular job leading to prolonged unemployment of the worker affected. Thus, if small firms were to lose their contact to the skilled labour market due to an increasing importance of firm-specific qualifications, it would not only jeopardize employment security for employees, but also undermine another important basis of small firms' 'active flexibility' (see Sengenberger 1987).

To sum up, the intensification of the inter-company division of labour induced by large enterprises confronts small supplier firms with the question whether they will be included in the process at all and, if so, whether their financial and organizational means will allow them to carry out the necessary adaptation measures. Additionally, if they adapt to the demands with short-sighted organizational changes oriented to the established patterns of large industrial enterprises, they endanger their special potential for active flexibility. All in all, it is still unclear, whether small firms will be able to compensate for this loss without achieving the ability possessed by large enterprises for long-term planning and larger manufacturing volume, i.e. without growing. In any case, it seems certain that small firms that want to have a chance in direct supply to large-scale producers, still might be called 'small' as far as their number of employees is concerned; in terms of their qualitative attributes, however, they will no longer have anything to do with the traditional understanding of a small firm.

SUMMARY AND CONCLUDING REMARKS

In view of developments to date and the inner logic of outsourcing within large inter-company manufacturing groups, it is questionable whether small firms will succeed in holding their position or will have to resort to positions downstream in the supply chain. It can be assumed, however, that the increased performance demands of large purchasing companies will not remain restricted to the first rank of suppliers and that their extension is likely to close rather than open the door in terms of the amount of parts and services ordered from small supplier firms. Therefore, in future, we are likely to be witnessing an even stronger concentration of supply in fewer companies – possibly organized in small units – with these enterprises seeking to use their new performance potential to get a competitive edge on other supply markets where somewhat lower standards of efficiency, flexibility and reliability are (still) valid.

Similar developments are observable with regard to supply markets not included in this analysis, i.e. those for capital goods and overhead services. Thus, the pressure which large-scale production enterprises are exerting on their first-rank parts suppliers is not only cascading down the supply chain, but it might also spread out to the entire supply market, without any indication of where and how this process will end and how great the pressure on companies occupying different positions along the supply chain will be. Already today, this process is affecting a number of small supplier firms – companies which, as of now, are not directly part of any ambitious integration schemes – in such a way that they are forced to mobilize their potential for passive flexibility, i.e. to pursue or accept an intensification of work, a worsening of working conditions, a squeeze on their profits and the wasting away of their working capital (see Deiß 1989).

The problem is that even when small supplier firms try to adjust productively to the increased requirements of the subcontracting market this may jeopardize their long-term market position: this is to be feared when adjustment and adaptation measures do not give due consideration to the specific advantages and conditions that small firms possess, and instead follow the patterns set down by large enterprises. Technologies and methods that enhance flexibility in larger enterprises can have a detrimental effect when implemented in small firms. Nevertheless, hardware and software suppliers seem to be concentrating on optimizing integration, without regard to the long-term effects of their technology on personnel policies and company perspectives in small firms.

It is to be feared that small firms that do have the chance to be incorporated in inter-company manufacturing systems and carry out the adaptation measures demanded of them will be confronted with increasing risks and demands on flexibility, while at the same time they will lose their traditional potential for 'active flexibility' (in terms of skills, qualifications, organization, etc.), without any sufficient (technological, economic etc.) compensation. Thus, from a long-term perspective, a short-term successful adaptation could prove a rather superficial achievement. In fact, it could turn out to be the nucleus of a grave threat to the small company's very existence.

All in all, the dangers which the expansion of outsourcing entails for small firms would seem to outweigh its opportunities. Yet, it is not quite clear that an exclusion of traditional small firms from the inter-company division of labour would be in the interest of the large purchasing enterprises. Rather, it is quite probable that burying the small firm's potential for 'active flexibility' would mean a loss in flexibility for the entire system.

This risk should encourage the parties involved to seek cooperative strategies for adapting to changing demands and for securing this flexibility that goes beyond short-term interests.

Part VII

New Technologies and Industrial Relations

23 Unions' Policies Towards New Technologies in the 1980s – An Example from the Metal Industry[1]

Norbert Altmann

INTRODUCTION

The thesis that the development of technology and organization – and therefore the utilization of manpower – can be directed, shaped and designed may be well known. In the everyday practice of instituting rationalization measures, however, this concept is hardly perceptible. Instead, technological determinism prevails among engineers, management, and even within unions and works councils. Concepts dealing with controlling and directing technological and organizational developments are only gradually gaining ground and still remain primarily influenced by interests of capital and manpower utilization. However, the shape and form of tomorrow's working life will be determined by the way in which concepts of rationalization and a humane design of work can be attuned to each other. Thus the conflict of interest between employers and employees and their representatives will be a central theme of the future.

In tackling this issue, the following chapter limits itself to the sector of manufacturing, mainly that of mass production within the metal-processing industry (including the electrical, machine-building, precision-mechanics and optics industries). However, the findings have a broader significance since much of the experience gained there is also valid in other areas. Taking company rationalization concepts and union design postulates as of 1988, the aim is to show certain tendencies which are conceivable as future developments in the FRG. Definite prognoses are not possible and not intended. Also, the new problems arising from unification with East Germany are not dealt with here.

RATIONALIZATION AND HUMANIZATION

Developments up to the end of the 1970s[2]

Up until the beginning of the 1960s, economic growth in the Federal Republic of Germany was based on an expanding industrial sector and a growing supply of labour (largely from the continuing stream of refugees from East Germany). Since then, however, growth has only been possible by intensifying production for several reasons. First, the labour market reserves were depleted and second, the overall volume of working hours was decreasing due to the age structure of the work-force, reduction in working hours, and extended education and training periods. One solution was to integrate workers with no industrial experience, particularly women and foreign workers, into the work process. In the 1960s, this increased the pressure towards job simplification and work organization with a high degree of division of labour, particularly in mass production. In manufacturing sectors where the average demands on skills were higher, in the machine-building industry for example, a strategy of polarized utilization of skilled workers on the one hand, and unskilled and semi-skilled workers on the other was pursued in order to overcome manpower shortages.

Tayloristic principles of work organization continued to spread, entailing short-cycle, monotonous work with unvarying work tasks performed at a fast pace. At the same time, demands made on performance were intensified, based on refined methods of time and motion studies (MTM), as well as performance-related payment schemes (incentive wages). Dequalification processes and processes of polarization became widespread, even in areas where skilled workers were employed.

During the post-war years, in the 1950s and 1960s, compensation of increased work loads by rising real wages met with employee acceptance. However, the transition to the 1970s, a boom period with extreme manpower shortages, saw a change in expectations and behaviour of industrial workers, skilled as well as unskilled. Tayloristically organized jobs became unattractive, particularly for German workers.

This change brought with it an increased awareness of physical work strains and mental stress. In many instances, long periods of overwork resulted in reduced working ability and in an increase in latent and manifest forms of performance refusal: high rates of absenteeism (rates of 20 per cent were no rarity in assembly processes of the electrical industry), high labour turnover (up to 30 per cent), rising reject rates and quality defects. The wildcat strikes of 1969 and 1973 were also manifestations of a rejection of extended demands on performance (Schumann *et al.* 1971).

In the first half of the 1970s, companies were faced with the question of whether Tayloristic work organization was still rational and efficient. The reasons for this, however, were not deficient employee performance or resistance, but the altered situation on the world market. Demands on flexibility and quality increased, and could not be met by rigidly organized Tayloristic work forms. Manpower with a limited range of skills could not be utilized flexibly, nor was manpower of this kind oriented to product quality. The companies' problems were aggravated by negative feedback effects such as absenteeism, high labour turnover and increased difficulties in recruiting skilled personnel for job setting, quality control etc. In the mid-1970s these problems were discussed under the keyword 'the crisis of Taylorism'.

During the phase from the beginning to the middle of the 1970s, several scientific concepts of work organization offering alternatives to Taylorism by showing ways out of rigid forms of work organization, unattractive work, and subjective loss of motivation gained considerable significance. Representative of such concepts are the well-known forms of work structuring and semi-autonomous work which have been developed in particular by the socio-technical approach of the Tavistock school: the optimization of technological and organizational design through the integration of social factors. This concept gained considerable influence in the Federal Republic through its application and extension in the large-scale Scandinavian experiments of 'industrial democracy'. The approach emerging from these experiments emphasized a high degree of autonomy for individual workers and work groups, a considerable amount of worker control over their own work and the involvement of employee representatives in the design of technology and organization. German industry, particularly the areas of mass production in the electrical industry, fine mechanics, and optics industry etc. began to take an active interest in these issues.

The programme for 'Humanization of Working Life' (*Bundesminister für Arbeit und Sozialordnung* 1974) initiated by the social democratic-liberal government in 1974 played a crucial role in the discussion concerning a reduction of Tayloristic work forms and in generally bringing the topic of improvement of work design to public attention. The programme's aims and objectives were:

1 the establishment of a scientific basis for a humane and productive design of work;
2 the development of models for a humane design of technology and organization; and
3 the dissemination of pertinent knowledge in the companies.

The programme was started primarily with the objective of gathering knowledge relevant to the practical application of laws which were passed at the beginning of the 1970s, the Work Constitution Act of 1972 among others. The law prescribes a ' humane' design of work in accordance with scientific findings (§ 90, 91). Other factors were certainly indirectly responsible, such as the previously mentioned problems of Tayloristic forms of industrial work organization, social democratic-liberal concepts for a modernization of the economic system, and strong union pressure for better working conditions during the economic boom period.

This programme was by no means merely concerned with the narrow aspects of ergonomic or organizational work design or on job safety and health-related factors, but also aimed at 'establishing the basic values and rights belonging to a democratic, social, and constitutional state within the world of work' (yearly report of the Project Management Humanization of Working Life 1976).

The research and development work carried out within the framework of this programme had been financed with approximately 1.4 billion DM from 1974 through 1989. A considerable amount of knowledge concerning the possibilities of humane work design was gained as well as a great deal about the limits imposed by the companies' performance policies. In 1989 the Humanization of Work programme was reformulated, and as of 1990 transformed into a research and promotion programme under the name 'Work and Technology'. Now, the accent is more strongly placed on mastering innovation in the production process, including that of inter-company integration, and on supporting small and middle-sized enterprises. Support, however, as before, comprises the humanization aspect with a particular accent on work and health protection and further training. Also, the issue of the design of work and technology remains a central focus. The yearly budget for this promotion programme is (from 1990) about 100 million DM.

From the outset, the employers underlined the necessity of connecting humanization and efficiency, and it was, in fact, the aspect of efficiency which prompted the reduction of Tayloristic work during the beginning of the 1970s when new demands were made on flexibility and quality. For the employers, improvement of working conditions goes along with the adaptation of technical progress to market requirements, in which the establishment of new co-determination rights are not naturally included.

From the employers' point of view, it was the effects that flexible work organization, such as job-enrichment or semi-autonomous groups, had on employee performance that were paramount. With such work forms, a wider level of skills could be attained, or skills previously blocked by a strict division of labour could be used more efficiently, thereby enabling a more flexible utilization of manpower. This, in turn, allowed staff reductions and savings due to more intensive and efficient utilization of personnel, as in the case of being able to transfer personnel to fill vacancies. Tighter staffing could also serve the purpose of gradually selecting certain workers and thereby raise the average skill level within the company. The humanization effect achieved by more meaningful work, a greater scope for decision-making and the general design of work was more a side effect of efficiency measures than anything else.

Originally the unions regarded the government programme as being just a continuation of their traditional struggle for better working conditions, co-determination and protection against rationalization. Initially the purely ergonomic and physiological aspects – such as acoustic insulation, dust control and others – played an important role in controlling the observance of legal regulations. By participating in this programme, the unions were not seeking a policy of reconciliation and harmonization of opposing interests within the wider framework of humanization of work, nor did they initially regard it as an instrument towards attaining a stronger control over the design of work within the companies. Some time later, however, the research findings and design activities organized by the programme became more and more oriented to effects on individual and collective performance at work. Then the unions became increasingly sceptical about the German version of the socio-technical approach to work design.

Since the beginning of the 1970s, the unions' approach towards technological and organizational 'progress' and their concepts for collective agreements and regulations had undergone a fundamental change: technological and organizational progress was no longer regarded as a 'natural' development following its own inherent laws, but as an area of political activity. The main objective was no longer protection against the effects of technological development ('protection against rationalization') or wage compensation for negative effects of rationalization (extra pay for work burdens or general wage increases). Instead unions wanted to influence and shape technological developments and thus the goal to influence work design in general became decisive. (Naturally the former objectives of protection, for instance, dismissals, still remained part of union policy.) The quantitative, i.e. wage-oriented, collective bargaining policy began to turn gradually towards a qualitative one and has been so ever since. According to the new union policies, the design of work should take all of man's existential needs into consideration, while technology should fulfil the requirements of 'social compatibility'.

The 'Collective Agreement on Wage Conditions II' in the metal-processing industry in North Württemberg/North Baden, which received much attention when it was formulated in 1973, had two main objectives: first, a precise monitoring of the demands made on performance (by fixing the norms for the determination of standard times in performance-related pay systems). In this respect, traditional orientations were adhered to, namely protecting the employees from the negative effects of a seemingly 'unalterable' form of job design (IG Metall 1977).

Second, however, the agreement laid down work pauses, certain criteria of humane work design and the minimum duration of cycles in assembly line work. This was the first example of direct intervention in the employers' work design rights which exceeded the traditional protective regulations concerning work and health. It should be pointed out, however, that actual enforcement of the agreement on the company level left much to be desired.

Conditions in the 1980s

During the transition from the 1970s to the 1980s three new conditions appeared which had an influence on the question of shaping technology, organization and humane work design:

1 The economic crisis, which set in around the end of the 1970s, and of course the development of new, computer-based technologies brought the impact on employment to the forefront. Employment protection, or at least a cushioning of dismissal effects by social plans and other similar measures, government employment policies, i.e. employment programmes and above all reduction in working hours became the focal points of union interest (Sengenberger 1984).

2 The transition from a social democratic-liberal government to a conservative-liberal government in 1983 resulted in government measures which amounted to a de-regulation of the existing employee status in order to create incentives for employment. Toward this end, employment conditions and contracts were made, as the key words indicate, to be more 'flexible' and 'individualized'. This was to be achieved particularly by the Employment Promotion Law of 1985, which facilitated the utilization of temporary workers and part-time workers, suspended small and new enterprises from employment security regulations and set higher requirements for the realization of social plans (Rosenfelder 1985).The unions were confronted with the problem of protecting the uniform employee status and the effectiveness of general social-political settlements.

3 The rationalization wave based on computer-aided so-called 'new technologies' (from CAD/CAM to FMS, particularly the integration of production planning, control and the manufacturing process itself (CIM) etc.) in German industry was a crucial factor (for a more detailed discussion of this issue see Parts II and III of this volume; IG Metall 1983).

Basic concepts of rationalization at the beginning of the 1980s

What consequences does the wave of rationalization occurring since the beginning of the 1980s on the basis of computer-aided technologies have for a humane design of work? (See e.g. Altmann *et al.* 1982a; Jürgens, Naschold 1984; Kern, Schumann 1984; Baethge, Oberbeck 1986; Schultz-Wild *et al.* 1986.) The discussions in the Federal Republic among managers point to two fundamentally different concepts of development. On the one hand, there is the extensively computer-integrated manufacturing process (CIM) up to the unmanned automatic factory; on the other, there are decentrally organized work processes, which are under the autonomous control of highly qualified workers with a homogenous level of skills who have access to central production process data by means of interactive communication systems.

The design objectives of centralistic automation are: minimal staffing, elimination of employee intervention, total irrelevance of working environment (since so few workers are necessary), elimination of all tasks requiring skills and qualification and the concentration of the latter in the development, planning and control departments outside the area of production. While this design concept certainly appears utopic, it shows a line of technological and organizational thinking which is a product of technocratic engineering science and which is sometimes called 'technocentric' (Brödner 1985).

In contrast to this 'technocentric' approach, there are numerous investigations pointing toward a different course of development which could be called manpower-oriented or 'anthropocentric'. According to this concept, manpower is regarded as an important potential for productivity because implementation processes encounter a considerable number of problems and automatic manufacturing processes do not run at optimal levels. A broad use of employee skills and qualification is recommended in order to avoid inefficiencies in the work process (see Part III in this volume for a discussion of CIM and organization of work).

While present research does not allow any definite prognoses, there are a number of signs pointing to a predominance of technocratic and centralistic concepts being undertaken; this fact, however, is overshadowed by the broad discussions of so-called 'new management concepts' (Kern, Schuman 1984) whose realization in practice is doubtful.

On the basis of data-technological integration a 'new type of systemic rationalization' is establishing itself (see chapter 4 of this volume for a description of this concept). This type of rationalization introduces completely new problems which were not considered at all in the research programmes or the policies of the unions in the 1980s.

THE DISPUTE BETWEEN EMPLOYERS AND UNIONS ON WORK DESIGN

This section deals with the far-reaching conflicts and differences between unions and employers concerning problems of work arising in the course of the introduction of new technologies (the metal industry and the metal workers unions – IG Metall remain the focus). The particular points to be addressed are:

1 new patterns of stress;
2 work content, scope for self-regulation and skills;
3 wages and work performance;
4 self-determination and control;
5 flexibilization of working hours and employee status;
6 co-determination at the workplace; and
7 employment protection.[3]

Problems of work design[4]

Work design and health risks

In the FRG it is generally assumed that the utilization of new technologies will cause a shift from – decreasing, but still existing – physical work burdens to mental and social stress. Innumerable aspects of work design have been subjected to regulation including ones initiated by the unions at the parliamentary level. Many of them, however, are becoming obsolete due to this shift in patterns of stress. This will make the works councils' task of monitoring and implementing these rules at the company level more difficult.

The German Works Constitution Act from 1972 grants the works councils the right to extensive information prior to rationalization measures, but no right of co-determination (§§ 80, 90). The latter only comes into effect (§ 91) if the works council can prove that the form of work design being introduced is not in accordance 'with demonstrated scientific findings of a humane design of work', the design of work 'clearly' contradicts these findings and that the worker is negatively effected 'to a notable extent'. It is difficult to furnish this proof, and thus worker representatives have few chances to intervene in the design of work. When new technologies are introduced, health risks may be closely connected to work organization, as in the case of stress caused by isolation on the job or number of hours working at video work stations with radiation. However, in these cases it is almost impossible to meet the requirements of the law for intervening in the design of work.

The disputes over the introduction of video terminals are an example. Because of the anticipated, and in part scientifically proven, health risks due to radiation exposure at video work stations, the unions have been seeking to establish ergonomic workplace design, limited working periods at video terminals, compulsory eye checkups, etc. Recently these concepts have been expanded in the form of a design of work organization which allows diversified work tasks. While such alterations of work design could not have been established on the basis of the existing co-determination rights, numerous agreements of a similar kind were achieved at the company level mainly due to political pressure brought forth by the works councils.

The basic structure of union activities in this area concentrate on two main areas:

1 to shape and design technology and organization, instead of just protecting against its undesirable consequences in the case of scientifically proven negative effects. Human forms of work design and not mere corrections of problems are to be established. This is not only a question of knowledge (and research) and models of design, but also one of the investment potential of the enterprises concerned and last but not least, a question of power in negotiation processes.
2 The works councils should not be burdened with the nearly impossible task of having to prove that certain company design measures do not meet the present standards of health, but instead, the companies themselves should have to prove that the new technologies they are introducing are in accordance with the latest standards of humane work design. Unless this problem is solved, it will be impossible for the works councils to deal with this issue in the future.

Work organization and training

It can be assumed – as previously outlined – that there are still strong leanings towards Tayloristic concepts of work organization with a high division of labour and polarization of skills.

According to union concepts the utilization of new technologies should be designed in such a manner that two basic forms of work organization are realized:

1 Diversified work tasks: an individual job should encompass self-regulation in the execution of work, operations scheduling, executing and monitoring tasks and functions.
2 Group work: in this case broad employee skills, and ideally a high, homogeneous qualification enable all employees to execute all tasks

within a certain area and to determine workplace rotation themselves. Qualified teamwork of this kind is regarded as a way to reduce work burdens and stress, attain a high level of skills and encourage solidarity.

In both models the intention is to avoid unqualified and stressful residual work which arises in the loading and unloading areas of automated systems.

These demands, which are in accordance with the course of manpower-oriented work design outlined above, coincide with somewhat similar managerial concepts in a number of companies. Examples where manpower-oriented work design are actually implemented can be found in the machine tool industry, where this kind of work has a long tradition, but also in the formation of teams for the monitoring, controlling and maintenance of flexible manufacturing systems and even in highly automated sections in the automotive industry. This type of personnel policy is generally restricted to circumscribed areas within the companies, however. Numerous residual jobs requiring unqualified workers remain (loading tasks, for example), and in many areas strong centralization tendencies are apparent.

No matter what form of work organization gets introduced, the question of training remains crucial. In principle, unions as well as employers reckon with a growing demand for qualified workers and the existence of a current shortage of skilled workers. This is demonstrated by the fact that company expenditures on further training have multiplied over the last decade, although these training measures have been essentially oriented to company-specific adaptation to new technologies.

The dispute between employers and unions on training concentrates on four points:

1 company specific training, management's general objective, as opposed to training of a more general nature which can be of long-term use to the individual on the labour market;
2 the restriction of access to further training by means of employer selection, as opposed to free access for all employees within the company wanting to pursue further vocational training;
3 the attempts on the part of management to use further training measures to secure employee acceptance of rationalization measures without considering alternative forms of work design; and
4 management attempts to circumvent the works councils' co-determination rights concerning further training by carrying out in-plant further training measures in forms for which no such co-determination rights exist as of yet, such as small-group activities.

The dispute leads back to two opposing training objectives: the companies are trying to achieve a higher level of skill in order to implement flexible manpower utilization and fast adaptation to changes within the company, primarily by building up the core work-force; additional goals are to gain the acceptance of rationalization measures and avoid co-determination.

The unions are trying to establish forms of further training to give the employees qualifications which will be of general use on the labour market in order to reduce employee risks in case of redundancy. Moreover, the unions calculate that a greater availability of skilled workers will serve to promote strongly those forms of work organization which can make use of these qualifications, secure or expand employee control over the work process, and allow room for individual development.

As far as this question is concerned, the unions primarily rely on the works councils' capacity to achieve agreements within the individual companies to realize these goals. The works councils have a co-determination right concerning company further training measures (§ 98 Works Constitution Act) which does not come into effect, however, when training measures are exclusively connected with company-specific requirements. In fact, this entire area does not belong to the works councils' traditional scope of activities. In those instances where the companies' 'training offensives' are not connected with general questions of personnel planning and wage agreements – as is the case in some large firms – the works councils have hardly any influence.

In the metal industry, a breakthrough occurred at the collective bargaining level with the 'new' general wage and salary agreement (I) in North-Württemberg/North-Baden in 1988. The agreement aims, among other things, at the 'maintenance and broadening' of qualifications and is supposed to make it possible for individual employees to carry out a diverse range of functions. The agreement stipulates that the employer is obligated to provide information on (long-term) training needs due to technological and organizational changes and to discuss the issue once a year with the works council. In this way employee interests can be considered and possibly influence implementation of measures. The agreement does not give the works council co-decision-making rights, nor does it include an obligatory training programme for all employees in a company. Also there is no guarantee for a job reflecting an employee's qualification level, or providing a corresponding job classification or remuneration. In any case, a more active role, one can even say politicized role, of the works council in a field that was previously neglected has, however, been achieved through the agreement.

Wages and work design

In the Federal Republic of Germany the grading of employees into wage groups is generally carried out on the basis of the requirements of the actual workplace (using a method of analytical workplace evaluation). In the case of rationalization measures, this can have the effects described below.

In those instances where the introduction of new technologies simplifies jobs – computer-aided typesetting methods in the printing industry for example – or when traditional, wage-compensated work burdens such as heat or noise no longer exist – downgrading (i.e. wage losses) may occur. Since traditional protection agreements only secure wages for a limited period of time, establishing jobs that make greater demands on skills and qualifications is a more lasting way of securing income.

Therefore, unions have attempted to negotiate collective wage agreements in a way that encourages management interest in a design of work processes that make greater demands on skills.

The agreement on wage differentiation at VW (a company wage agreement) is a first tentative step in this direction. Wage grouping is not determined by the actual jobs, but on a work system in the sense of diversified work tasks and group work. Wherever an employee works within this system, he earns a fixed wage for being able to meet system demands and requirements. This enables the company to make flexible use of personnel within the work process, with a minimum of administrative expenditures and complicated wage-calculation and without any problems for supervisors when employees have to rotate within the system.

The union is ambivalent with regard to the attempt to link wages to individual qualifications (rather than work requirements). On the one hand, downgrading due to technological and organizational changes would not be possible; on the other, differences in social status would be rigidified. The Industrial Metalworkers Union tried to resolve this contradiction in a model wage agreement in one company (Vögele Wage Agreement 1982).

In the Vögele case, wages are determined exclusively by individual qualifications (not by the actual job requirements), uniformly for all members of staff, including white-collar workers. Downgrading and resulting income losses due to decreasing work requirements caused by rationalization measures are therefore no longer possible. The agreement also lays down the conditions of technological and organizational design as well as work content: accordingly, employees are to be encouraged and given the chance to expand their know-how, a versatile use is to be made of their skills and work requirements and conditions which could lead to a reduction in performance and dequalification are to be avoided. This agreement can certainly be regarded as a considerable achievement in establishing a humane design of industrial work. However, the union was never able to establish similar collective wage agreements on a broader scale. Indeed, the German association of employers reacted to this settlement by expelling the company from its ranks. It is apparent, however, that the company has succeeded in increasing its staff's qualification level, motivation and orientation towards quality as well as achieving a high degree of flexibility as far as manpower utilization and adaptation is concerned.

Also on this issue, the 'new' collective agreement from North-Württemberg/North-Baden in 1988 discussed above represented a broadening of former practice. Significant improvements were achieved by applying the (still used) analytical job evaluation at workplaces in which new technologies and new organizational forms were introduced. Moreover, the remuneration systems were adjusted to flexible and differing tasks, carried out by an employee within a single work area. The job classification was commensurate with the highest job requirements which were dealt with in a half-year period.

Adapting the wage differentiation system in the Federal Republic to production methods with new technologies is no easy task. Both unions and companies find themselves confronted with contradictions with regard to this issue:

1 Employers try to maintain wage differentiation as a way of controlling performance, a measure which at the same time impedes flexible utilization of personnel and cooperation within a given work system.
2 The unions are basically trying to reduce wage differentials, but are forced to adhere to the principle of performance related wage systems, because only in the latter case are they granted co-determination rights in wage design (according to the Work Constitution Act).

The fact that an objective, analytical evaluation of individual performance is not possible in computer-aided, integrated technological systems and that certain tasks are decoupled from direct results in the work process has caused a trend towards payment systems artificially linked to performance and premium pay, etc. The unions aim to make these performance and working conditions part of the collective wage agreements. At present, the works councils can hardly make any legally based investigations on the connection between wages, performance and working conditions.

Computer-aided monitoring of employee performance and behaviour

Computer monitoring is a highly significant area of conflict in the utilization of new technologies. The urgency of this problem is increasing in proportion to the new technologies which are being used to record and integrate production processes and personnel data.

Up until the mid-1970s, personnel information systems were primarily being used for statistical and administrative purposes (recording of working hours and wage accounting), whereas now they are increasingly deployed for planning purposes (personnel planning, for example). These systems can be used to monitor employee behaviour and performance when highly differentiated data concerning skills and qualifications, absenteeism, willingness to work overtime, etc. are linked with information concerning breakdowns, machine use, product quality, etc. (workshop-data processing-systems).

The unions criticize the development of a systematic evaluation using both data systems for its negative effects on employees, the most important being reduction of the scope for self-regulation on the shop-floor, selection of persons with under-average performance for expulsion from the core work-force, intensification of work by recognition of breaks in the work process and other similar factors.

Basically the works councils do have co-determination rights regarding the introduction of technical equipment for monitoring behaviour and performance (§ 87, 1(6) Works Constitution Act). It is from this legal basis that the unions derive the works councils' right to co-determine the design of data systems.

The unions are concerned with two principles in influencing work design:

1 Retention of a minimum of individual and collective scope for taking autonomous action on the job and also the preservation of employee production knowledge, thus securing a certain position of influence and power for the employee.
2 Transparency of performance should not be exclusively a management privilege, otherwise the works councils' potential for negotiation and their strategies would be severely restricted.

The severity of conflicts over these developments is due to the fact that central issues of company hierarchy are at stake. The interlinkage of personnel data and operational data enables the companies to pursue a form of personnel policy which is strongly anticipatory, disciplinarian and oriented to selection and which is not at the disposal of worker representatives. At present the works councils are decidedly unable to cope with these complicated problems of technology and planning. The unions are trying to cope with this problem by offering works council members instruction in dealing with this new area, by defining the legality and illegality of instances of technologically based monitoring by court order, and also by developing model-plant-agreements. Work disputes have already taken place in this area and employee awareness of these problems is particularly high, even though these control systems have not yet been widely implemented.

Flexibilization of working hours and employee status

The disputes over new working hours policies in the Federal Republic are not only concerned with flexibilization or the effects of reduced working hours on employment. They also dealt with the employees basic right to have the power to dispose freely of 'working' and 'leisure' time.

The dissemination of new technologies has made the flexibilization of working hours an issue of great importance; new technologies allow increasing technological and organizational flexibility due to the possibilities of decoupling worker tasks from the production process; market developments and new logistic systems are impelling the companies towards flexible personnel policies; and lastly, the unemployment situation is demanding new concepts for regulating working hours.

The employers are trying to use the separation of working hours and operational times to expand their companies' utilization periods. From the employer's viewpoint, a reduction of regulations concerning working hours will facilitate the following: adaptation of personnel volume to market fluctuations; expansion of operating times for machines or plant equipment; the individualization of working hours; a rise in employment due to a larger number of part-time jobs (particularly by converting full-time jobs into part-time jobs); and lastly, faster recruiting of new personnel due to less binding forms of employment contracts. For all of these reasons, flexibilization represents a new employer approach to working time policy.

The unions are confronted with a two-sided problem: on the one hand, their goal is to create employment incentives by a reduction of working hours; within the context of the labour disputes in 1984 over the 38.5-hour week they had to concede the possibility of company specific solutions, thereby opening the door to flexible working hours. On the other hand, however, they must try and secure the so-called 'normal employment relationship', i.e. the permanent employment of individual employees. Only this form of employment provides an adequate individual income for a family, and also an adequate old-age pension on the basis of conventional legal and social security regulations. Finally, it is the 'normal employment relationship', i.e. the permanent integration of personnel as employees, on which the unions' own power rests.

Within the Federal Republic of Germany, the deregulation of the 'normal employment relationship' (particularly encouraged by the Employment Promotion Law 1985) has been resulting in an increasing tendency towards utilization of part-time work and temporary work, which are displacing full-time jobs.

There are also contradictory interpretations of the employment incentive resulting from reductions in working hours. Employers give economic growth rates the credit for new recruitments or slower rates of staff reduction, whereas the unions point to the effects of reduced working hours. The German Institute for Employment Research (IAB) states that the reduction of working hours by 1.5 hours for 4 million employees in 1985 was responsible for 80,000 of the 250,000 persons newly employed (resulting in an employment effect of approximately 50 per cent in the areas directly affected by working hours' reduction). The metal workers union estimated, for 1987, (the year in which the 37.5-hour work week was introduced in the metal industry) 60 per cent of the reduced working time to have positive employment effects.

The collective wage agreements of 1984 focused on the introduction of the 38.5-hour work week in the metal-processing and printing industries and in parts of the wood and plastic-processing industry. The employers, however, succeeded in establishing a flexibilization of working time, i.e. the hours set by the collective agreement only had to be met on average for all of the workers in an enterprise. Different groups of workers could have differing working hours ranging between 36 and 40 hours per week. This flexibilization (in 1987) was utilized by only 10 per cent of all companies, primarily large companies.

The employers are continuing to pursue their aim of establishing more flexible working hours and they are now increasing incentives by offering tangible wage rises as an alternative to reductions in weekly working hours.

While the metal workers union was originally reluctant to participate in the discussion on flexibilization, it could not avoid having to deal with this issue in the second half of the 1980s. What the union was and is demanding for the employees is currently being termed as 'autonomy of working time' or 'sovereignty of working time'.

The most important points in the mid-1980s were: a maximum of eight working hours daily; a maximum of 40 working hours per week, and compensation in free time for any deviations in weekly working hours within two months; two work-free days on the weekends, which means particularly that Saturdays must be kept work-free; a maximum of 10 hours of overtime per month, and compensation for this with free time.

The issues of variable shift times, variable forms of daily work and flexible working hours, etc. remain virulent. In 1990 the metal industry agreed to the introduction of the 35-hour week beginning in 1994. The agreement stipulates that work intensification through the shortening of work times should be prevented through special restrictions, and that performance should be defined and agreed upon at the enterprise level. Flexibility in working times was secured in that as much as 18 per cent of the employees in an enterprise can voluntarily extend their normal working time to 40 hours for which they can receive compensation in the form of free time or extra pay. Overtime and the right to a free weekend remain regulated as described above. However, exceptions do exist and are increasing through special agreements at the enterprise level which has led to some conflicts between unions and work councils. The debate between worker representatives and employers continues to centre around these issues and on the relationship between working time and the company's operating time.

These developments have had considerable consequences for work design and the quality of working life because flexible working hours and employee status are closely interrelated. For example, flexibilization tends to encourage an increase in marginal work-forces. This puts the works councils in an increasingly difficult position because flexibilization may secure positive working conditions for the core work-force, while reducing the works councils' potential to achieve a design of work which corresponds to the unions' concepts for a reduction in the division of labour, and which also secures the employees' legal status in the sense of a 'normal employment relationship'.

At least within areas of mass production the flexibilization of working hours seems to add momentum to Neo-Tayloristic and centralistic work designs and to undermine the foundations of worker representation.

Co-determination and work design

According to the dual system of employee representation, the unions cannot participate in the companies' design of technology and organization except within the framework of very general decisions that lie in the context of the 'Co-determination Law' of 1976. The previously mentioned collective agreements contain only limited possibilities for influencing the design of work. For years the unions have been trying to establish co-determination rights pertaining to the introduction of new technologies within the context of the Works Constitution Act. However, at the moment the chances for this are rather poor.

The works councils' opportunities can only be briefly sketched here. As far as central issues of work design are concerned, they only have rights of information and consultation. It is only in cases of work design with severe negative effects on employees that their co-determination rights come into play. In any case, the works councils' resources and information concerning technological and organizational changes have been rather deficient up to now. This is all the more so in the case of introduction of new technologies (see chapter 24).

Co-determination at the workplace, with the direct participation of employees, has traditionally played a subordinate role in the official policies of the German unions (while having caused much union-internal controversy). The objective of the Federation of German Trade Unions' (DGB) 'Co-determination Initiative' of 1984, however, was direct employee participation in bringing about a humane design of their own work. Of course, their participation was not intended to be independent of the institutionalized forms of worker representation, but to augment them. The models for direct employee participation are still in the discussion phase.

The management of numerous companies have also initiated forms of employee participation along the principles of small-group activities. These have been met with considerable interest from employees who regard these participation forms as opportunities for making use of their skills and know-how to co-design their own work and cope with areas and tasks normally excluded from their work. Opposition arises (particularly in areas of performance and pay) where rationalization effects, due to work in quality circles for example, directly result in work intensification or staff cuts.

A number of investigations have backed up union critiques that work in these small groups serves to promote strongly further rationalization and acts as a way and means of attaining relatively criticism-free acceptance of new technologies. These activities become particularly dubious as an instrument of participation in the case of complex, interconnected technological and organizational systems whose positive or negative effects on employees in other areas cannot be foreseen.

Generally, the unions' position, particularly that of the metal workers union, is ambivalent on this issue. The use of small-group activities to establish other forms of participation apart from institutionalized worker representation is strictly rejected. On the other hand, the works councils' participation in these activities is regarded as a chance to achieve a certain degree of co-determination at the workplace and to harness employee motivation for union interests. Attempts are being made to use collective agreements within the company to influence the goals, participants and tasks of small-group activities, and these have already been partially successful.

Employment protection and work design

Employment protection is certainly the focal point of union policies and activities. The question is: can the issue of humane work design remain a priority despite the pressure the unions are under to first and foremost prevent personnel reductions?

In the case of staff cuts, the measures taken by worker representatives are mainly defensive in nature and concerned with cushioning the impact on employees. There are numerous highly complex forms of combined socio-political and collective agreements and regulations. Their effects centre on the 'balance of interests' anchored in the Works Constitution Act (§ 112), which then usually result in a social plan regulating severance pay, selection of persons to be dismissed etc. An influence on the actual design of work is usually not found in this context. The settlements within the steel and printing industry concerning the manning of jobs are an exception and rest on a long tradition.

New forms of social planning are taking shape which could exert considerable influence on training, manpower utilization and participation, i.e. the creation of employment plans rather than social plans. In this regard, the staff reductions which took place at one of the largest German companies in consumer electronics in the mid-1980s, was one of the first examples and still attracts considerable attention.

The unions accepted the necessity of dismissing approximately 3,000 employees from this electronics company in Southern Germany. With the participation of local union representatives, company management and the works council worked out a social plan, according to which:

1 Each worker affected by rationalization is given the opportunity for further vocational training at company expense.
2 Dismissed employees receive government re-training support which is supplemented by company payments for a period of two years.
3 Two equally staffed commissions were established with two objectives: one commission ('new production lines') analyzes new products and new markets in order to develop new employment possibilities and to secure locations of the plants. The second commission ('work design and training') is assigned with the task of determining the content of the re-training programmes and promoting the regional development of qualified jobs with humane working conditions.
4 Employees over the age of 55 and those choosing to leave the workforce are granted early retirement payments or severance pay according to conventional regulations.

This case demonstrated the need for intervention into regional political structures to implement such far-reaching employment programmes. In this instance, new regional political activities involving local communities and authorities were initiated. These particular efforts have served to make the perspectives of controlling, shaping and designing technology and organization beyond the individual company level as well as within particular companies, and the importance of humane job design stand out as opposed to traditional protection measures against rationalization.

Union instruments for influencing technology and organization

According to the concepts traditionally held by German unions, work is a process in which the individual should be able to find self-fulfilment and which should create social solidarity. Next to the protection against factors threatening basic living conditions, the shaping of technology and organization using the criteria of humane work design is presently (the end of the 1980s), at least by virtue of this definition, an issue of nearly equal significance.

Naturally the unions are active in all areas of social reform and social politics, and policies concerning technology are a central issue in our day and age. In addition to humane work design and co-determination, there are other demands being made on technology such as that of environmental compatibility, sensible use of natural resources, as well as disarmament. Aside from political campaigns and influencing the parties in parliament, the unions in the Federal Republic presently have little chance of influencing the design of technology. The numerous technology development programmes – promoting micro-electronics, information and communication technology, manufacturing technology – are all being carried out with minimal or no union participation at all (the Programme for Humanization of Working Life and the programme initiated in 1989, 'Work and Technology' being exceptions to this).

Collective agreements remain the central instrument of union influence on the companies and thereby also on work design. In the Federal Republic collective agreements are usually valid for an entire industrial branch and for large regions (with the exception of special company collective agreements, particularly in large-scale enterprises). Although a large number of collective agreements exist – in 1989 there were 32,000 valid collective agreements, 8,000 of which were company agreements – the basic tendency, at least within individual branches, is towards standardization of working conditions and employment status. This implies two things:

1 An orientation towards what weaker companies can achieve and thereby a determination of the lower limits of the standards aimed for; and
2 the realization, and, circumstances permitting, the improvement of general standards within the framework of the dual system of interest representation by the works councils.

By their professional and strongly legalistically oriented courses of action, the works councils have so far been successful – at least to a certain extent – in realizing the second factor as far as wages and safety norms are concerned. The objective of influencing the design of technology, organization and working conditions, however, confronts the works councils with totally new and unaccustomed tasks, for which their members are not at all qualified. Above all, this task is impossible to attain within the framework of traditional forms of participation and monitoring of adherence to given norms.

In this difficult situation the unions are pursuing three strategies:

1 They seek the conclusion of appropriate collective agreements. A model was presented in one area of the metal-processing industry in South Germany, which, however, was not realized (model of a collective Wage Framework Agreement for South Wüttemberg/Hohenzollern and South Baden 1982). This model agreement attempted to achieve a comprehensive regulation of the conditions of work and labour performance according to the criteria of a 'humane design of work' and contained numerous suggestions for regulations, such as job rotation and group work, company obligation to train all employees, fixing the relationship between wages and performance, etc. This model had up to 1990 never been implemented.

2 The unions are also trying to instruct the members of the works councils in how to influence the design of work by making use of systematic combinations of existing legal regulations. The unions are also utilizing the research findings of the Programme for the Humanization of Working Life to present models of humane work design, to demonstrate their economic viability, and they are attempting to put these models into company practice via the works councils. The development of various forms of supportive material for the works councils is an essential task towards this end. The fact that the unions presently have few resources for these service tasks, is a crucial problem.

3 Various forms of cooperation above the individual company level have come about, which, however, are still very heterogeneous and vary in scope. For example: union-organized consulting centres for the implementation of new technologies; collaboration with universities or research groups from the areas of social science and engineering science; collaboration of works councils from various companies on a regional level, so-called technology work groups or automation groups, particularly with the participation of union-oriented engineers, and other activities.

CONCLUSIONS – PROBLEMS WITHIN THE DUAL SYSTEM OF EMPLOYEE REPRESENTATION IN ESTABLISHING HUMANE WORKING CONDITIONS

The conclusions presented here represent possibilities and should not be viewed as prognoses. They concern the unions' role in the establishment of humane working conditions within the framework of the dual system of employee representation in the Federal Republic of Germany. Another important consideration that cannot be dealt with here is the effect of the unification of the two German states.

The present wave of rationalization is by no means uniform. The type of new technologies, their scope and the period of time in which they are implemented depend on the economic and innovative potential of a particular company. This, in turn, differs according to industrial branches, regions, in-house personnel structures, the amount of qualified manpower on the labour market, etc.

Basically, it can be expected that the working conditions of employees within the companies and also the conditions prevailing between the firms will become more heterogeneous in the next several years. Standardization of working conditions, to be brought about by the instrument of collective agreements valid for entire industrial branches and regions, is hardly realizable. In this way, the works councils' tasks will undergo a change, through the necessity of having to convert general perspectives of technological and organizational design into company specific implementations.

The usual conclusion in industrial relations research is that this implies a loss in union functions and power and a concomitant increase in the works councils' influence and power.[5]

This need not necessarily be so, however. While it may be the case within large-scale enterprises, in the automotive industry or the chemical industry for example, the majority of works councils in medium and small-sized companies are increasingly unable to cope with these new tasks. New forms of union instruction and consulting and new union activities concerned with the co-design of technology and organization are the only way out of this impasse.

Influencing the design of work necessitates a new way of asserting demands. It is no longer adequate for the works councils to concentrate solely on negotiating working conditions within the framework of collective agreements or that of the works constitution, or to 'enforce' regulations and see that they are adhered to. Influencing the design of work and technology means putting forth alternative concepts and not merely attempting to ameliorate negative effects. If this policy is carried out, it could become a new union instrument to mobilize and politicize the works councils, even when these concepts cannot be fully realized.

In fact, initial steps in this direction can already be discerned. To the extent it is successful it would also serve to counteract the process of integration of the works councils into the interest sphere of their own companies, which some observers presume will take place.

There can be no doubt that the unions are thereby challenged to increase their consultation and service activities. This not only requires a redistribution of union resources, but also new strategies of involving union members. There are tendencies pointing to the possibility that concepts and instruction measures could be developed on a regional level, which would focus directly on actual local problems. While this results in a certain amount of narrowing of the larger picture, a concentration of organizational resources and a process of union-internal democratization becomes possible. This process could serve to release new forces for mobilization, for example by reducing the communication deficiencies between the works council members of various companies within a region, by integrating regional policies concerning infrastructure, location of plants etc. and also through the increased transparency of local union activities for the particular workers affected. These developments are more likely to strengthen the unions than to weaken them.

In this context, the following is the most significant aspect: if the heterogenity of personnel within the companies causes the works councils to disintegrate by overburdening them with the task of having to represent widely divergent interests, and if the works councils do not succeed in integrating the groups of engineers and white-collar workers whose numbers continue to grow relative to blue-collar workers, then it will be most difficult to make the works councils effective. All attempts so far of involving engineers in union-oriented design programmes have met with limited success.

The new technologies themselves have the potential to lend new power to the unions. The intra- and inter-company integration of logistic and distributive processes being brought about on the basis of the rapid development of information and communication technology make new strategies of labour disputes possible: spot strikes or conflict measures can result in extensive, far-reaching problems for the companies. The fact that works councils are not allowed to carry out strikes, points to the potential for strengthening unions rather than weakening them.

The German unions cannot retreat from their traditional orientation to problems concerning society as a whole. Their moderate stand on social and wage policies was thus far pursued out of political consideration for the limited, but nevertheless significant co-determination and participation rights. In the 1980s, a process of social polarization took place within the Federal Republic of Germany which touched on the problem of unemployment and which caused this 'social contract' to crumble. New problems are arising due to the unification with East Germany. It not only strains the resources of the unions which, already in the 1980s, were too limited to solve the problems discussed here; more damaging, it once more brings to the surface the issues of unemployment, regional differentiation, and the standardization of working conditions for all those employed. This leads to an important question: are the unions in danger of increasing this process of polarization by fighting for a humane design of work which primarily benefits those already employed, and then perhaps only members of the core work-force (and in the western part of the FRG)?

Seen from the perspectives this paper has presented, the question can be answered without any reference to actual requirements of employment or social policies: the unions can only succeed in maintaining their own recruitment chances and the support and solidarity of their members, who are mainly skilled workers, if they manage to exert decisive influence on the working conditions of the core workers. This is the precondition for their power to improve the working conditions of the marginal workers, as well as influence the overall design of technology and work organization and thus, the employment situation as a whole.

Influencing working conditions is a political concept and contains the vision of an active, purposeful design for the quality of working life as opposed to the rather vague compensation of negative social effects brought about by rationalization.

24 Rationalization Strategies and Representation of Worker Interests[1]

Norbert Altmann

INTRODUCTION

This chapter focuses on new rationalization strategies in industry and their effects on the representation of employee interests on the shop-floor level in the Federal Republic of Germany. The three points that are dealt with are: which direction rationalization strategies will take; what opportunities employee representatives will have in influencing the design of technology and work organization; the problems faced by employee representatives in the FRG.

The discussion confines itself to employee representation at the company level, that is on the activities of the works council, and concentrates on the manufacturing area within companies, primarily in the metal industry.

Examples from the large enterprises in the automotive industry as well as smaller companies are given to elucidate the three points listed above. Companies of differing size and market dominance with strong and weak works councils are considered to point out the differences in work design and co-determination that can occur within companies, and the actual problems that arise in representing employee interests.

RATIONALIZATION STRATEGIES

'A spectre is haunting Europe – the spectre of Post-Taylorism.' Saturated sales markets and increasing international competition have made rapid product changes, the availability of a wide range of different models and types, high quality and short delivery periods, etc. a necessity. These are conditions, which could not and cannot be met by the traditional forms of Tayloristic and Fordistic work design.

The question remains as to whether a new utilization of 'human resources' to meet these conditions is really a central focus of new rationalization strategies.

In the Federal Republic of Germany the concepts and ideas in work design which will play a central role in the future are often oriented to the so-called 'New Production Concepts' put forward by Horst Kern and Michael Schumann (1984). In the opinion of the authors, these concepts, propagated by a so-called 'enlightened' management, are an expression of company interest in achieving a better utilization of human resources. They are based on labour relations which, from the standpoint of management, are no longer regarded as being determined by a conflict of interest, but are viewed more in terms of 'mutual trust'. According to the 'New Production Concepts' there is a necessity for and a tendency towards re-professionalizing manufacturing work for reasons of flexibility and quality assurance, as well as the objective of achieving maximum utilization of expensive manufacturing facilities, etc.

This interpretation of new developments in work design is well accepted by employers in the FRG, as numerous official statements indicate. However, the assumptions concerning the segmentation of the labour force that derive from these concepts, which have also been described by the authors, are largely ignored. The unions are sceptical, but as they have placed great hopes on the development of more skilled work, particularly on skilled group work as the basis of a humane work design and greater chances for participation, the discussion going on within them is ambivalent.

The topic is also highly controversial within the field of social sciences in Germany (Malsch, Seltz 1987; Düll 1985). Rather than enter into this debate here, it seems more fruitful to point out actual developments in rationalization strategies based on investigations carried out at the ISF (Döhl *et al.* 1989; Deiß *et al.* 1989; Altmann, Sauer 1989).

The point of departure of this research is the fact that within German industry a high degree of division of labour and central manufacturing control are still predominant organizational concepts. On the one hand, these organizational concepts are based on measures taken by management which are of a structurally conservative nature, and on the other hand, on the traditional attitudes prevailing among engineers. The latter tend to regard their own work designs in a rather exclusively technologically oriented manner, namely from the standpoint of pursuing what they consider to be the 'one best way', which usually means that of maximum control of the manufacturing process by means of control technology. This approach always involves the elimination of human intervention – the prime source of mistakes in the opinion of engineers – to the greatest extent possible. Moreover, this basic attitude of designers of technology goes hand in hand with the widespread assumption that workers will somehow find a way to adapt to the machinery once it has been installed. In their opinion, the problem of training can be solved by simplifying work.

The decisive factor, however, is not this structurally conservative and technology-centred point of view, and it is also not the concepts held by management, whether enlightened or not. Rather, the decisive factor is the strategic direction rationalization measures are taking which can be observed empirically and objectively and interpreted using capital utilization criteria. Under the present conditions of competition on the sales markets and the potential presented by information technology, rationalization is evidently taking a new direction, one which we have termed 'systemic rationalization' (for a more detailed discussion of this concept, see chapter 4).

This type of rationalization can be briefly characterized and summarized by the following three aspects:

1 Rationalization strategies are no longer primarily oriented toward the performance of individual manufacturing processes or the efficiency and utilization rates of individual facilities, but instead toward optimal integration of all in-plant processes. New technologies tend to permit a more systemic kind of rationalization, which may not be pre-planned or consciously pursued in a systematic manner, yet which nevertheless effect the entire company and all areas of its production process. Concrete examples are the different ways and means that 'computer-integrated manufacturing' (CIM) is applied.
2 Besides this, the new type of rationalization strategically integrates supply, processing and distribution processes that take place outside the company, thus changing the division of labour between companies as well as the inter-company relationships which previously were determined by market and competitive forces and were regulated by contracts. With the help of information technology it has become possible to directly interlink company–external processes with in-plant operations. The companies' relationship to supply and sales markets as a whole is thereby subjected to rationalization measures. Examples of these developments are a number of new and different logistics systems, JIT-systems, etc. and new criteria of MOB-decisions.
3 The new type of rationalization is characterized by what are seemingly contradictory objectives: on the one hand, flexibilization is intended to meet the new sales market requirements, on the other hand, in view of stiff competition, companies are seeking ways to reduce the cost of production at the same time. These conflicting objectives could not be attained through the use of conventional technology and organization. However, now it seems that the use of integrative organization and control technologies makes both objectives obtainable. This means that the flexible potential of human labour is no longer regarded as being the central issue of flexibilization and quality assurance, etc., for now the

> potential for flexibility is contained in technology instead. The prime objective is to make the entire manufacturing system – from procurement of parts and materials to the distribution of the finished product (including transport processes involved) – more productive and efficient over and beyond individual processes and plants; and also to reduce the amount of capital fixed in the system.

This means that rationalization strategies are not specifically directed towards labour as a resource and therefore we cannot say that the end of Taylorism is in sight. What current rationalization strategies aim for is the utilization of flexible technology and the integration of all manufacturing, administrative and distributive sub-processes with the help of information technology: thereby perpetuating the division between planning and execution.

It is quite clear that this does not mean that these developments eliminate the problem of human labour altogether or aim to establish 'full automation' or 'unmanned factories'. The practical problems are obvious, namely the considerable limits of control technology, the problems arising in cases of partial or insular implementation, interface problems between man and machine and between different systems, the lack of skilled workers in key positions or bottleneck areas, the diverging interests of key groups within the company, and last but not least, the limited experience of management in this field leading to exaggerated expectations based on the promises made by system manufactures and software houses.

The potential implied by new technologies, however, is also evident: the tendency to achieve even greater integration which is generated by control systems once they are installed and the pressure to organize inter-company relationships in such a way that an integrated manufacturing process is brought about. The result could be called computer-based Neo-Taylorism with a new polarization of labour – not only within companies, but between them as well. It would be misleading to suggest that this new type of systemic rationalization is already widely established. However, it is definitely a direction in which strategies have been developing.

THE REPRESENTATION OF EMPLOYEE INTERESTS AND THE CONTROL OF TECHNOLOGY AND WORK[2]

Before dealing with the problems of employee representatives in coping with systemic rationalization, some remarks on the existing institutional structure of industrial relations in the Federal Republic of Germany may be useful.

1 A characteristic feature of industrial relations in Germany is the strict division between industrial unions active on a supra-company level,

encompassing both region and industrial branch, and the internal company organs representing the employees, namely the works councils. This dual structure was regulated by the Works Constitution Act in 1972. Collective agreements represent the unions' central instrument in their dealings with the employers' associations. These agreements contain general provisions regulating the conditions of employment, the basic factors on which wages are based and the reference figures for wages.

2 The works councils are entrusted with the task of giving concrete form to the collective agreements at the individual company level. Toward this purpose, an independent level of negotiation takes place between the works council and company management. The results of these negotiations are generally formulated by company agreements. There are – besides others – three characteristic aspects of this level of negotiation: firstly, the strong orientation to regulations defined by Federal law and by collective agreements; secondly the commitment to achieving consensus stipulated by the works constitution (the spirit of mutual trust and ban of industrial actions based on works councils' decisions); and thirdly the cooperative and company-oriented attitude held by the works councils.

3 In contrast to union organization in Great Britain and the United States, it was possible to cope with the social effects of the conventional forms of rationalization which have taken place step-by-step, point-for-point within this structural framework. The principles involved in this process have traditionally been the stabilization or increase of wages, employment security, and the reduction of work loads and stress or at least their material compensation.

Up to now, one of the weaknesses of the German system of employee representation lay in the fact that the instruments of negotiation as defined by law and collective agreements did not permit any substantial intervention in the technological and organizational design of work processes. The initial attempts toward achieving a so-called 'qualitative' collective bargaining policy (as opposed to a quantitative or wage-based policy), in effect since the beginning of the 1970s, which were geared toward the design of technology and work organization, have made few inroads so far.

The basic demands centre on an abstract idea of 'humane design' of technology and work organization. Primarily this means the reduction of division of labour, particularly between planning and execution.

Apart from eliminating or reducing stress, work loads, monotonous assembly-line work, and so on, the concrete demands focus on three points:

1 the creation of meaningful and comprehensive work content granting the individual worker a broader scope for taking action and making

decisions, and allowing the integration of mental and physical demands made on workers, in other words, diversified jobs;

2 the development of group work, for which numerous individual aspects are defined, such as the number of group members, the wage structure, the exact definition of tasks, etc.; and
3 all aspects of these demands for the design of work forms and work organization include co-designing training for such jobs.

All in all, the unions' primary goal is not the pursuit of protectionist and compensatory policies, although these are still necessary and are also undertaken, but rather that of gaining some kind of control over technological/organizational developments, and of obtaining a certain amount of influence on the design of technology and work organization at the company level (see chapter 23).

If it were really true that the general thrust of rationalization strategies were going in the direction of re-professionalization of production work, and the division of labour were dissolving, then the demands which the works councils have made on work design in terms of job diversification, group work and training would be welcomed with open arms.

This is, however, clearly not the case. If we look at the automotive industry, for example, we cannot speak of an end of Taylorism. Of course it cannot be ignored that numerous pilot projects and measures have been carried out in this industrial branch which were influenced by Japanese management concepts, experiments in the United States or in Scandinavia (at Volvo, for instance) and which in the FRG were also brought about due to the support of strong works councils.

Nevertheless, the picture that emerges from the findings of a number of different investigations and studies is that no long-lasting or revolutionary changes have come to pass as the following examples indicate (Jürgens *et al.* 1989; Klebe, Roth 1987; Roth, Kohl 1988; in the sense of 'new production concepts' Kern, Schumann 1984). The largest share of tasks still involve direct manual labour (in parts assembly and final assembly, for example) which is mostly simple, semi-skilled work. In assembly processes we find forms of work design that were basically developed back in the 1970s under the key word 'work-structuring' (job-enlargement for various tasks requiring the same level of skill; job rotation and decoupling from rigid machine cycles, etc.) These measures, however, amount mainly to a new performance policy (Altmann *et al.* 1982). They serve to exploit existing tacit skill potential which was previously blocked by a rigid division of labour, and also enable a more flexible utilization of labour. This, in turn, allows tighter staffing and cost savings, due to the fact that personnel can be utilized more intensively. Flexible utilization of labour makes possible the reallocation of individual workers to similar workplaces where they are able to replace absent colleagues. This tighter manning can also be used for a step-by-step selective elimination of personnel, thereby enhancing the average skill level. In this sector, meaningful job tasks and a greater scope for self-determination on the job are more side effects than anything else.

In areas with a higher degree of automation (in the pressroom or metal-working departments, for example) there is an increased utilization of skilled workers, although their numbers are still limited. These workers have to deal with tasks relating to maintenance, quality assurance, program adjustment and production control. There are numerous ongoing experiments strongly geared to a general utilization of skilled workers. However, there are diverging opinions about whether these forms of work will actually be established on a broader scale. In any case, it is not possible to conclude that an increase of average skill level within individual work processes will result in a requalification of manufacturing work. These higher average skill levels are the result of the quantitative reduction of lesser skilled tasks, such as loading, and the integration of maintenance tasks into manufacturing work.

Accordingly, this development has resulted in a qualitative and quantitative erosion of previously independent tasks and areas, such as maintenance and repair. All in all, this amounts to a restructuring of the work-force at the plant level in which skilled workers perform manufacturing work and lesser skilled workers are eliminated. In general semi-skilled workers have little chance of assuming the role of skilled worker through company training measures or on the basis of their experience.

The monitoring of complex equipment by so-called system leaders does not typify the future of work, but instead documents the fact that a few key positions will arise at the critical interface areas between central control and fine tuning on the shop-floor level. While the range of tasks to be coped with in these positions is broad and diversified, they do tend to increase segmentation of work as a whole.

Group work does exist in the sense of so-called system supervision teams which have a higher and more homogeneous level of skill and cope with larger facilities and manufacturing systems. Experience gained to date on the effects of group work indicates the following: first, far reaching training measures for so-called 'polyvalent hybrid skilled'[3] workers are reduced, because these workers cannot really make use of their broad range of skills in everyday work; second, the self-determination of work within the group results in new forms of division of labour and specialization, particularly in order to meet the demands for a rapid attendance to break-downs and failures; while third, self-selection processes are triggered off by which those group members who are substandard in terms of perfor- mance and skills are pushed out of the groups, thus forming the basis for a gradual selection of certain personnel and eventually, personnel reduction.

Naturally training measures play a considerable role here and all automobile manufacturers invest substantial sums and didactic know-how in such training. It must be emphasized that these training measures are strongly geared to meet current, company specific requirements, and that the selection of trainees is primarily decided by members of medium and lower management. By no means can it be said that this training is accessible to all interested persons.

The possibilities for comparing operational data and performance data give rise to new forms of monitoring, or at least a new dimension to supervision, even in view of the fact that direct performance monitoring by technical means is not permitted without co-determination of the Works Council (§ 87,6 Works Constitution Act). In large-scale companies in the automobile sector numerous company agreements have been concluded with the aim of restricting the connection between personnel information systems and production data acquisition (PDA).

These forms of work readily portray a type of rationalization which is technology-centred and geared toward central control which only builds up skilled and engaged personnel in certain key positions.

Today, the decisions made at the headquarters of large automobile manufacturers not only affect their own manufacturing sites, performance and output targets, but directly or indirectly determine the suppliers' products, manufacturing methods, scheduling, and ultimately the conditions of work in supplying firms. Initial CAD/CAM integration between automobile manufacturers and suppliers on the basis of standardized, manufacturer-unspecific data transfer systems, just-in-time (JIT) connections with flexible delivery calling, detailed specifications for methods of quality assurance and control, etc. represent considerable demands being made on the suppliers' capacity to adapt in terms of technology and organization, quite apart from the pressure of international competition to which these companies are exposed.

Systemic rationalization forms are directed toward the entire context of the manufacturing process, from material and particularly the parts supply, all the way through to distribution of the finished product. In what way will these forms of rationalization actually affect the suppliers?

The specific situation of suppliers is too complex to go into here in detail, especially in light of the diversity in companies that would have to be considered (for a discussion of suppliers problems see chapter 22; Altmann, Sauer 1989). The decreasing number of parts and components manufactured by the automobile manufacturers themselves and increasing product complexity certainly offer opportunities for suppliers. Yet, at the same time, as the German magazine *Manager* put it, the supply companies have the automobile manufacturers 'breathing down their neck'. The general trends are well known: the increasing dualism of suppliers with a capacity for development and innovation on the one hand, and those who are no more than an extended work bench on the other; reduction of direct supply and single sourcing, the pressure to change locality in order to adapt to the manufacturers; far-reaching specifications concerning the purchase of material, manufacturing methods and quality assurance; the trend toward component assembly at the supplier level; initial CAD/CAM integration; manufacturing and delivery on call; integration into just-in-time systems, etc.

How do these developments, which are an important part of the rationalization measures taken by the automobile manufacturers, affect the employees in supplier companies? Some factors are:

1 Relocation to the proximity of automobile manufacturers in order to secure just-in-time delivery may have considerable effects on employment and unemployment in the original locations.
2 Production synchronous manufacturing and just-in-time systems often result in structures of working hours in supplier companies which threaten to become increasingly difficult to control, leading to overtime, short-time work, flexible working hours, weekend work and the utilization of temporary work and fixed-term work contracts.
3 In cases in which suppliers fulfil the function of extended work benches, more unskilled and repetitive forms of work will be the result.

All in all, the integration of suppliers into the logistics chain makes them increasingly dependent – with the exception of large-scale enterprises in the electronics industry and other similar big suppliers. Management within the dependent companies becomes less capable to manoeuvre as far as the design of technology and work is concerned.

Highly significant is the fact that, apart from the automotive industry, we have been able to discern similar developments in other industries, for example in a branch of the consumer goods industry (i.e. furniture manufacturing; see Döhl *et al.* 1989; Deiß *et al.* 1989). Here the new forms of company internal, integrative rationalization as well as the inter-company division of labour and supply have brought about even more problematic effects on employees than those occurring within the auto industry. One such effect is the industrialization of small crafts businesses, which came about through the concentration of production of parts and components in series, and resulted in a considerable trend toward dequalification for the employees. Another example is the shifting of work processes involving dangerous materials into the small business sector, where an adequate know-how in dealing with such materials is lacking.

THE PROBLEMS OF REPRESENTING EMPLOYEE INTERESTS IN COMPANIES

It is clear that the representatives of employee interests can hardly ride on a wave of mutual interest with management as far as the reduction of division of labour and the end of Taylorism are concerned. In fact, shifts in work design generate new problems.

One must realize that union programmes addressing possible future forms of work design place a further burden on employee representatives at the firm level in coming to terms with the new, systemic-oriented rationalization forms. This is the case because collective agreements formulate their demands in an increasingly general form. It is up to the works councils to give these demands a concrete form. This is the reason why some scientists and unionists refer to a shift of employee interest representation to the company level: meaning that the power of the works councils on the company level is on the rise, while the role of the unions is being reduced to the function of a service institution (see Müller-Jentsch 1988a).

This is a rather lopsided picture, however. Large and strong works councils, and this is particularly the case in the automotive industry, are capable, under certain circumstances, of concretizing the skeletal specifications of programmes without union help. However, the works councils in economically weak industrial branches or enterprises, or in small firms, might not succeed in this task even with the help of union counselling, let alone without it, especially in industrial branches outside of the metal industry (with regard to automobile suppliers see Doleschal 1989). An important aspect of this issue is the basic question of what is involved when union policies try to introduce their demands on the design of work and technology at a political level.

It is true that the works councils in the automotive industry have succeeded in achieving some spectacular co-design measures. But they are also proof of the fact that even strong works councils have difficulty in breaking through the technology-oriented design concepts of systemic rationalization. And when they do succeed, it is usually in limited areas. These few cases serve as models for alternative and humane forms of work design.

One example is group work, which has been acclaimed worldwide as representing a humane and also productive form of work. However, group work in the German automotive industry is to be found only in mechanized and automated processes such as pressrooms, foundries, metalworking shops, body assembly and paint shops. In these areas, as mentioned above, what is primarily involved is the monitoring of complex manufacturing facilities and the integration of control tasks, maintenance and repair, trouble-shooting and sometimes also the fine tuning of control programmes. Mainly, when we talk about group work at present, we are talking about pilot and experimental programmes. All in all, there is only a small number of mostly skilled workers employed in some type of group work in the automotive industry. (For example, of programmes initiated by the works councils, about 7,000 workers out of about 450,000 employees in the automobile industry were involved in some form of group work in 1988). What is more, apart from system supervision teams, several work groups only came about due to the pressure exerted by works councils. In these instances, the conflicts concentrated firstly on the integration of lesser skilled employees (instead of selective processes to eliminate them) and secondly on the introduction of wage systems for these groups, which would dampen individual competition and which would compensate for the increased demands made by the rotating job assignments within a given system (Brumlop 1986). The works councils have only succeeded in achieving these aims to a limited extent.

What are the problems which the many, yet in most cases, weaker works councils face outside of the automotive industry, in supplier companies or in those industrial branches which are economically weaker and less organized in terms of employee representation?

The following points touch on some problems which have their objective basis in the systemic character of company internal integration and in the structure of conventional interest representation (Altmann, Düll 1987). As a scenario we have chosen the gradual, but still difficult implementation of individual components of CIM – i.e. CAD, CAM or production planning and control systems (PPC) etc. – which tend to be oriented to centralized control.

The effects resulting from *integration based on data technology* within the company (as caused by CAD/CAM, for example) and between companies (between client and supply companies, for example) are difficult for the works councils to assess. Past experience is no longer applicable. Once decisions have been made in favour of comprehensive systems which are open for further expansion, and the latter are installed, it is almost impossible to correct the effects. The works councils are understaffed and lack the required know-how to cope with this situation.

The *problem of obtaining information* undergoes a complete change. Up to now it was primarily a question of prompting management to provide timely and comprehensive information (Böhle 1986). But how is the information provided by management on manpower-relevant aspects of technology implementation going to be judged? Even for management, especially in smaller companies, it is difficult to objectively assess the implications of introducing various new technologies and control systems, as well as their possible extensions. Thus the required implementation period and the necessary skills are generally underestimated. At the same time, the works councils are no longer able to draw on traditional sources of information such as persons responsible for the manufacturing area or members of personnel management. Information channels to those promoting computer-based integration measures, such as project groups of engineers and planners do not exist, and, in any case, communication between the two parties on technical subjects is nearly impossible. Communication from the bottom upward, from the work-force, is also negligible, due to the fact that they have little or no experience with the new rationalization measures.

In the Federal Republic the traditional provisions of *protection against rationalization* only come into effect when a direct connection can be proved to exist between the rationalization measures and their effects on personnel or an individual person. In the case of integration measures it is difficult to prove such a connection. The reasons for rationalization and the effects on personnel are usually not connected to each other: the redundancies or transferrals of employees occurring in such cases often take place in upstream or downstream company areas, or in supply companies or downstream firms, such as repair workshops. The same thing holds true for the demands made on altered structures of working hours or for demands made on skills, etc. This means that it becomes increasingly difficult to apply the corresponding laws, collective agreements and company agreements. The interaction between management and works councils is beset with mutual uncertainty and lack of calculability and the conventional forms of solving conflicts lose their effectiveness (Köhler *et al.* 1987; Deiß 1988).

It remains difficult to estimate in what way the *structures of personnel and skills* within the companies will change; it is virtually impossible to make any generally valid predictions. Any statements made must do justice to the particular company and other industrial branch-related aspects. It is certain, however, that the weight given to various employee groups will shift and thus the basis of employee interest representation within the company. Competing interests among the employees, particularly those of the rationalization winners and the rationalization losers, will have an effect on the works councils and will make it difficult to achieve a uniform definition of problems, negotiation concepts and design objectives. It will be difficult to standardize negotiation results attained within a given company in other companies or throughout an entire industrial branch. Thus a pillar of solidarity and labour mobilization will break apart.

Employee training, or at least their adaptation to new technologies, will become more important for the companies; of course, the same also holds true for employees. Traditionally, the sectors of training and re-training have not been a central focus of negotiation for the works council, although a limited right of co-determination does exist. Both sides have a growing interest in training and it will certainly become an area of conflict due to contradictory interests: for one, re-training is connected with the selection of individual workers instead of ensuring comprehensive training for all; furthermore, some forms of re-training, for example in quality circles, are carried out in a way that primarily ensures the acceptance of company measures and thereby undercuts the co-determination rights of the works councils; and lastly, it will be difficult to translate the newly acquired skills into higher wages, as they do not fit into the traditional job evaluation criteria.

New technologies offer new and considerably enlarged potential for *centralized monitoring and control of performance and employee behaviour*. This has often been paraphrased as the development toward the 'transparent employee'. Of course, this is currently more a matter of discussion than actual reality – apart from those exceptions which must nevertheless be taken seriously. Management's increasing potential for acquiring information concerning the work situation and employee behaviour deprives the works council of its up to now 'exclusive, hands-on' knowledge and thus its basis for demands, negotiation leeway and mobilization.

All of this is hard to reconcile with the *unions' demand for a 'humane design' of technology and organization*. Works councils in small and medium-sized companies hardly have any chance of receiving direct advice from their unions when new technologies are introduced, or in those cases where such technologies are forced upon their company from the outside within the context of integration measures. Here, the classic forms of union education prove inadequate. By no means is it possible to turn works councils into broad-based experts; however, they must at least be capable of making well-founded and viable demands on company experts and on management if they wish to achieve anything at all. Whether they will be able to count on the help of union-oriented engineers, computer scientists or similar employee groups must remain totally open at the present time.

A second scenario of systemic rationalization deals with the inter-company division of labour and integration. The effects that rationalization measures have on upstream and downstream companies are not considered at all by the works councils of those companies from which they originate. Traditionally, there has been virtually no contact between the works councils of companies which are part of a logistics chain. (Incidentally, this is quite in contrast to the many efforts made by the management of supplier companies to keep in touch with each other and find a joint approach to at least some of their problems.) This occurs all the more because the supply, service and trade relations of these companies extend beyond regional boundaries and the limits of industrial branches and thus go beyond the scope of the industrial unions.

The works councils in upstream and downstream companies cannot count on receiving any information or help from their colleagues. In so far as we are talking about works councils of small and medium-sized suppliers, they are in a much weaker position from the start. In such companies, management's capacity for devising strategies is more or less limited, particularly when the company is part of a technological and organizational integration concept, via delivery systems, for example. Such companies not only have to cope with pressure exerted on prices by the purchasing companies, but must handle extensive demands on the technological and organizational design of their own manufacturing processes. It is decisive here that it is not only a matter of the company's own machinery and facilities; integration into a wider system as we have shown – directly affects the skills structure of the employees, the time structure of work, etc. The extension of integration measures requires negotiations and agreements with the superordinate companies, and possibly with external advisors, in which the works councils cannot take part according to present works constitution and co-determination laws. Management of supplier companies may make tactical use of their limited capacity for devising strategies by pressuring the works councils to accept certain measures lest the company's existence be threatened.

In sum: in most cases there is no forthcoming information or cooperation whatsoever from the works councils of the superordinate companies which could provide the works councils of dependent companies with an indication of how external rationalization measures might affect staff within their own companies.

The basic question of the out-sourcing or the re-integration of production is also a special problem for the works councils of the large, dominant companies: make-or-buy decisions are no longer solely determined by economic and market criteria, but also by whether or not suppliers can be integrated into the overall logistic chain by means of data transfer technology. In such instances the works councils have virtually no chance of influencing MOB decisions. Even in cases where at least some rights to check on out-sourcing decisions have been granted, as was the case in a large-scale company in the Federal Republic at the end of the 1980s, there is no feedback at all on the effects on employment in the given supplier companies.

This 'non-relationship' between the works councils even extends into subsidiaries of one company. In instances where there are formal relationships, via the general enterprise works council, for example, or through direct personal contacts, interplant relations are often deliberately blocked off and information withheld when it is a matter of employment security at one's own location. This 'non-relationship' is even more pronounced where the connections extend beyond the boundary of a given industrial branch, i.e. in instances where the respective union cannot be called upon. In this context, one should keep in mind that when regional union representatives are summoned by the works council, in principle they are only concerned with solving the problems of the given company. These problems are beginning to get recognized and at least within the automotive industry, a change in attitude is beginning to occur at the regional level.

All in all, the unions are faced with a critical and completely unsolved organizational problem, namely the organization of employee interest representation throughout the logistics chain. This is all the more a problem in that concepts for coordinating the works councils must not only extend beyond the individual companies, but also beyond entire industrial branches. Global sourcing and internationalization of the firm itself (for example, in connection with the European Common Market) will intensify this problem – not to mention 'new problems' caused by the integration of newly installed works councils in the region of the former GDR. Its solution is an urgent task which the industrial unions will not be able to ignore much longer in view of systemic rationalization in the form of inter-company integration and internationalization.

Notes

1 NO END IN SIGHT – CURRENT DEBATES ON THE FUTURE OF INDUSTRIAL WORK

1 There are a number of works in German on the development of German industrial sociology. A few examples are: Lutz, Schmidt 1977, from the perspective of ISF; Schmidt 1974, in the context of European and American industrial sociology; Schmidt *et al.* 1982; and Pries *et al.* 1989; 1990, on the 1980s; Brandt 1990, with regard to Marx's influence on German industrial sociology. Foreign language analyses on this subject can be found in Lutz 1978; Altmann 1987.

2 This is one source of the difficulty in translating the texts in this volume. References to social theory and specific theoretical traditions oblige the use of special terms even in detailed analyses of particular technological and work systems (see chapter 2). See Lutz, Schmidt 1977 for a more detailed discussion of this issue.

3 A good overview of the German discussion on social scientific research methods and an extensive bibliography can be found in Kern 1982.

4 A number of methods are used simultaneously or successively in the case study: expert interviews, intensive interviews, group discussions, various forms of analyzing jobs and work processes, analyses of company materials, company-related analyses of the regional labour market, and local conditions surrounding the company. More in-depth questioning of the labour force generally proceeds with partially standardized interviews that are mostly based on the experiences gathered from the qualitative methods.

5 See the comprehensive literature in the Federal government's research project 'Humanization of Working Life'; on the international discussion Coriat 1982; Hirschhorn 1984; Knights, Willmat 1985; Blackburn *et al.* 1985. Important contributions on this subject are: Mickler 1981; Dohse *et al.* 1984. From ISF: Asendorf-Krings 1979; Altmann *et al.* 1982a; Drexel 1981; 1981a; 1982; Drexel, Nuber 1979; Köhler, Schultz-Wild 1985; Lutz 1976, 1981a, 1982a.

6 See Brandt 1981, 1986, 1990; Benz-Overhage *et al.* 1982; Bergmann *et al.* 1986; Malsch, Seltz 1987; Jürgens *et al.* 1989; Pries *et al.* 1989, 1990; and many others.

7 The work of ISF is based on theoretical considerations, which cannot be presented in detail here, known as the 'Munich *Betriebsansatz* (strategy

approach),' which achieved wide recognition in German industrial sociology. See Altmann, Bechtle 1971; Altmann *et al.* 1978; Bechtle 1980.

2 STRANGER IN PARADISE – AN AMERICAN'S PERSPECTIVE ON GERMAN INDUSTRIAL SOCIOLOGY

1 This paper is based on several years' experience working in German research institutes, discussions with other strangers grappling with the intricacies of German social structures and institutions, and explanations from willing German colleagues about their society and their research.

2 Interestingly enough, quantitative data analysis has gained momentum in German academic circles in recent years. Influences from abroad, pressure from companies and government agencies, the increased attempt to 'market' results, and desires to be legitimated through the use of numbers, even questionable ones, all have led to this development. Ironically, in the United States, historical sociology and more qualitatively directed research are gaining ground, probably for the same reasons that the opposite trend is occurring in Germany.

3 Having an 'Abitur' does not rule out the vocational education route. Several training programmes, particularly in the white-collar sector, prefer their participants to have completed *gymnasium*.

4 Union structure is unusual in German institutional forms in the way boundaries between worker classifications and employee types are crossed. This came about because, after the Second World War, unions decided that they should not compete with one another, as politically based unions do, and decreed that only one union could exist in a particular industry (*Industriegewerkschaft*) and, above all, for all employees (both blue- and white-collar) in one plant (*Einheitsgewerkschaft*).

Unlike the United States, in Germany, a particular industrial union (from the German Federation of Trade Unions, DGB) has the right to represent all of the employees in that industry, and all firms belonging to the employer association are bound by its agreements. (The situation for small general unions such as the Federation of the Christian Unions (CGB) or White Collar Worker's Union (DAG) is somewhat different. They can represent employees across industrial lines, but have little influence.) Employees themselves decide whether they personally want to be union members. In 'closed or union shops' in the United States, the employees vote for or against unionization and a certain union. If the majority votes pro-union, all employees covered by union agreement must be members.

5 This is a complicated issue since the relationship between skill, job classification and wage differs by industry and private and public sector in Germany. Here we are speaking only about the manufacturing sector such as the auto industry, metalworking industry, electrical industry, etc.

3 THE CONTRADICTIONS OF POST-TAYLORISTIC RATIONALIZATION AND THE UNCERTAIN FUTURE OF INDUSTRIAL WORK

1 This chapter was written in the context of the ISF project 'B2' on post-Tayloristic rationalization in conjunction with the special research unit SFB

333. It is based in part on the results of the research presented in Part III of this volume.

2 This path refers to Peter F. Drucker, an American specialist on organization, author of *Concept of the Corporation* (1972b), who has written extensively on humanistic forms of work organization.

4 SYSTEMIC RATIONALIZATION AND INTER-COMPANY DIVISIONS OF LABOUR

1 This article is based on empirical research on rationalization in the German furniture-making industry, carried out between 1983 and 1985, and on research dealing with the establishment of new relationships between manufacturers and suppliers in the automobile industry, which consisted of intensive case studies in different branches of that industry. The most important publications coming out of this research are: Altmann *et al.* 1986; Döhl *et al.* 1989; Deiß *et al.* 1989; Altmann, Sauer 1989, and Deiß *et al.* 1990.

2 It has become increasingly difficult to make a conceptual distinction between technology and organization at the company level in developed capitalist countries. Hierarchical organizational constraints are built into the organization of information and communication technologies and are inseperably interconnected within organization technology. Thus, organization technology's function in the process of rationalization – mediated by production techniques and labour – is to bring about and control the entire organizational network or, in other words, to coordinate all productive work (see Brandt *et al.* 1978).

3 As a result, new dependencies as well as new sectoral and industrial structures develop (for more on this issue, see the contributions in: Altmann, Sauer 1989 and Bieber, Sauer 1991).

4 This is not contradicted by the fact that the application of systemic rationalization measures has significant effects upon workers. We would like to point out here, for instance, that tendencies toward extensive utilization of technological equipment have led to an enormous increase of work at night and on Saturdays (as well as to a debate on the extension of work to Sundays). In line with the often discernible reduction of in-house manufacturing in the centres of capital accumulation (see pp. 56–9), this extension of de facto work hours also effects upstream production stages.

5 ON THE HISTORY OF NC-TECHNOLOGY – DIFFERENT PATHS OF DEVELOPMENT

1 The issues raised here represent preliminary hypotheses and findings coming out of an ongoing research project at the ISF in Munich (see Hirsch-Kreinsen 1989). The empirical basis of the following arguments are the extensive material laid out by David Noble (Noble 1979; Noble 1984) concerning NC-development in the United States, and about a dozen detailed conversations conducted mainly in 1988 with experts who had a large role in the development of NC-technology in Germany. The undoubtedly important role of Japan in the general issue of NC-development is not dealt with here due to the preliminary nature of the work presently underway.

6 THE ROLE OF MANUFACTURING TECHNOLOGY MARKETS IN SYSTEMIC RATIONALIZATION PROCESSES

1 This chapter is based on part of the results of an extensive empirical investigation which was carried out on the subject of changing rationalization strategies in the furniture industry, the machine tool industry and the foundry industry. The study was completed in 1987; the findings on the furniture industry have since been published (Deiß *et al.* 1989; Döhl *et al.* 1989); see also chapter 15. A less detailed version of this paper was the basis for a presentation, later published in Lutz (ed.) 1989.

2 Two points should be made here: (1) These two trends represent widespread marketing strategies, which can be plausibly argued at a theoretical level; they are, however, ideal-typical forms, and thus not completely generalizable. (2) We view firms who belong to a widespread 'middle group', not dealt with here, who try to pursue both market strategies or try to market products somewhere between exclusive and mass-produced, in danger of disappearing from the market. At the same time, if firms from the two types described above manage to obtain an economic and technological-organizational foothold, they are in a good position to eventually pursue both strategies. In this context, a selection and centralization process takes place, which could conceivably cause a convergence of the two market strategies. This process is too complex to go into detail here.

3 At this point, it should be pointed out that the existing (or potential) network of supplier relations effects the solution to these problems (see Deiß 1989; Döhl 1989, and chapter 22 of this volume).

7 THINKING MACHINES, DREAMING ENGINEERS? TOWARDS APPLIED EXPERT SYSTEM TECHNOLOGY

1 This chapter was written in conjunction with the German Senate's Investigatory Commission on Technology Assessment. An earlier version, under the title, 'Evaluation of Expert System Applications in Industrial and Service Environments', appeared in: T. Bernold, U. Hillenkamp (eds): *Expert Systems in Production and Services II*, Elsevier Science Publishers, Amsterdam/Oxford/New York/Tokyo 1989, pp. 130–146.

2 In the third case the utilization depends on the source of the shortage, that is the company interpretation of the source, whether this is personnel policy, the labour market or any other factor.

3 Government-defined vocational training of skilled workers according to specified occupations, usually a 3 1/2-year apprenticeship, with alternate training in companies and education in public vocational school.

4 International comparative investigations have shown that countries with a lesser developed system of vocational training, such as Great Britain and the United States, show a higher degree of in-plant division of labour under similar conditions (for example Lutz 1976; Sorge *et al.* 1983).

5 In several enterprises I had the curious experience when dealing with managers and system engineers, in which the very same people who vehemently opposed socialist economic planning pushed totally planned systems in their own enterprises. The vision of total calculability, rejected on a social level, seems to be very attractive at the enterprise level.

6 This replacement may occur in a number of different ways: replacing parts and aspects of expert work in order to accelerate work speed; dividing expert work into development of knowledge and a larger number of application functions; and in some rare cases, replacing experts altogether.

8 COMPUTERIZED MANUFACTURING AND SENSORY PERCEPTION – NEW DEMANDS ON THE ANALYSIS OF WORK

1 This chapter is based on two research projects, whose results are published in: F. Böhle, B. Milkau: *Vom Handrad zum Bildschirm – Eine Untersuchung zur sinnlichen Erfahrung im Arbeitsprozeß*, Campus Verlag, Frankfurt/New York 1988, and H. Rose (ed.): *Programmieren in der Werkstatt – Perspektiven für Facharbeit mit CNC-Maschinen*, Campus Verlag, Frankfurt/New York 1990.
2 For more detail see Hacker 1987: 355
3 See in this regard Prinz 1984 and the critical discussion of psychological theories of perception in Hoffmann-Axthelm 1984, as a review of theories of perception.
4 In its original meaning in the study by Popitz, Bahrdt *et. al.*, this refers to the ability 'to perform a complicated technical operation in such a way as to be able to assimilate it immediately in one's own motions and to continue doing it'. It is described in more detail as follows: 'This is neither an intellectual achievement nor is it a mechanical or automatic form of response. The word 'intuitiveness' comes to mind, because it nicely describes the state between the two.' (Popitz, Bahrdt *et. al.* 1957: 197.)
5 Individual approaches to this (e.g. Popitz, Bahrdt *et al.* 1957) have not been developed systematically. Only in studies on 'emotional work' in the area of personal and support services or in coming to terms with 'housework' have phenomena such as emotionally guided action, and subjective involvement been included as important components of job skills. Extensions of the concept apply, however, only to dealings with people, and even then just to particular aspects of work-related actions.
6 Studies on new kinds of dangers in the work process deal, for example, with such phenomena as: the destruction of holistic, emotional perception and the ability to think intuitively. Critics of artificial intelligence see man's superiority as lying in his capacity for intuitive action, associative thought and physical sensory perception. In our opinion, these studies indicate important aspects of dealing with new technologies; but here, too, reference is made mainly to human skills whose significance for the work process has yet to be defined. See critical discussions of new technologies e.g. Weizenbaum 1978; Brod 1984; Turkle 1984; Rose 1982; Roszak 1986; Mettler-Meibom 1987; Volpert 1985. For critical discussions on the limits of artificial intelligence, good examples are Dreyfus 1985 and Dreyfus, Dreyfus 1986.
7 Our approach differs in this sense from primarily psychological theories (of perception) which do not explicitly take into account and reflect this relationship. Thus the emphasis is not on intra-psychic events and physiological processes, but on the function of sensory perception in practical action.
8 The focus here is on the identification of differences between man and machine, i.e. the computer (see Volpert 1987).

9 This largely corresponds to the term emotion as used in psychological theories; a good example is Izard 1981.

10 See the description and critical appraisal of the concept of feelings which predominates in science and practice in Ulich 1982.

11 Particular mention should be given to studies and concepts of sensory perception based on phenomenological approaches and Gestalt psychology (Straus 1956; Merleau-Ponty 1966, etc.); approaches in psychological research derived from the criticism of the 'cognitive transition' in psychology and in which feelings are seen as being more than merely subordinate or disturbing components of action and whose significance in terms of practical action are examined (Ulich 1982; Mandl, Huber 1983 etc.); studies of the systematic quality of eidetic and associative thought and intuitive cognition (Goldberg 1985; Watzlawick 1982 etc.); investigations into the social function and meaning of non-rational forms of behaviour which are significant in terms of cultural psychology (Boesch 1980; Boesch 1983; Lorenzer 1981) and, finally, studies within the framework of philosophical epistemologies in which the contemporary concept of rationality is modified and extended. See, in particular, the pathbreaking work of Langer 1965/84.

12 For the concept of the body referred to here see particularly Böhme, Böhme 1985.

13 In this context it should be pointed out, for example, that within the framework of a philosophical theory of cognition even as late as the nineteenth century, 'identification' was labeled an important principle of cognition (Böhme, Böhme 1985).

14 Excerpts from interviews with experts which now follow are based on a presentation of our findings in Böhle, Milkau 1988. We refer to these without quoting them explicitly.

15 The phenomena described here are mentioned in existing industrial sociological studies. However, they have never been systematically related to one another, but have been treated as isolated phenomena. Studies oriented to social-psychological factors and concerned with the way in which workers 'experience' their work represent a more productive approach. The results of these studies – for instance on skilled workers' relationships to their machines – closely resemble our findings (see Volmerg *et al.* 1986).

16 It should be pointed out here that this does not fully describe the activities of skilled workers. We have accentuated only those aspects which are of particular interest for our topic. Moreover, the following analysis is limited to skilled workers' activities in small and medium-sized series manufacturing, i.e. work and production structures that are not, or only slightly, affected by Tayloristic principles of work organization.

17 See, for example, the presentation in Bergmann *et al.* 1986; and the chapters in Part III.

18 See as an example the work of Kern, Schumann 1984.

19 Compare, for example, the study on development in the machine-building industry by Dörr 1985.

20 We use the term 'strain' here – in contrast to its use in work sciences – in the sense of demands which are experienced by the workers as basically negative and unmanageable.

9 DIFFUSION OF CIM-TECHNOLOGIES – DYNAMIC DISSEMINATION AND ALTERNATIVE PATHS OF INNOVATION

1 Portions of this paper, which is an extended version of one first published in *Computer-Integrated Manufacturing Systems (CIMS)*, vol. 2, no. 4, Nov. 1989, under the title 'On the Threshold of Computer-Integrated Manufacturing – Diffusion Trends of CIM Technologies in West German Industries', pp. 240–8, are reprinted by permission of the publishers Butterworth-Heinemann Ltd. The paper deals with results from a study on diffusion trends of CIM-technologies which ran from 1986 to 1989. The empirical basis of the project was a postal survey of standardized questionnaires in Autumn/Winter 1986–7 answered by about 1,100 firms from the German capital goods industries. As a follow-up to the postal survey, short case studies were undertaken in 60 firms of the sample which demonstrated advanced stages of computer integration. The results of the project are summarized in Schultz-Wild *et al.* 1989 and Hirsch-Kreinsen *et al.* 1990. Besides the author, members of the project were Christoph Nuber, Frank Rehberg and Klaus Schmierl.

2 Compare, for example, the illustrative descriptions of successful cases of computer-based industrial automation in Jorissen *et al.* 1986.

3 Especially with regard to the dynamics of the increasing application of flexible manufacturing systems and cells compare Fix-Sterz *et al.* 1987 and 1990. For a more detailed account of the possibilities and effects of this line of technology on employment policies compare chapter 11 in this volume and Schultz-Wild *et al.* 1986.

4 Refer to AWF 1985 for an attempt to clarify the abbreviations and terminology dealing with computer technology.

5 The following statements are based on 1,096 questionnaires filled out by West German companies in the capital goods industry. The results of the questionnaires were weighted by size of company and industrial branch according to their proportion of employment in the West German capital goods industry. The survey, carried out by ISF Munich, occured in the context of the RKW-Project A 161 and was conducted during the autumn and winter of 1986–7. For more details of the survey findings see Schultz-Wild *et al.* 1989.

6 Computer-aided production planning and control systems, also termed manufacturing requirements planning (MRP).

7 Computer-aided process planning and programming.

8 Computer-aided design (and engineering).

9 Computerized production data acquisition.

10 According to our survey findings skilled workers are the largest personnel group in machine-building companies; on the average their share is 41 per cent compared to 28 per cent in the other sectors of the capital goods industries. For blue-collar workers only, there is a ratio of about 1.7 skilled to 1 semi-/unskilled worker in machine-building companies whereas in the rest of the capital goods industry the ratio of about 1.2 semi-/unskilled to 1 skilled worker holds true. For further details see chapter 10 and Hirsch-Kreinsen *et al.* 1990: 40 ff.

11 Please note that in this case and also all others, specifications only refer to whether an individual company is using or plans to utilize a given technology in the near future. This says nothing about the number of systems or aggregates per company or the extent of application. For example there may be a considerable

future increase in the total number of CNC-machine tools even if there are hardly any new companies expected as first time users and a sort of saturation limit can be said to have been reached in this respect.

12 CAQ-systems predominate in the electrical industry, where they are far more widely utilized in companies of all sizes compared with mechanical engineering.

13 Only within the electrical industry, with its prevalence of large-scale enterprises, are higher diffusion rates in evidence.

14 The following paragraphs are based on contributions of my colleague Hartmut Hirsch-Kreinsen; compare our collective papers Hirsch-Kreinsen, Schultz-Wild 1990 and 1990a, and for a more detailed analysis and examplification Hirsch-Kreinsen *et al.* 1990.

15 For a more extensive and systematic analysis, compare Hirsch-Kreinsen *et al.* 1990, esp. chapter VI, from p. 139.

16 This point of view is also gaining increasing attention in the primarily technologically oriented discussions within engineering science; see for example Martin, Kivinen *et al.* 1990; Martin, Ulich, Warnecke 1990.

10 TECHNOLOGICAL INNOVATION – ORGANIZATIONAL CONSERVATISM?

1 The empirical basis of this chapter are two projects on technological and organizational change in the mechanical engineering industries, carried out in the years 1985–90. Both quantitative and qualitative methods were utilized. The results of the quantitative survey are presented in chapter 9. The analyses of organizational developments – the focus of this article – are based on 60 case studies in companies with a high level of CIM-applications. A summary of the two projects' research results can be found in Schultz-Wild *et al.* 1989 and Hirsch-Kreinsen *et al.* 1990. Members of the project group were – apart from the authors of this article – Marhild von Behr, Hartmut Hirsch-Kreinsen, Christoph Nuber, and Rainer Schultz-Wild.

2 The 60 case studies (between one and five days per plant) focused on manufacturing departments with a high rate of computer utilization. The main points of reference were the development of division of labour according to functions and skills and the concepts for rationalization put forward by the managers interviewed; for more details see Schultz-Wild *et al.* 1989.

3 On the difference between centralistic and workshop-oriented CIM-technologies, see chapters 5, 6, 9 and 18.

4 Compare chapter 11 for a detailed description of such a strategy.

5 The labour supply certainly is of high importance for the development of job structures. This becomes particularly evident in international comparisons. Compare Sengenberger in chapters 14, 16, 17; see also Düll 1989; Köhler, Grüner 1990.

11 FLEXIBLE MANUFACTURING SYSTEMS AND WORK ORGANIZATION

1 This article presents results from a project on flexible manufacturing systems, job structures and training policies conducted in the years 1978–85. The

conceptual background was developed in the the projects 'C3' and 'C4' of the special research unit 101 (see foreword and Part I). The analysis is based on a longitudinal case study dealing with the entire process of planning, designing and implementing a flexible manufacturing system (FMS) in a large West German machine-building plant. This process required approximately seven years (1977–84) and was sponsored by the German government. The project group participated as researchers in the implementation process. A summary of the findings is published in Schultz-Wild *et al.* 1986. Members of the project group – apart from the authors of this article – were Inge Asendorf and Burkart Lutz.

12 BRINGING SKILLS BACK TO THE PROCESS

1 This is the revised version of a paper based on studies commissioned by the CEC-FAST programme in cooperation with the European Centre for the Development of Vocational Training (CEDEFOP) entitled 'New Production Systems' (see Brödner 1987). Some parts of the paper were published in *Computer Integrated Manufacturing Systems*, vol. 1, no. 2, May 1988, pp. 82–8, and later in Warner *et al.* (eds) 1990, pp. 87–99 and are reprinted by permission of the publishers Butterworth-Heinemann Ltd. It is based on results of several research projects on computer-integrated manufacturing (CIM) and the development of skill structures in West German metalworking industries, results of which are presented in the preceding chapters (see also Hirsch-Kreinsen *et al.* 1990). The author would like to thank his team colleagues from the ISF, particularly Hartmut Hirsch-Kreinsen, Christoph Köhler, and Christoph Nuber, for their helpful suggestions and feedback.

2 For a discussion of the advantages of the use of qualified skilled workers in complex manufacturing systems see chapter 11.

3 Compare, for example, the paper by Peter R. Everitt 1985: 'The Manufacturing Cell: 'A Plant within a Plant''.

4 This was how an executive of a German machine-building company at Systec '86 in Munich paraphrased the demand presently being voiced to give more consideration to those affected by changes in the work situation when planning and introducing new computer-aided systems (Hummel 1986).

5 Compare individual CAM-components: CNC, DNC, FMC, FMS etc. Unfortunately the use of terminology for CIM-components is by no means uniform (see Lay 1987).

6 See also Braczyk's argument about 'prescriptions' on work organization and skills embodied in production technologies (1987).

7 For examples see Hirsch-Kreinsen, Schultz-Wild 1990a, Hirsch-Kreinsen *et al.* 1990, and chapter 9 in this volume.

8 It is interesting to note that recently there are indications that systems which started out with different basic concepts are beginning to converge. However, this is likely to prove to be a very long-term developmental process. An example of this is in the development of NC-programming methods. It can be assumed that due to the further development of interactive-graphic programming methods a standardization of operator modi will take place between systems originally based on office programming and those oriented towards workshop programming (see chapter 5). Generally, the concepts of open-system

architectures seem to better suit the diverse technical and organizational conditions of different users and open up market opportunities, especially for small supplier companies.

9 For a more detailed report of the relevant constellations and interests in the course of company implementation processes see Hirsch-Kreinsen, Schultz-Wild 1986a; compare also Hirsch-Kreinsen *et al.* 1990 and chapter 10 in this volume.

10 See the recommendations by the working group (organized by FAST and CEDEFOP 1986/87) to the European Community concerning initial and continuing training for new manufacturing systems, Sellin 1987. Compare also Sorge 1990.

11 A systematic, comparative analysis of the different European educational and vocational training systems is naturally not the intent of this chapter; we will restrict ourselves in the following to pointing out several problem constellations which are important for the question of future factory structures.

12 The dual system is the combination of in-plant training and vocational training in public schools in the FRG (see chapters 2 and 16).

13 A comparison of the utilization of flexible manufacturing systems in France and in Germany, for example, indicates that certain variations in system-concepts and layout (greater significance of the central computer and software programs in France; stronger concentration on processing machines, handling facilities and transport systems in Germany) relate to different labour supply and professional orientations of engineers. See chapter 18.

14 See, for example, the case where semi-skilled workers were successfully trained to become skilled operators in a complex flexible manufacturing system (chapter 11); for more details see Schultz-Wild, Asendorf *et al.* 1986.

13 WORK STRUCTURING AND COMPANY PERFORMANCE POLICIES

1 This chapter is based on surveys on the implementation of 'New Forms of Work Organization' in the late 1970s and the beginning of the 1980s in the FRG as well as in Italy and France (Altmann, Düll 1979; Düll *et al.* 1983), and comprehensive empirical studies (1976–80), which made use of case studies, process analyses, expert interviews and group discussions (nine long-term and 12 short-term case studies in companies, mainly in mass-production industries, including 17 cases of new work forms, compared with nine traditional ones). The results are published in Altmann *et al.* 1981; 1982a. This contribution is a revised and edited version of an article published in 1982 in German (Altmann *et al.* 1982).

2 By sequential processes, we mean forms of work organization in which the individual operations within a section of manufacturing or assembly follow one another both in terms of steps and time. In contrast to assembly-line work, however, manual and automatic stations, work stations with and without cycles, connected and unconnected to technical systems, divided and not divided by buffers, are connected to one another. Sequential processes differ from the job shop principle, production islands, and automatic processes, etc.

3 For an investigation of the works councils' position on these issues see Altmann *et al.* 1981.

14 THE FUTURE OF THE MASS PRODUCTION WORKER

1 This chapter is based on joint international research conducted by the ISF and the Paris-based Groupe de Sociologie du Travail, CNRS. The study was entitled 'Assembly automation as an element of company-wide rationalization'. German members were – besides the authors – N. Altmann and M. Moldaschl; French contributors were D. Chave and G. Lemaître. Research was undertaken between 1986 and 1990 in German, French and Italian production sites of a (European) multinational company in the consumer electronics field. Of central concern was the relationship between company rationalization strategies and personnel policy. The method used was case studies relating to the production and assembly of televisions. Research results will be published in 1991.

2 In answering this question we will not specifically consider an important socio-economic dimension of the mass production model's crisis, namely the labour market and the gradual diminution of that labour reservoir, out of which the 'mass production worker' was previously recruited (Lutz 1982b).

3 Here, we will not take into consideration the fact that Germany has, of course, always witnessed contradictory interests between labour and its organizations of collective representation concerning the societal regulation of performance intensity and wage levels (see Part VII).

4 Discussions with experts were carried out on all levels of the company organization (a total of 150 discussions); they ranged from the company management to the foremen, and included technical offices directly dealing with production, members of the work council, and union representatives (from outside the company). The collection of material was supplemented by a total of 40 very detailed work-psychological analyses and 150 interviews with assembly workers.

5 CAP = *Certificat d'aptitude professionelle*. In France, this certificate is a basic substitute for initial training; similar to the German craft certificate, it is generally recognized in the labour market. Significant differences between the two do exist, both with regard to the training content and the transferability of the respective certificates.

6 Both in the Federal Republic and in France, social plans serve as a 'settlement' between the company and its employees in cases of dismissals; such plans usually entail a financial settlement. In the Federal Republic, the work council has the right to co-determine the social plan's design. 'Alternative social plans' (in French: *reconversion*) attempt to substitute financial settlements with measures aimed at a reintegration into the labour market. Besides further training, these measures include placement assistance, creation of 'employment companies' along the lines of technology parks, etc. Despite several spectacular cases (e.g. in the French steel industry), 'alternative' social plans have not really got off the ground. For the situation in the Federal Republic see Part VII.

7 Directly connected with the group model is the master craftman's loss of function. He is taken out of the hierarchy and is given a kind of directing function as a 'technical advisor' of several groups ('area technician'). The groups are only responsible to the area supervisor, who is responsible for a total of about 300 employees.

8 For a discussion of the historical development of union positions regarding the model of Tayloristic work organization in France and in the Federal Republic, see Linhart *et al.* 1989.

9 Regarding the system of 'norm-based negotiation' in the Federal Republic see Düll, Bechtle 1988.

15 TOWARDS A POLARIZATION OF SKILL STRUCTURES

1 This chapter is based on the findings of a comprehensive empirical investigation concerned with the implementation of new technologies in the woodworking industry in the Federal Republic of Germany. Particular attention was paid to the furniture-manufacturing industry. The investigation was carried out between 1983 and 1987 and consisted mainly of case studies and interviews conducted with experts, which were performed at numerous furniture manufacturers (firms utilizing new technologies), as well as analyses carried out at supplier firms and the manufacturers of wood processing machinery and new organization and control technologies. For a detailed account of the findings see Döhl *et al.* 1989, and Deiß *et al.* 1989.

2 Moreover, it should be pointed out that the (often occurring) initial increase of auxiliary tasks such as the feeding, piling and transport of workpieces, is followed by the step-by-step automation of transport and handling areas; this, in turn, invariably results in the reduction of many unskilled and simple semi-skilled jobs.

3 Concerning the development of the purchasing firm's demands and the changes on the supplier markets in this branch, see Deiß 1989, and Döhl 1989a.

16 VOCATIONAL TRAINING, JOB STRUCTURES AND THE LABOUR MARKET – AN INTERNATIONAL PERSPECTIVE

1 This chapter arose from the ISF project 'B 5' on 'Labour Market Structures and the Small Firm Sector' carried out in conjunction with the special research unit 333. It summarizes research results from several studies on company job structures and labour market structures. For a more comprehensive discussion of these issues, see W. Sengenberger: *Struktur und Funktionsweise von Arbeitsmärkten – Die Bundesrepublik Deutschland im internationalen Vergleich*, Frankfurt/New York 1987.

2 During the 1980s re-definition and restructuring of occupational classification took place. In the metalworking sector, for example, 40 different occupations were consolidated into six major categories.

3 This is not to say that the overall representation of worker interests and worker concerns is necessarily inferior in Germany. It suggests merely that, by international comparison, job control through restrictive rules has not been very advanced in West Germany.

17 EDUCATION AND JOB HIERARCHIES – CONTRASTING EVIDENCE FROM FRANCE AND GERMANY

1 This chapter came out of a presentation given at an international seminar of the UNESCO Institute for Education in 1978 and a 1981 published collection of the proceedings. The information is based on a detailed presentation and analyses of results from an investigation of seven matching pairs of French and German

factories carried out by ISF, Munich and LEST (Labouratory for the Economics and Sociology of work, Aix-en-Provence) described in Lutz 1976. 'Politique d'éducation et organisation industrielle en France et Allemagne' (Maurice *et al.* 1982; in English 'The Social Foundations of Industrial Power – A Comparison of France and Germany', Cambridge, Massachusetts 1984) is also based on material gathered from this study.

2 On this problem and the significance of the former higher technical institutions see Lutz, Kammerer 1975.

18 MECHANICAL ENGINEER VERSUS COMPUTER SCIENTIST – DIFFERENT ROADS TO CIM IN FRANCE AND GERMANY

1 This chapter is a product of the project group 'B2' ('Influences and paths of development of Post-Tayloristic rationalization strategies' of the special research unit 333). It is based on a wide range of research projects whose findings are presented elsewhere in this volume (see chapters 3, 5).

19 NEW PRODUCTION STRUCTURES À L'ITALIANO? – SIMILARITIES AND DIFFERENCES IN THE WEST GERMAN AND ITALIAN STEEL INDUSTRIES

1 This chapter is based on a presentation given for employee representatives and experts in the Italian Steel Industry in Sesto San Giovanni. It was first published in Italian and then – in an expanded version – in German (Drexel 1983; Drexel 1985). The information contained is partly based on an extensive investigation carried out in the West German steel industry (Drexel, Nuber 1979; Drexel 1982), and partly on secondary analyses of Italian studies (Butera 1979; Micheli *et al.* 1976; Bechtle 1982; Barisi 1982; Villa 1983), supported by discussions with Italian experts.

2 The investigations carried out by Drexel and Nuber in various continuous process industries in West Germany, particularly in the chemical and steel branches, centred on changes in the skill and training of production workers: the introduction of state-regulated and -controlled training schemes for skilled workers in areas, which up until then only had unskilled and semi-skilled personnel trained on the job (including the foreman [*Meister*]). These momentous changes observed in enterprise qualification structures and policy were revealed to be an aspect of a transformation of the entire structure of production work which was subject to the same logic in all branches of the continuous process industries included in our investigations: rationalization through a strong reduction in the quantity of labour, along with an intensified utilization of the quality of labour – thus, the same logic which is central to the rationalization strategies characterized by Kern, Schumann as 'new production concepts' (see Drexel 1980 for a summary).

3 The majority of these case studies are analyses and documentations of experiments with new forms of work organization, which were set in motion by companies and carried out by scientists. Thus the following analysis represents what companies and scientists considered the central problems; the opinions of

workers and their representatives are only indirectly represented, but will be dealt with below in the context of union demands.

4 Also for reasons of available material, the chapter is limited to the period up to 1981.

5 We must not, however, overrate the importance of the collective perspective implied in the *unità operative* as it is always limited to the individual team. Also, the workers belonging to wage group 6 are excluded from them.

20 THE DEVELOPMENT AND STRUCTURE OF SMALL-SCALE FIRMS

1 This chapter is based on a research project within the framework of the 1986 programme 'New Industrial Organization' administered by the International Institute for Labour Studies, Geneva (see Sengenberger *et al.* 1990). It is a revised version of a paper presented at the Congress of the Comitato Regionale Dell' Emilia Romagna of the Confederazione Nazionale Dell'Artigianato 1989.

2 There are no strict regulations or clear criteria, such as number of employees, turnover, or the existence of a particular production structure, in the Federal Republic governing the classification of a firm as a craft enterprise; it is simply a legal-normative definition. If a firm wishes to be classified as a craft enterprise, a master craftman's certificate is required in order to manage the firm.

21 SMALL SUPPLIER FIRMS AT THE CROSSROADS – MEETING NEW CHALLENGES THROUGH COOPERATIVE RETRAINING SCHEMES

1 This chapter is based on a research project on new demands on suppliers in large and small company alliances and the challenge for training policies; the research was conducted by the authors in 1989, in the course of which experts were interviewed from several supplier sectors with small company structures (hardware production, plastics processing, tool and mould-making), in large purchasing companies (from the automobile, electrical and mechanical engineering industries), and in further training institutions and associations. The German report was published in 1990 under the title: S. Weimer; H.G. Mendius: Neue Anforderungen an Zulieferbetriebe im Groß-Kleinbetriebsverbund – Eine Herausforderung für die Qualifizierungspolitik' in the series: Working Papers of the working group SAMF (Social Science Labour Market Research), Universität, Gesamthochschule Paderborn, November 1990.

2 It seems that small and medium-sized companies employ more unskilled and semi-skilled workers than was previously assumed; see Paulsen 1987.

3 Issues of finance, business economics and production engineering predominate in the consulting services offered by the statutory associations and by private business consultants.

4 In the Federal Republic of Germany, further training measures are offered by several institutions, such as the chambers of industry and commerce, craft associations, employers' associations and unions, state institutions and commercial further training enterprises, in addition to the companies themselves.

5 In view of the special types of qualification requirements (key qualifications)

connected with the new technologies, it is assumed that on-the-job learning processes, and hence the company as a place of learning, will become more important in the future.

6 According to one survey, 6 per cent of craft businesses and 23 per cent of industrial companies have already made use of training associations or are interested in them (see Meyer, Schwiedrzik 1987).

7 Experiences with primary vocational training associations reveal that the commitment of an individual institution played a central role in the creation of nearly all the associations (see Meyer, Schwiedrzik 1987).

8 This is true, for example, of certain basic qualifications in the area of CNC, CAD, SPC, etc.

9 For the sake of simplicity we speak of 'the suppliers'. Obviously, the situation is very complex and depends on the subcontracted product actually supplied (e.g. a relatively low-value serial part or a highly complex tool), the position on the product market in question, the share of turnover earned with the respective customer, and many other factors. For reasons of space, this article is unable to elaborate on most of the differences. Arguments refer generally to suppliers of serial parts (see Altmann, Sauer 1989).

10 Of course, there are enormous variations in the interest of a cooperation on the customer side, depending on the type of supplier and the products bought. Here it is important whether the customer is actually or potentially able to make the parts in question himself, whether he has the right production know-how for the job, whether there are many, few or no other roughly equivalent alternative suppliers, and how innovative the supplier is held to be, etc.

11 A further factor is that many subcontractors look on the staff in these departments of their customers first and foremost as being those persons who conduct the often highly unpopular 'audits', who reproach them for real or nonexistent defects, who impose conditions with more or less consideration, who classify them as A, B, C or D suppliers, etc. All these functions do not make it easy for the customer's agent to be viewed primarily as someone with whom you can confidently discuss problems and develop solutions.

12 The customers' further training departments have an additional specific and possibly very vital interest: cooperation on further training with subcontractors can help to raise their status within their own company. First in what tends to be a 'qualitative' dimension, i.e. when others perceive that they have contributed at least indirectly to improving the quality of the end product (a benchmark of every modern corporate philosophy); and second (and far more profanely), by showing that the cooperation on training with suppliers serves either to improve the utilization of (otherwise threatened) existing capacities or to catalyst the installation of additional capacities.

13 It is obvious that (even with extensive adaptation to the suppliers' needs) just a part of the measures organized by the customers is suitable for suppliers and that, vice versa, for most aspects of the suppliers' training needs the customers are irrelevant. In many cases, however, the provision of instructional materials and places of learning could be an important aid for initiating measures to be carried out by suppliers or further training institutions.

14 The fact that this could well provide a starting point for cooperation on further training is clearly shown by a workshop we conducted in the course of the project, at which we managed to assemble a number of representatives of the above-mentioned departments from several large purchaser companies,

executives from small supplier firms, representatives from associations and educational organizations, and scientists.

15 This would certainly be natural if all the businesses involved were suppliers of the large company participating in the training association. In our talks there was also occasional willingness to offer training beyond the group of companies for which business relations exist, e.g. to potential suppliers as well.

16 It should be noted that a worker with a fundamental, broadly based vocational qualification – in contrast, say, to a semi-skilled worker – is equipped with the preconditions for obtaining additional qualifications which, although highly oriented to actual conditions or company-specific problems, can be used in other areas than the one in which they were acquired. An occupational qualification provides good opportunities for productive integration of company-specific elements – a basis lacking even in those unskilled and semi-skilled workers with lengthy experience.

17 The role played by worker representatives in company further training is not discussed here. The question as to whether 'further training is a sphere of activity for works councils' has been up for debate for at least 15 years (see Maase *et al.* 1975). The question has always been answered in the affirmative by the unions; recently there have also been collective agreements with regulations to this effect and more are being striven for. It is also true, however, that in those small businesses where there is a works council, further training is, as a rule, not one of the works council's priorities, in contrast again to initial vocational training. Forward-looking further training concepts still depend on the employee and on the cooperation of the of company representatives. Some deliberations on this subject are to be found in: Mendius, Wendeling-Schröder (1990).

18 The 'training' focal point of the special labour market programme in 1979 can be taken as an example. Some of its initiators also looked on it as a sounding board for the promotion and regulation of further training within the framework of the Employment Promotion Act. Not least for this reason it was the centre of massive political disputes, but at the same time it was also accompanied and evaluated scientifically with relative intensity. There was – sometimes massive – criticism of the programme's details and of substantial shortcomings in management, resulting in considerable cashing in. On the whole, however, the promotion of company training was classified generally as a sensible instrument of labour market and training policy, which needed to be refined and selectively applied (see for example, Mendius *et al.* 1983; Bosch *et al.* 1983; Sauter 1982; Scharpf *et al.* 1982). With the change of government in 1982, critics of the political components of the special programme took over the functions, and since then the findings do not seem to have been used systematically, nor have the medium-term effects been examined.

19 Consideration could be given, for example, to the supplier circles which – originating in Baden-Württemberg – already exist in a number of the German states. Medium-sized businesses from various supply sectors have joined forces in these circles, mainly so as to work together in the procurement of orders, in purchasing, in developing contracts and in sales.

20 In other words, the company and inter-company worker representatives should be included along with the participating suppliers and purchasing companies, professional associations and training organizations.

22 SMALL FIRMS IN BIG SUBCONTRACTING

1 In addition to the quoted sources, this chapter is based on a number of interviews with experts, particularly with individuals from trade associations and several large German automobile manufacturers; these interviews were conducted within the framework of a project sponsored by the Federal Ministry of Reseach and Technology and carried out together with H.G. Mendius and S. Weimer on the potential of small firms for taking strategic action. The contribution is a revised version of K. Semlinger, 'Stellung und Probleme kleinbetrieblicher Zulieferer im Verhältnis zu großen Abnehmern', in N. Altmann; D. Sauer (eds) *Systemische Rationalisierung und Zulieferindustrie*, Frankfurt/New York 1989.

2 For example, the German Automotive Industry Federation estimates that 280 firms account for 80 per cent of the parts and components supplied to the automotive industry by subcontractors.

23 UNIONS' POLICIES TOWARDS NEW TECHNOLOGIES IN THE 1980s – AN EXAMPLE FROM THE METAL INDUSTRY

1 This chapter is based on a synthesis of results from diverse studies on work design, carried out by the ISF München and which also included union policies and activities. It is a revised version of a paper presented to the Nordic Institute of Advanced Occupational Environment Studies (Espoo, Finland, 1987).

2 See Altmann, Düll 1979 for an extensive bibliography.

3 For a further discussion of the issue of inter-company integration and its consequences for the unions, see chapter 24.

4 Much of the information in this section is based on union and employer periodicals, in particular: WSI-Mitteilungen; Die Mitbestimmung; Kittner, Gewerkschaftsjahrbuch (yearly); Zimmermann 1982; Institut für Angewandte Arbeitswissenschaft Köln: Mitteilungen; Bundesvereinigung der Deutschen Arbeitgeberverbände 1983.

5 Ongoing discussions in the United States and also those in the Federal Republic generally foresee a decline of the unions (see Müller-Jentsch 1988). The following factors are held responsible for this: the differences between old and new industries; the increasing use of qualified manpower (engineers, white-collar workers) without any orientation, ties or loyalty to the unions; polarization of staff according to rationalization 'winners' and 'losers'; decreasing company-size and extinction of the 'mass worker'; company specific works council orientation; decreasing chances for a standardization of working conditions; restriction of union activities to the area of service functions.

In the Federal Republic of Germany, however, there are trends moving in the opposite direction which should be pointed out:

1 apparent willingness of union members to actively seek better working conditions and their solidarity even in times of economic crisis (strike for 35-hour week);

2 altered priority of values on the part of the employees, implying new goals in life and less blind acceptance of 'technical progress';

3 new social movements with effects on the unions (women's liberation movements, the German Environmental Parties, i.e. the Green Party);
4 young workers who disapprove of the unions' bureaucratic organization while accepting their political objectives;
5 the continued recognition of the unions as the sole representative of employees, regardless of their actual integration;
6 initial fears of the effects of rationalization among engineers and white-collar workers and their need for representation; and
7 the employers' own interest in functioning unions, due to the fact that the traditional forms of industrial relations ensure calculability and relatively limited conflicts.

These aspects point away from the thesis of the decline of the unions in the FRG.

24 RATIONALIZATION STRATEGIES AND REPRESENTATION OF WORKER INTERESTS

1 This chapter extends the analysis provided in the previous contribution on Work Design and Unions in that it addresses the problems of the works councils and thus the 'dual structure' of interest representation. It is based on the same study as the preceding chapter as well as the investigation on systematic rationalization (see Döhl *et al.* 1989; Deiß *et al.* 1989; Altmann, Sauer 1989).
2 We deal with the effects of rationalization using the example of the automotive industry.
3 This is a German concept of a type of skilled worker which derived from some companies' attempts to train skilled workers so that they could master all components of a modern production process (electronics, mechanics, hydraulics, pneumatics, etc.). This allows a group of workers who are overseeing a production facility full autonomy and flexibility in carrying out their work assignments. In practice, this concept was never realized.

References

Acs, Z.J.; Audretsch, D.B.: 'Entrepreneurial Strategy, Entry Deterrence, and the Presence of Small Firms in Manufacturing'. Discussion Paper IIM/IP 86–28, Wissenschaftszentrum Berlin für Sozialforschung, Berlin 1986.

Aiginger, K.; Tichy, G.: *Entwicklungschancen der Klein- und Mittelbetriebe in den achtziger Jahren.* Politische Studien, Sonderheft 1, München 1982.

Altmann, N.: 'Company Performance Policies and the Role of the Works Council'. In: S. Tokunaga; J. Bergmann (eds): *Industrial Relations in Transition*, University of Tokyo Press, Tokyo 1984, pp. 255–80.

Altmann, N.: 'Shin Gijutsu to Rodo Kumiai – Sangyo Reberu ni okeru Gorika to Rodo no Ningenka – (Qualität des Arbeitslebens, Rationalisierung und industrielle Beziehungen in der Bundesrepublik Deutschland)'. In: N. Altmann; M. Nomura (eds), *Nishidoitsu no Gijutsu Kakushin to Shakai Hendo (Neue Technologie und Strukturwandel der deutschen Gesellschaft)* Daiichi-Shorin, Tokyo 1987.

Altmann, N.; Bechtle, G.: *Betriebliche Herrschaftsstruktur und industrielle Gesellschaft*, Carl Hanser Verlag, München 1971.

Altmann, N.; Bechtle, G.; Lutz, B.: *Betrieb – Technik – Arbeit – Elemente einer soziologischen Analytik technisch-organisatorischer Veränderungen*, Campus Verlag, Frankfurt/München 1978.

Altmann, N.; Bechtle, G.: 'Gijutsu Kakushin to Rodo – 1970 Nendai, 80 Nendai no Nishi-doitsu Sangyo Shakaigaku – (Technischer Wandel und Arbeit. Zu einigen Aspekten deutscher Industriesoziologie in den 70er und 80er Jahren)'. In: N. Altmann; M. Nomura (eds), *Nishidoitsu no Gijutsu Kakushin to Shakai Hendo (Neue Technologie und Strukturwandel der deutschen Gesellschaft)* Daiichi-Shorin, Tokyo 1987.

Altmann, N.; Binkelmann, P.; Düll, K.; Mendolia, R.; Stück, H.: *Bedingungen und Probleme betrieblich initiierter Humanisierungsmaßnahmen, 4 Bde.*, Fachinformationszentrum Karlsruhe, Karlsruhe 1981.

Altmann, N.; Binkelmann, P.; Düll, K.: 'Neue Arbeitsformen, betriebliche Leistungspolitik und Interessen der Beschäftigten'. In: *Soziale Welt*, Heft 3/4, 33. Jg., 1982, S. 440–65.

Altmann, N.; Binkelmann, P.; Düll, K.; Stück, H.: *Grenzen neuer Arbeitsformen – Betriebliche Arbeitsstrukturierung, Einschätzung durch Industriearbeiter, Beteiligung der Betriebsräte*, Campus Verlag, Frankfurt/New York 1982a.

Altmann, N.; Deiß, M.; Döhl, V.; Sauer, D.: 'Ein "Neuer Rationalisierungstyp" – neue Anforderungen an die Industriesoziologie'. In: *Soziale Welt*, Heft 2/3, 37. Jg., 1986, S. 191–206.

Altmann, N.; Düll, K.: *New Forms of Work Organization in the European Community – Germany*. (European Foundation for the Improvement of Living and Working Conditions, ed.), Dublin 1979.

Altmann, N.; Düll, K.: 'The Role of the Parties Involved in the Introduction of New Technology – Company Strategies and Participation', mimeograph, Dublin 1985.

Altmann, N.; Düll, K.: 'Rationalisierung und neue Verhandlungsprobleme im Betrieb'. In: *WSI-Mitteilungen*, Heft 5, 40. Jg., 1987, S. 261–9.

Altmann, N.; Nomura, M. (eds): *Nishidoitsu no Gijutsu Kakushin to Shakai Hendo (Neue Technologie und Strukturwandel der deutschen Gesellschaft)*, Daiichi-Shorin, Tokyo 1987.

Altmann, N.; Sauer, D. (Hrsg.): *Systemische Rationalisierung und Zulieferindustrie – Sozialwissenschaftliche Aspekte zwischenbetrieblicher Arbeitsteilung*, Campus Verlag, Frankfurt/New York 1989.

Anglo-German Foundation for the Study of Industrial Societies (ed.): *New Technology Based Firms in Britain and Germany, Part 1*, London 1988.

Arnold, U.: *Strategische Beschaffungspolitik – Steuerung und Kontrolle strategischer Beschaffungssysteme von Unternehmen*, Europ. Hochschulschriften, Reihe 5, Volks- und Betriebswirtschaft, Bd. 380, Bern 1982.

Asendorf, I.: *Facharbeiter und Rationalisierung – Das Beispiel der großbetrieblichen Instandhaltung*, Campus Verlag, Frankfurt/New York 1979.

Asendorf-Krings, I.: 'Facharbeiterqualifikation und Rationalisierung'. In: *Gewerkschaftliche Monatshefte*, Heft 7, 1979a.

Asendorf-Krings, I.; Drexel, I.; Kammerer, G.; Lutz, B.; Nuber, Ch.: *Reform ohne Ziel? – Zur Funktion weiterführender beruflicher Schulen – Eine bildungspolitische Analyse auf empirisch-statistischer Grundlage*, Europäische Verlagsanstalt, Frankfurt/Köln 1975.

AWF (Ausschuß für Wirtschaftliche Fertigung e.V.): *Integrierter EDV-Einsatz in der Produktion – CIM (Computer Integrated Manufacturing), Begriffe, Definitionen, Funktionszuordnungen*, Eschborn 1985.

Bade, F.-J.: 'The Economic Importance of Small and Medium-Sized Firms in the Federal Republic of Germany'. In: D. Keeble; E. Wever (eds): *New Firms and Regional Development in Europe*, London/ Sydney/Dover 1986, pp. 256–74.

Baethge, M.; Oberbeck, H.: *Zukunft der Angestellten – Neue Technologien und berufliche Perspektiven in Büro und Verwaltung*, Suhrkamp Verlag, Frankfurt 1986.

Bardeleben, R. v.; Böll, G.; Uppenkamp, E.; Walden, G.: 'Weiterbildungsaktivitäten von Klein- und Mittelbetrieben im Vergleich zu Großbetrieben'. In: *Berufsbildung in Wissenschaft und Praxis*, Heft 11, 1989.

Barisi, G.: *Relations collectives de travail et stratégies des acteurs sociaux dans la sidérurgie à participation d'Etat en Italie (1968–82)*, Paris 1982.

Bechtle, G.: 'Enterprise, procès de production, utilisation de la force de travail'. In: *Sociologie du travail*, Heft 1, 1978.

Bechtle, G.: *Betrieb als Strategie – Theoretische Vorarbeiten zu einem industriesoziologischen Konzept*, Campus Verlag, Frankfurt/München 1980.

Bechtle, G.: *Arbeitsorganisation als Verhandlungsfeld zwischen Betrieb und Gewerkschaften – Eine empirische Untersuchung in der italienischen Stahlindustrie*, Campus Verlag, Frankfurt/New York 1982.

Bechtle, G.: 'Ristrutturazione aziendale e nuove forme della rappresentanza sindacale; il caso tedesco dell'industria del automobile (VW ed Opel)', *FLM Notizie*, Turin No. 9, 1986.

Bechtle, G.; Heiner, S.: 'Die Schwierigkeiten einer Klassenpolitik – Der Fall der italienischen Gewerkschaften 1968–78'. In: J. Bergmann (Hrsg.): *Beiträge zur Soziologie der Gewerkschaften*, Frankfurt 1979.

Bechtle, G.; Lutz, B.: 'Die Unbestimmtheit post-tayloristischer Rationalisierungsstrategie und die ungewisse Zukunft industrieller Arbeit – Überlegungen zur Be gründung eines Forschungsprogramms'. In: K. Düll; B. Lutz (Hrsg.): *Technikentwicklung und Arbeitsteilung im internationalen Vergleich*, Frankfurt/New York 1989, S. 9–91.

Beck, J.; Hohwieler E.; Potthast A.: 'Stand, Probleme und Zielsetzungen einer Vernetzung zwischen CAD und einer CNC-Werkzeugmaschine'. In: M. v. Behr; Ch. Köhler (Hrsg.): *Werkstattoffene CIM-Konzepte*, KfK-PFT 157, Karlsruhe 1990, S. 23–50.

Becker, W.; Weber, J.: 'Strategische Beschaffung als Schlüssel zum Einkäufermarkt'. In: *Beschaffung aktuell*, Heft 4, 1986, S. 36–8.

Behr, M. v.; Hirsch-Kreinsen, H.: 'Qualifizierte Produktionsarbeit und CAD/CAM-Integration – Erste Befunde und Hypothesen'. In: *VDI-Z*, Bd. 129, Nr. 1, 1987, S. 18–23.

Behr, M. v.; Köhler, Ch. (Hrsg.): *Werkstattoffene CIM-Konzepte – Alternativen für CAD/CAM und Fertigungssteuerung*, KfK-PFT 157, Karlsruhe 1990.

Benz-Overhage, K.; Brumlop, E.; Freyberg, T. v.; Papadimitriou, Z.: *Neue Technologien und alternative Arbeitsgestaltung – Auswirkungen des Computereinsatzes der industriellen Produktion*, Campus Verlag, Frankfurt/New York 1982.

Benz-Overhage, K.; Brandt, G.; Papadimitriou, Z.: *Computertechnologien im industriellen Arbeitsprozeß. In: Kölner Zeitschrift für Soziologie und Sozialpsychologie*, Sonderheft 24, Westdeutscher Verlag, Opladen 1982, S. 84–104.

Bergmann, J.: 'Industriesoziologie – eine unpraktische Wissenschaft?' In: U. Beck (Hrsg.): *Soziologie und Praxis, Soziale Welt, Sonderband 1*, Verlag Otto Schwartz & Co., Göttingen 1982, S. 397–416.

Bergmann, J.: 'Die Fragmentierung der Lohnarbeiterklasse in Japan'. In: *Leviathan*, Heft 1, 1983.

Bergmann, J.; Hirsch-Kreinsen, H.; Springer, R.; Wolf, H.: *Rationalisierung, Technisierung und Kontrolle des Arbeitsprozesses – Die Einführung der CNC-Technologie in Betrieben des Maschinenbaus*, Campus Verlag, Frankfurt/New York 1986.

Bernold, T.; Hillenkamp, U. (eds): *Expert Systems in Production and Services – Impact on Qualifications and Working Life*, Elsevier Science Publishers B.V., North-Holland/Amsterdam/New York/Oxford/Tokyo 1988.

BIBB (Bundesinstitut für Berufsbildung Berlin)/Federal Institute for Vocational Training (ed.): *The New Industrial Metalworking Occupations*, BIBB, Berlin 1988.

BIBB/IAB (Bundesinstitut für Berufsbildung; Institut für Arbeitsmarkt- und Berufsforschung) (Hrsg.): *Neue Technologien – Verbreitungsgrad, Qualifikation und Arbeitsbedingungen*, Beiträge zur Arbeitsmarkt- und Berufsforschung Nr. 118, Nürnberg 1987.

Bieber, D.; Sauer, D.: 'Kontrolle ist gut! Ist Vertrauen besser? – "Autonomie" und "eherrschung" in Abnehmer-Zulieferbeziehungen. In: H.G. Mendius; U. Wendeling-Schröder (Hrsg.): *Zulieferer im Netz – Zwischen Abhängigkeit und Partnerschaft*, Köln 1990.

Blackburn, P.; Coombs, R.; Green, K.: *Technology, Economic Growth, and the Labour Process*, London 1985.

Boden, M.: Training and Employment Patterns in the Age of Expert Systems. In: T. Bernold; U. Hillenkamp (eds): *Expert Systems in Production and Services*, North-Holland/Amsterdam/New York/Oxford/Tokyo 1988, pp. 191–8.

Boesch, E.E.: *Kultur und Handlung – Einführung in die Kulturpsychologie*, Hans Huber Verlag, Bern/Stuttgart/Wien 1980.

Boesch, E.E.: *Das Magische und das Schöne – Zur Symbolik von Objekten und Handlungen*, Verlag Friedrich Frommann/Günther Holzboog GmbH & Co., Stuttgart-Bad Cannstatt 1983.

Bosch, G.; Seifert; H.; Spies, B.: *Arbeitsmarktpolitik und Gewerkschaftliche Interessenvertretung*, Endbericht, DGB-Bundesvorstand und Wirtschafts- und Sozialwissenschaftliches Institut des DGB, Düsseldorf 1983.

Boyer, R.: *Regulation Theory – A Critical Assessment*, Columbia University Press, New York 1989.

Böhle, F.: *Strategien betrieblicher Informationspolitik – Eine systematische Darstellung für Betriebsräte und Vertrauensleute*, Bund-Verlag, Köln 1986.

Böhle, F.; Milkau, B.: *Vom Handrad zum Bildschirm – Eine Untersuchung zur sinnlichen Erfahrung im Arbeitsprozeß*, Campus Verlag, Frankfurt/New York 1988.

Böhme, H.; Böhme, G.: *Das Andere der Vernunft – Zur Entwicklung von Rationalitätsstrukturen am Beispiel Kants*, Suhrkamp Verlag, Frankfurt 1985.

Braczyk, H.-J.: 'Human Centred Technology, Strategic Options and User's Consent'. In: P. Brödner (ed.): *Strategic Options for 'New Production Systems'*, Brussels 1987, pp. 143–54.

Braczyk, H.-J.; Grüneberg, U.; Schmidt, G.: *Produktionsmodernisierung und soziale Risiken – Eine Fallstudie zur Gießereiarbeit*, Campus Verlag, Frankfurt/New York 1982.

Brandt, G.: 'Einleitende Bemerkungen zum Thema "Technologieentwicklung, Rationalisierung und Humanisierung". In: G. Brandt u.a. (Hrsg.): *Technologieent-wicklung, Rationalisierung und Humanisierung*, BeitrAB 53, IAB, Nürnberg 1981, S. 1–20.

Brandt, G.: 'Das Ende der Massenproduktion – wirklich?' In: R. Erd u.a. (Hrsg.): *Strukturwandel in der Industriegesellschaft*, Frankfurt/New York 1986, S. 103–22.

Brandt, G.: *Arbeit, Technik und gesellschaftliche Entwicklung. Transformationsprozesse des modernen Kapitalismus – Aufsätze 1971–87*, hrsg. v. D. Bieber; W. Schumm, Suhrkamp Verlag, Frankfurt 1990.

Brandt, G.; Kündig, B.; Papadimitriou, Z.; Thomae, J.: *Computer und Arbeitsprozeß – Eine arbeitssoziologische Untersuchung der Auswirkungen des Computereinsatzes in ausgewählten Betriebsabteilungen der Stahlindustrie und des Bankgewerbes*, Campus Verlag, Frankfurt 1978.

Brod, C.: Technostress – *The Human Cost of the Computer Revolution*, Addison-Wesley Publishing Company, Reading, Massachusetts etc. 1984.

Brose, P.: 'Hemmende und fördernde Faktoren im Innovationsprozeß von Klein- und Mittelbetrieben'. In: *Internationales Gewerbearchiv*, Heft 1, 32. Jg., 1984, S. 21–37.

Brödner, P.: *Fabrik 2000 – Alternative Entwicklungspfade in die Zukunft der Fabrik, edition sigma*, Berlin 1985.

Brödner, P. (ed.): 'Strategic Options for "New Production Systems" – CHIM: Computer and Human Integrated Manufacturing', FAST Occasional Papers No. 150, EEC, Brussels 1987.

Brumlop, E.: *Veränderungen der Arbeitsbewertung bei flexiblem Personaleinsatz*, Campus Verlag, Frankfurt/New York 1986.

Bullinger, H.-J.: 'Technische Entwicklung und Veränderung der Qualifikationsanforderungen, Referat auf der Fachtagung des IAO zu "Innovation und Qualifikation, berufliche Bildung und neue Technologien im Handwerk"' in Bonn am 15./16. Oktober 1987.

Bundesminister für Arbeit und Sozialordnung (Hrsg.): 'Forschung zur Humanisierung des Arbeitslebens, Aktionsprogramm des Bundesministers für Arbeit und Sozialordnung und des Bundesministers für Forschung und Technologie'. In: *Sozialpolitische Informationen*, Sonderausgabe, Jahrgang VIII, Bonn/Duisdorf 1974.

Bundesminister für Arbeit und Sozialordnung (Hrsg.): *Übersicht über die soziale Sicherheit – Übersicht über das Recht und die Arbeit*, Bonn 1990.

Bundesvereinigigung der Deutschen Arbeitgeberverbände: *Unternehmerische Personapolitik*, Köln 1983.

Buschhaus, D.; Gärtner, O.; Goldgräbe, A.; Hoch, H.-D.; Krischok, D.: 'Neue Berufsstruktur für die industriellen Metallberufe. In: Berufsbildung in Wissenschaft und Praxis, Heft 6, 1984.

Butera, F.: *Lavoro umano e prodotto technico – Una ricerca sulle acciaiere di Terni*, Einaudi, Turin 1979.

Clemens, R.; Friede, C., unter Mitarbeit von Dahrenmöller, A.: *Existenzgründungen in der Bundesrepublik Deutschland – Grundlagen einer Existenzgründungsstatistik*, Poeschel Verlag, Stuttgart 1986.

Cooley, M.J.E.: *Architect or Bee: The Human Price of Technology*, Hogarth Press, London 1987.

Corbett, J.M.: 'Prospective Work Design of a Human-centred CNC Lathe'. In: *Behaviour and Information Technology*, No. 3, Vol. 4, 1985, pp. 201–14.

Corbett, J.M.: 'Human Work Design Criteria and the Design Process – The Devil in the Detail'. In: P. Brödner (ed.): *Skill Based Automated Manufacturing*, IFAC, Karlsruhe 1986.

Coriat, B.: 'Differenzierung und Segmentierung des Gesamtarbeiters (Arbeiterschaft) in Industrien mit Prozeßfertigung'. In: *Kölner Zeitschrift für Soziologie und Sozialpsychologie*, Sonderheft 24, Opladen 1982, S. 185–202.

Cramer, U.: 'Klein- und Mittelbetriebe: Hoffnungsträger der Beschäftigungspolitik?' In: *MittAB* (Mitteilungen aus der Arbeitsmarkt- und Berufsforschung), Heft 1, 1987, S. 15–29.

D'Iribarne, A.: 'Nouvelles technologies et structures de qualification'. In: O. Diettrich; J. Morley (eds): *Relations between Technology, Capital and Labor* – Proceedings of the First Community Symposium in Social Sciences, held in Pont-a-Mousson 1981.

Deiß, M.: 'Arbeitsschutz und neue Rationalisierungsformen'. In: *WSI-Mitteilungen*, Heft 7, 41. Jg., 1988, S. 412–20.

Deiß, M.: 'La rationalisation systemique – Quelques remarques à propos de ses caractéristiques et de son rôle dans l'intégration inter-entreprises'. In: P. Cohendet (eds): *L'après-Taylorisme. Nouvelles formes de rationalisation dans l'entreprise en France et en Allemagne*, Paris 1988a, p. 97–106.

Deiß, M.: 'Entwicklung der Arbeitsbedingungen in den Zulieferbetrieben der Möbelindustrie'. In: N. Altmann; D. Sauer (Hrsg.): *Systemische Rationalisierung und Zulieferindustrie*, Frankfurt/New York 1989, S. 53–88.

Deiß, M.; Altmann, N.; Döhl, V.; Sauer, D.: *Neue Rationalisierungsstrategien in der Möbelindustrie II – Folgen für die Beschäftigten*, Campus Verlag, Frankfurt/New York 1989.

Deiß, M.; Döhl, V.; Sauer, D., unter Mitarbeit von Altmann, N.: *Technikherstellung und Technikanwendung im Werkzeugmaschinenbau – Automatisierte Werkstückhandhabung und ihre Folgen für die Arbeit*, Campus Verlag, Frankfurt/New York 1990.

Deppe, R.; Hoß, D.: *Arbeitspolitik im Staatssozialismus – Zwei Varianten: DDR und Ungarn*, Campus Verlag, Frankfurt/New York 1989.

Deutsche Bundesbank: 'Der Dienstleistungssektor in der Bundesrepublik Deutschland als Träger des wirtschaftlichen Wachstums', Monatsbericht August, Frankfurt 1988.

Deutschmann, C.: 'Industrial and Enterprise Unionism – Organizational Structures of Unions and Labor Market Sources of Union Power in West Germany and Japan in a Historical Perspective'. In: *East Asia*, No. 2, 1983.

DFVLR/Projektträgerschaft 'Humanisierung des Arbeitslebens' (Hrsg.): *Schriftenreihe 'Humanisierung des Arbeitslebens', 1981-present*, Campus Verlag, Frankfurt/New York.

Die Mitbestimmung (Hans-Böckler-Stiftung, Düsseldorf): published yearly.

DiFilippo, A.: *Military Spending and Industrial Decline – A Study of the American Machine Tool Industry*, Greenwood Press, New York/Westport/London 1986.

Döhl, V.: Die Rolle von Technikanbietern im Prozeß systemischer Rationalisierung. In: B. Lutz (Hrsg.): *Technik in Alltag und Arbeit*, Berlin 1989, S. 147–66.

Döhl, V.: 'Rationalisierungsstrategien von Abnehmerbetrieben und Anforderungen an die Zulieferer – Das Beispiel Möbelindustrie'. In: N. Altmann; D. Sauer (Hrsg.), Frankfurt/New York 1989a, S. 29–51.

Döhl, V.; Altmann, N.; Deiß, M.; Sauer, D.: *Neue Rationalisierungsstrategien in der Möbelindustrie I – Markt und Technikeinsatz*, Campus Verlag, Frankfurt/New York 1989.

Dohse, K.; Jürgens, U.; Malsch, T.: 'Reorganisation der Arbeit in der Automobilindustrie – Konzepte, Regelungen, Veränderungstendenzen in den USA, Großbritannien und der Bundesrepublik Deutschland – Ein Materialbericht', Internatio nal Institute for Comparative Social Research, Labor Policy, Discussion Paper No. IIVG 84–220, Berlin 1984.

Dohse, K.; Jürgens, U.; Malsch, T.: 'Umstrukturierung in der Automobilarbeit – Zwischenbefunde aus einer internationalen Vergleichsstudie', mimeograph, Berlin 1984a.

Doleschal, R.: 'Just-in-time-Strategien und betriebliche Interessenvertretung in der deutschen Automobilindustrie'. In: N. Altmann; D. Sauer (Hrsg.), Frankfurt/New York 1989.

Dörr, G.: 'Schranken betrieblicher Transparenz, Schranken betrieblicher Rigidität – eine Problemskizze zum Umbruch der Arbeitsformen im Maschinenbau'. In: F. Naschold (Hrsg.): *Arbeit und Politik*, Frankfurt/New York 1985.

Drexel, I.: 'Die Krise der Anlernung im Arbeitsprozeß – Betriebliche und gesellschaftliche Ursachen der Trennung von Qualifizierung und Produktion'. In: *Soziale Welt*, Heft 3, 31. Jg., 1980, S. 368–95.

Drexel, I.: 'Ausbildung für Angelernte – Zweiter Bildungsweg zum Facharbeiter?' In: *Das Mitbestimmungsgespräch*, Heft 11, 1981.

Drexel, I.: 'Anlernung im Produktionsprozeß oder formalisierte Berufsausbildung?' In: *Rationalisierung*, Heft 12, 32. Jg., 1981a, S. 275–8.

Drexel, I.: *Belegschaftsstrukturen zwischen Veränderungsdruck und Beharrung – Zur Durchsetzung neuer Ausbildungsberufe gegen bestehende Qualifikations- und Lohnstrukturen*, Campus Verlag, Frankfurt/New York 1982.

Drexel, I.: 'Problemi di qualificazione e politiche aziendali nell'industria siderurgica della Repubblica Federale Tedesca ed in quella italiana'. In: A. Pichierri (ed.): *Lavoro operaio e strutture organizzative nella evoluzione della siderurgia in Italia*, Città di Sesto San Giovanni 1983, pp. 17–27.

Drexel, I.: 'Neue Produktionsstrukturen auf Italienisch? Veränderungen von Qualifikation, Arbeitsorganisation und Entlohnung in der westdeutschen und der italienischen Stahlindustrie – Parallelen und Unterschiede'. In: *Soziale Welt*, Heft 1, 36. Jg., 1985, S. 106–27.

Drexel, I.; Nuber, C.: *Qualifizierung für Industriearbeit im Umbruch – Die Ablösung von Anlernung durch Ausbildung in Großbetrieben von Stahl und Chemie*, Campus Verlag, Frankfurt/München 1979.

Dreyfus, H.L.: *Die Grenzen künstlicher Intelligenz – Was Computer nicht können*, Verlag Athenäum, Königstein/Ts. 1985.

Dreyfus, H.L.; Dreyfus, St. E.: *Mind over Machine – The Power of Human Intuition and Expertise in the Era of the Computer*, Basil Blackwell, Oxford 1986.

Düll, K.: 'Que signifie est la "stratégie d'autonomie" de l'entreprise?' In: *Sociologie du travail*, Heft 1, 1978.

Düll, K.: 'Gesellschaftliche Modernisierungspolitik durch neue "Produktionskonzepte?"' In: *WdSI-Mitteilungen*, Heft 3, 38. Jg., 1985, S. 141–5.

Düll, K.: 'Einige kritische Überlegungen zum Ende der Arbeitsteilung'. In: B. Lutz (Hrsg.): *Soziologie und gesellschaftliche Entwicklung*, Frankfurt/New York 1985a, S. 398–403.

Düll, K.: 'The Analysis of New Forms of Work Organization – Results of Case Studies from the Federal Republic of Germany, France and Italy'. In: P. Grootings *et al.* (eds): *New Forms of Work Organization and their Social and Economic Environment*, Budapest 1986, pp. 197–231.

Düll, K.: 'New Forms of Work Organization: Case Studies from the Federal Republic of Germany, France, and Italy'. In: P. Grootings *et al.* (eds): New *Forms of Work Organization in Europe*, Transaction Publishers, New Brunswick/Oxford 1989, pp. 193–216.

Düll, K., in collaboration with Altmann, N.; Chave, D.; Barisi, G.; Dadoy, M.; Bechtle, G.; Boffo, S.: 'European Foundation Guidelines – International Comparison of New Forms of Work Organization', Consolidated Report, München 1981/Dublin 1983 (also available in German, French, Italian, Danish, Dutch, Greek).

Düll, K.; Bechtle, G.: 'Die Krise des normierten Verhandlungssystems – Rationalisierungsstrategien und industrielle Beziehungen im Betrieb'. In: K.M. Bolte (Hrsg.): *Mensch, Arbeit und Betrieb*, Weinheim 1988, S. 215–44.

Erbe, H.-H.; Schaefer, B.; Schloemer, R.: 'Qualifying for Computer Based Skilled Work in SME's', a paper presented at the IFAC Symposium 'Skill Based Automated Production', Wien, November 1989.

Ernst, A.: 'Japans unvollkommene Vollbeschäftigung'. In: *Mitteilungen des Instituts für Asienkunde*, Nr.115, Hamburg 1980.

Ernst, A.: *Dauerbeschäftigung und Flexibilität in Japan – Beschäftigungspolitik japanischer Unternehmen in Rationalisierungs- und Krisenphasen*, Campus Verlag, Frankfurt/New York 1988.

Everitt, P.: 'The Manufacturing Cell "A Plant Within a Plant"'. SME Technical Paper MS 85 – 901, Dearborn 1985.

Faunce, W.A. (ed.): *Readings in Industrial Sociology*, Appleton Century Crofts, New York 1967.

Feuerbaum, E.: 'Die Zubringer – und die Montageindustrie'. Abhandlungen aus dem Industrieseminar der Universität zu Köln, Heft 2, Berlin 1956.

Fix-Sterz, J.; Lay, G.; Schultz-Wild, R.; Wengel, J.: 'Flexible Manufacturing Systems and Cells in the Scope of New Production Systems in Germany', FAST Occasional Paper No. 135, EEC, Brussels 1987.

Fix-Sterz, J.; Lay, G.; Schultz-Wild, R.; Wengel, J.: 'Flexible Manufacturing Systems and Cells in the Federal Republic of Germany'. In: M. Warner *et al.* (eds): *New Technology and Manufacturing Management*, Chichester/New York/Brisbane/ Toronto/Singapore 1990, pp. 191–206.

Francis, A.; Grootings, P. (eds): *New Technologies and Work – Capitalist and Socialist Perspectives*, Routledge, London/New York 1989.

Gelder, E.: 'Innovative Endprodukte erfordern neuartige Zusammenarbeit mit den Vorlieferanten – nicht nur in der Technik'. In: *Beschaffung aktuell*, Heft 8, 1986, S. 34–5.

GEPRO (Gesellschaft für Produktionstechnik und Organization mbH): 'Entwicklung von Hilfsmitteln und Methoden zur Marktsicherung und Markterweiterung von Zulieferunternehmen', Aachen 1985.

Giebel, U.J.: *Sozialleistungen und Unternehmensgrößenstruktur. Analyse unter nehmensgrößenspezifischer Belastungswirkungen von gesetzlichen, tarifvertraglichen und freiwilligen Regelungen im Sozialbereich.* Untersuchungen zur Wirtschaftspolitik, Bd. 65, Institut für Wirtschaftspolitik der Universität Köln, Köln 1985.

Gill, K.S.: 'Expert Systems and Knowledge Transfer'. In: T. Bernold; U. Hillenkamp (eds): *Expert Systems in Production and Services*, North-Holland/ Amsterdam/New York/Oxford/Tokyo 1988, pp. 199–213.

Goldberg, P.: *The Intuitive Edge*, Crucible, Wellingborough 1985.

Gourevitch, P.; Martin, A.; Ross, G.; Bornstein, S.; Markovits, A.; Allen, C.: *Unions and Economic Crisis – Britain, West Germany, and Sweden*, George Allen and Unwin, London 1984.

Göranzon, B.; Josefson, I. (eds): *Knowledge, Skill and Artificial Intelligence*, Springer Verlag, London/Berlin/Heidelberg/New York/Paris/Tokyo 1988.

Grabher, G.: *De-Industrialisierung oder Neo-Industrialisierung? – Unternehmerische Reorganisation in einer traditionellen Industrieregion*, edition sigma, Berlin 1988.

Gramsci, A.: *Selections from the Prison Notebooks*, edited and translated by Q. Hoare and G.N. Smith, International Publishers, New York 1971.

Grupe, U.: *Programmiersprachen für die numerische Werkzeugmaschinensteuerung*, de Gruyter, Berlin/New York 1974.

Habermas, J.: *Theorie des kommunikativen Handelns*, Suhrkamp Verlag, Frankfurt 1981.

Hacker, W.: *Arbeitspsychologie – Psychische Regulation von Arbeitstätigkeiten*, Hans Huber Verlag, Bern/Stuttgart/Wien 1987.

Halder, K.; Sowa, R.; Ganz, W.; Schlaich, P.: *Unterstützung der Verbreitung neuer Technologien in Kleinbetrieben des Zulieferhandwerks, Forschungsbericht, Handwerkskammern Stuttgart und Ulm*, Stuttgart/Ulm 1988.

Harting, D.: Hauptsache billig? – 'Die LWA zeigt den anderen den Weg'. In: *Beschaffung aktuell*, Heft 4, 1986, S. 42–3.

Hirsch-Kreinsen, H.: 'Entwicklung einer Basistechnik – NC-Steuerungen von

Werkzeugmaschinen in den USA und der Bundesreuplik'. In: K. Düll; B. Lutz (Hrsg.): *Technikentwicklung und Arbeitsteilung im internationalen Vergleich*, Frankfurt/New York 1989, S. 161–211.
Hirsch-Kreinsen, H.; Schultz-Wild, R.; Köhler, Ch.; Behr, M. v.: *Einstieg in die rechnerintegrierte Produktion – Alternative Entwicklungspfade der Industriearbeit im Maschinenbau*, Campus Verlag, Frankfurt/New York 1990.
Hirsch-Kreinsen, H.; Köhler, Ch.; Moldaschl, M.; Schultz-Wild, R.: 'Technological Preconditions for Skilled Production Work in Computer Integrated Manufacturing, Expertise on behalf of the Institut Arbeit und Technik, Gelsenkirchen, as part of the FAST-project "Prospects of Anthropocentric Production Systems"', Munich, September 1990a.
Hirsch-Kreinsen, H.; Schultz-Wild, R. (Hrsg.): *Rechnerintegrierte Produktion – Zur Entwicklung von Technik und Arbeit in der Metallindustrie*, Campus Verlag, Frankfurt/München 1986.
Hirsch-Kreinsen, H.; Schultz-Wild, R.: 'Implementation Processes of New Technologies – Managerial Objectives and Interests'. In: P. Brödner (ed.): *Skill Based Automated Manufacturing*. IFAC Workshop, Karlsruhe, Federal Republic of Germany, September 3rd – 5th, 1986a.
Hirsch-Kreinsen, H.; Schultz-Wild, R.: 'Implementation Processes of New Technologies – Managerial Objectives and Interests'. In: *Automatica*, No. 2, Vol. 6, 1990, pp. 429–33.
Hirsch-Kreinsen, H.; Schultz-Wild, R.: 'Skilled and Cooperative Production Work in Computer Integrated Manufacturing'. Paper Prepared for the 11th IFAC World Congress – Automatic Control in the Service of Mankind, Vol. 12, Tallinn/Estonia/USSR, August 1990a, pp. 213–19.
Hirschhorn, L.: *Beyond Mechanization – Work and Technology in a Postindustrial Age*, MIT-Press, Cambridge, Massachusetts. 1984.
Hoffmann-Axthelm, D.: *Sinnesarbeit – Nachdenken über Wahrnehmung*, Campus Verlag, Frankfurt/New York 1984.
Hoppe, M.; Erbe, H.-H. (Hrsg.): 'Rechnergestützte Facharbeit', Reihe 'Berufliche Bildung', Band 7, Wetzlar 1986.
Hotz, B.: 'Productivity Differences and Industrial Relations Structures – Engineering Companies in the United Kingdom and the Federal Republic of Germany'. In: *Labour and Society*, No. 4, Vol. 7, 1982.
Hough, G. W.: *Technology Diffusion – Federal Programs & Procedures*, Lomond Books, Mt. Airy/Maryland 1975.
Hummel, R.: 'Computerintegrierte Fertigung am Beispiel einer Blechfertigung. Beitrag zum internationalen CIM-Kongreß zur Systec '86'. In: VDI-Gesellschaft Produktionstechnik (ADB) (Hrsg.): *Rechnerintegrierte Konstruktion und Produktion 1986*, Düsseldorf 1986, S. 127–51.
IG Metall (Hrsg.): *Werktage werden besser*, Europäische Verlagsanstalt, Köln/ Frankfurt 1977.
IG Metall (Hrsg.): *'Maschinen wollen sie – uns Menschen nicht' – Rationalisierung in der Metallindustrie*, Frankfurt 1983.
Institut für Angewandte Arbeitswissenschaft Köln: Mitteilungen, published yearly.
Irsch, N.: 'Kleine und mittlere Unternehmen im Strukturwandel'. In: *Kreditanstalt für Wiederaufbau*, Analysen-Meinungen-Perspektiven, Frankfurt 1985.
Isermann, R. (ed.): '10th IFAC World Congress on Automatic Control', Vol. 5, International Federation of Automatic Control, Munich, Federal Republic of Germany, July 27–31, 1987.

ISF München: 'Die Auswirkungen des technischen Fortschritts auf die Struktur und Ausbildung des Personals in den Walzwerken'. Hrsg. von der Hohen Behörde der EGKS, Luxemburg 1966.

ISF München: 'Die Veränderungen in der Struktur und Ausbildung der Arbeitskräfte der Eisen- und Stahlindustrie'. Hrsg. von der Kommission der Europäischen Gemeinschaften – EGKS (veröffentlicht in den Amtssprachen der Gemeinschaft), Luxemburg/Brüssel 1968.

Izard, C.E.: *Die Emotionen des Menschen*, Beltz Verlag, Weinheim/Basel 1981.

Jackson, S.; Browne, J.: 'An Interactive Scheduler for Production Activity Control'. In: *International Journal of Computer Integrated Manufacturing*, No. 1, Vol. 2, 1989, pp. 2–14.

Jones, B.: 'Division of Labor and Distribution of Tacit Knowledge in the Automation of Metal Machining. In: T. Martin (ed.): *Design of Work in Automated Manufacturing Systems*, Pergamon, Oxford 1984.

Jorissen, H.D.; Kämpfer, S.; Schulte, H.J.: *Die neue Fabrik – Chance und Risiko industrieller Automatisierung*, VDI-Verlag, Düsseldorf 1986.

Junne, G.: 'Das amerikanische Rüstungsprogramm – Ein Substitut für Industriepolitik'. In: *Leviathan*, Heft 1, 1985, S. 23–37.

Jürgens, U.; Malsch, T.; Dohse, K.: *Moderne Zeiten in der Automobilfabrik – Strategien der Produktionsmodernisierung im Länder- und Konzernvergleich*, Springer Verlag, Berlin/Heidelberg etc. 1989. (English edition forthcoming from Cambridge University Press, *Modern Times in the Automobile Factory*, 1991.)

Jürgens, U.; Naschold, F. (Hrsg.): 'Arbeitspolitik, Leviathan, Sonderheft 5/1983', Westdeutscher Verlag, Opladen 1984.

Kaiser, M.: 'Berufliche Flexibilität – Konzeption, Befunde, Schlußfolgerungen'. In: D. Mertens (Hrsg.): *Konzepte der Arbeitsmarkt- und Berufsforschung, BeitrAB 70*, Nürnberg 1982.

Kern, H.: *Empirische Sozialforschung – Ursprünge, Ansätze, Entwicklungslinien*, C.H. Beck Verlag, München 1982.

Kern, H.; Schumann, M.: *Das Ende der Arbeitsteilung? – Rationalisierung in der industriellen Produktion*, C.H. Beck Verlag, München 1984.

Kern, H.; Schumann, M.: 'Neue Produktionskonzepte haben Chancen – Erfahrungen und erste Befunde der Folgestudie "Industriearbeit und Arbeiterbewußtsein"'. In: *Soziale Welt*, Heft 1/2, 35. Jg., 1984a, S. 146–158.

Kern, H.; Schumann, M.: 'Industriearbeit im Umbruch – Versuch einer Voraussage'. In: B. Lutz (Hrsg.): *Soziologie und gesellschaftliche Entwicklung*, Frankfurt/New York 1985, S. 382–97.

Kern, H.; Schumann, M.: 'Limits of the Division of Labour – New Production and Employment Concepts in West German Industry'. In: *Economic and Industrial Democracy* (SAGE, London, Beverly Hills, Newbury Park and New Delhi), Vol. 8, 1987, pp. 151–70.

Kern, H.; Schumann, M.: 'New Concepts of Production in West German Plants'. In: P.J. Katzenstein (ed.): *Industry and Politics in West Germany – Toward the Third Republic*, Cornell University Press, London/New York 1989.

Kerr, C.: 'The Balkanization of Labor Markets'. In: E. Wright *et al.* (eds): *Labor Mobility and Economic Opportunity*, MIT-Press, Cambridge, Massachusetts. 1954.

Kittner, M. (Hrsg.) *Gewerkschaftsjahrbuch – Daten, Fakten, Analysen, Bund-Verlag*, Köln (yearly).

Klebe, T.; Roth, S. (Hrsg.): *Informationen ohne Grenzen – Computernetze und internationale Arbeitsteilung*, VSA-Verlag, Hamburg 1987.

Knights, D.H.; Willmat, D.: *Job Redesign: Critical Perspectives on the Labour Process*, Hamphshire, U.K. Gower, London 1985.

Koch, R.: 'Weiterbildung in Kleinbetrieben beim Einsatz neuer Techniken'. In: *Berufsbildung in Wissenschaft und Praxis*, Heft 4, 1988.

Köhler, C.: 'Nutzungsformen elektronischer Leitstände – Ergebnisse einer Anbieterbefragung'. In: M. v. Behr; Ch. Köhler (Hrsg.): *Werkstattoffene CIM-Konzepte*, Karlsruhe 1990, S. 79–100.

Köhler, C.; Grüner, H.: 'Foreign Workers – From the Necessary Evil to the Backbone of the Industry? The Case of the West-German Automobile Industry'. In: J. Fijalkowski (Hrsg.): *Transnationale Migranten in der Arbeitswelt*, Berlin 1990, S. 33–57.

Köhler, C.; Nuber, C.; Schultz-Wild, R.: 'Rationalisierungsprozesse mit verdeckten Folgen – Ansätze gewerkschaftlicher Politik'. In: *AFA-Informationen (Arbeits ausschuß für Arbeitsstudien)*, Heft 4, 37. Jg., 1987, S. 9–25.

Köhler, C.; Schultz-Wild, R.: 'Technischer Wandel und innerbetriebliche Mobilität – Mechanismen der Verdeckung von Rationalisierungsfolgen'. In: H. Knepel; R. Hujer (Hrsg.): *Mobilitätsprozesse auf dem Arbeitsmarkt*, Frankfurt/New York 1985, S. 329–50.

Köhler, C.; Sengenberger, W.: *Konjunktur und Personalanpassung – Betriebliche Beschäftigungspolitik in der deutschen und amerikanischen Automobilindustrie*, Campus Verlag, Frankfurt/New York 1983.

Kreibich, R.: *Die Wissenschaftsgesellschaft – Von Galilei zur High-Tech-Revolution*, Suhrkamp Verlag, Frankfurt 1986.

Langer, S.: *Philosophie auf neuem Weg – Das Symbol im Denken, im Ritus und in der Kunst*, Fischer Verlag, Frankfurt 1965/84.

Lay, G.: 'Strategic Options for CIM Integration'. In: M. Warner *et al.* (eds): *New Technology and Manufacturing Management*, Chichester/New York/Brisbane/Toronto/Singapore 1990, pp. 125–44.

Linhart, D.; Düll, K.; Bechtle, G.: 'Neue Technologien und industrielle Beziehungen im Betrieb – Erfahrungen aus der Bundesrepublik Deutschland und Frankreich'. In: K. Düll; B. Lutz (Hrsg.): *Technikentwicklung und Arbeitsteilung im internationalen Vergleich*, Frankfurt/New York 1989, S. 93–159.

Lorenzer, A.: *Das Konzil der Buchhalter – Die Zerstörung der Sinnlichkeit*, Europäische Verlagsanstalt, Frankfurt 1981.

Lutz, B.: 'Bildungssystem und Beschäftigungsstruktur in Deutschland und Frankreich – Zum Einfluß des Bildungssystems auf die Gestaltung betrieblicher Arbeitskräftestrukturen'. In: ISF München (Hrsg.): *Betrieb – Arbeitsmarkt – Qualifikation*, Frankfurt 1976, S. 83–151.

Lutz, B.: 'La renaissance de la sociologie industrielle en République Fédérale'. In: *Sociologie du travail*, Heft 1, 1978.

Lutz, B.: 'Social Endogeny of Technical Progress and Question of Development of Human Labour, 3–4 September'. In: O. Diettrich; J. Morley (eds): *Relations between Technology, Capital and Labour – Proceedings of the First Community Symposium in Social Sciences*, held in Pont-a-Mousson 1981.

Lutz, B.: 'Lernziel Facharbeiter – Materialien'. Hrsg. IG Chemie-Papier-Keramik, Hannover 1981a.

Lutz, B.: 'Einige gesellschaftliche Funktionen berufspraktischer Ausbildung'. In: D. Mertens; M. Rick (Hrsg.): *Berufsbildungsforschung, BeitrAB 66*, Nürnberg 1982.

Lutz, B.: 'Personalstrukturen bei automatisierter Fertigung'. In: B. Lutz; R. Schultz-Wild (Hrsg.): *Flexible Fertigungssysteme und Personalwirtschaft*, Frankfurt/New York 1982a, S. 85–101.

Lutz, B.: 'Kapitalismus ohne Reservearmee? – Zum Zusammenhang von Wirtschaftsentwicklung und Arbeitsmarktsegmentation in der europäischen Nachkriegszeit'. In: *Kölner Zeitschrift für Soziologie und Sozialpsychologie*, Sonderheft 24, West-deutscher Verlag, Opladen 1982b, S. 329–47.

Lutz, B.: *Der kurze Traum immerwährender Prosperität – Eine Neuinterpretation der industriell-kapitalistischen Entwicklung im Europa des 20. Jahrhunderts*, Campus Verlag, Frankfurt/New York 1984 (2. Auflage 1989).

Lutz, B. (Hrsg.): *Technik in Alltag und Arbeit – Beiträge der Tagung des Verbunds Sozialwissenschaftliche Technikforschung (Bonn, 29./30.5.1989)*, edition sigma, Berlin 1989.

Lutz, B.: *Le mirage de la croissance marchande – Essai de réinterprétation du développement du capitalisme industriel dans l'Europe du XXe siècle*, Ed. de la Maison des Sciences de l'Homme, Paris 1990.

Lutz, B.; Kammerer, G.: *Das Ende des graduierten Ingenieurs? – Eine empirische Analyse unerwarteter Nebenfolgen der Bildungsexpansion*, Europäische Verlagsanstalt, Frankfurt/Köln 1975.

Lutz, B; Moldaschl M.: *Expertensysteme und industrielle Facharbeit – Ein Gutachten über denkbare qualifikatorische Auswirkungen von Expertensystemen in der fertigenden Industrie*, Campus Verlag, Frankfurt/New York 1989.

Lutz, B.; Schmidt, G.: Industriesoziologie. In: R. König (Hrsg.): Handbuch der em-pirischen Sozialforschung, Bd. 8, Ferdinand Enke Verlag, Stuttgart 1977.

Lutz, B.; Schultz-Wild, R.: 'Aufklärung als Gestaltung – Zur Rolle der Sozialwissenschaften bei technisch-organisatorischen Innovationsvorhaben'. In: *WSI-Mitteilungen*, Heft 10, 1986, S. 669–78.

Lutz, B.; Sengenberger, W.: 'Arbeitsmarktstrukturen und öffentliche Arbeitsmarktpolitik – Eine kritische Analyse von Zielen und Instrumenten', Band 26 der Schriftenreihe der Kommission für wirtschaftlichen und sozialen Wandel, Verlag Schwartz & Co., Göttingen 1974.

Maase, M.; Sengenberger, W.; Weltz, F.: *Weiterbildung – Aktionsfeld für den Betriebsrat? – Eine Studie über Arbeitnehmerinteressen und betriebliche Politik*, 1. Auflage 1975, Europäische Verlagsanstalt, Köln/München 1975. (2. Auflage 1978, Campus Verlag, Frankfurt/New York 1978.)

Mahnkopf, B.: *Der gewendete Kapitalismus – Kritische Beiträge zur Theorie der Regulation*, Westfälisches Dampfboot, Münster 1988.

Maier, H.E.: 'Das Modell Baden-Württemberg – Über Institutionelle Voraussetzungen differenzierter Qualitätsproduktion – Eine Skizze', Discussion Paper IIM/LMP 87–10a, Wissenschaftszentrum Berlin für Sozialforschung, Berlin 1987.

Majchrzak, A.: *The Human Side of Factory Automation – Managerial and Human Resource Strategies for Making Automation Succeed*, Jossey-Bass Publishers, San Francisco/London 1988.

Malsch, T.: 'Die Informatisierung des betrieblichen Erfahrungswissens und der "Imperialismus der instrumentellen Vernunft"'. In: *Zeitschrift für Soziologie*, Heft 2, 16. Jg., 1987.

Malsch, T.; Seltz, R. (Hrsg.): *Die neuen Produktionskonzepte auf dem Prüfstand – Beiträge zur Entwicklung der Industriearbeit*, edition sigma, Berlin 1987.

Mandl, H.; Huber, G.L. (Hrsg.): *Emotion und Kognition*, Urban & Schwarzenberg, München/Wien/Baltimore 1983.
Männel, W.: *Eigenfertigung und Fremdbezug*, 2. Auflage, Poeschel Verlag, Stuttgart 1981.
Manske, F.: 'Alternative Strategies of Production Planning and Control (PPC)'. In: P. Brödner (ed.): *Strategic Options for "New Production Systems"*, Brussels 1987, pp. 131–40.
Marahrens, N.: 'Entwicklungschancen und Wachstumsgrenzen im Produzierenden Handwerk'. In: *Internationales Gewerbearchiv*, Heft 1/2, Berlin 1980, S. 35–44, 92–107.
Martin, T.; Kivinen, J.; Rijnsdorp, J.E.; Rodd, M.G.; Rouse, W.B.: 'Appropriate Automation – Integrating Technical, Human, Organizational, Economic, and Cultural Factors'. Paper prepared for the 11th IFAC World Congress – Automatic Control in the Service of Mankind, Vol. 1, Tallinn/Estonia/USSR, August 1990.
Martin, T.; Ulich, E.; Warnecke, H. J.: 'Appropriate Automation for Flexible Manufacturing'. In: R. Isermann (ed.): *10th World Congress on Automatic Control*, Vol. 5, 1987.
Martin, T.; Ulich, E.; Warnecke, H.J.: 'Appropriate Automation for Flexible Manufacturing'. In: *Automatica*, Vol. 26, No. 3, 1990, pp. 611–16.
Maurice, M.; Sellier, F.; Silvestre J.-J.: *Politique d'éducation et organisation industrielle en France et an Allemagne – Essai d'analyse sociétal*, Presses Universitaires de France, Paris 1982.
Maurice, M.; Sellier, F.; Silvestre J.-J.: *The Social Foundations of Industrial Power – A Comparison of France and Germany*, MIT Press, Cambridge, Massachusetts 1984.
Mayntz, R.; Hughes, T. P. (eds): *The Development of Large Technical Systems*, Campus Verlag, Frankfurt 1988.
Mendius, H. G.; Sengenberger, W.; Köhler, C.; Maase, M.: 'Qualifizierung im Betrieb als Instrument der Öffentlichen Arbeitsmarktpolitik – Begleitforschung zum Schwerpunkt 1 des Arbeitsmarktpolitischen Programms der Bundesregierung für Regionen mit besonderen Beschäftigungsproblemen, Forschungsbericht 89'. Hrsg. vom Bundesminister für Arbeit und Sozialordnung, Bonn 1983.
Mendius, H.-G.; Sengenberger, W.; Weimer S.: *Arbeitskräfteprobleme und Humanisierungspotentiale in Kleinbetrieben*, Campus Verlag, Frankfurt/New York 1987.
Mendius, H.G.; Wendeling-Schröder, U. (Hrsg.): *Zulieferer im Netz – Zwischen Abhängigkeit und Partnerschaft, Neustrukturierung der Logistik am Beispiel der Automobilzulieferung*, Bund-Verlag, Köln 1990.
Merleau-Ponty, M.: *Phänomenologie der Wahrnehmung*, de Gruyter, Berlin 1966.
Mertens, D.: 'Der unscharfe Arbeitsmarkt – Eine Zwischenbilanz der Flexibilitätsforschung'. In: *MittAB (Mitteilungen aus der Arbeitsmarkt- und Berufsforschung)*, Heft 4, 1973.
Mertens, D.; Rick, M. (Hrsg.): *Berufsbildungsforschung, BeitrAB 66 (Beiträge aus der Arbeitsmarkt- und Berufsforschung)*, Nürnberg 1982.
Mertens, P.; Borkowski, V.; Geis, W.: *Betriebliche Expertensystemanwendungen – Eine Materialsammlung*, Springer Verlag, Berlin/Heidelberg 1988.
Mettler-Meibom, B.: *Soziale Kosten in der Informationsgesellschaft – Überlegungen zu einer Kommunikationsökologie*, Fischer Verlag, Frankfurt 1987.

Meyer, K.; Schwiedrzik, B.: 'Der betriebliche Ausbildungsverbund – Praxis und Rechtsfragen'. In: *Berichte zur Beruflichen Bildung*, Heft 88, Berlin 1987.

Micheli, S.; Mighetto, F.; Montobbio, P.; Parodi, E.; Rouveri, L.: 'Esperienze di ricerca e intervento sull' organizzazione del lavoro in una grande azienda: discrezionalità soggetività e partecipazione'. In: *Quaderni di sociologia*, Vol. 2/3, 1976.

Mickler, O.: *Facharbeit im Wandel – Rationalisierung im industriellen Produktionsprozeß*, Campus Verlag, Frankfurt/New York 1981.

Moldaschl, M.: 'Grenzen der Mechanisierung geistiger Arbeit – Expertensysteme in der Produktion'. In: *Die Mitbestimmung*, Heft 9, 34. Jg., 1988, S. 481–5.

Moldaschl, M.: 'Mit "Potemkinschen Dörfern" in die Zukunft? – Personalplanung in der Montage'. In: *Technische Rundschau*, Heft 51/52, 81. Jg., 1989, S. 14–21.

Müller, K.: 'Zwischenbetriebliche Zusammenarbeit steigert die Leistungsfähigkeit'. In: *Management-Zeitschrift*, Nr. 5, Zürich 1987.

Müller-Jentsch, W.: *Zukunft der Gewerkschaften – Ein internationaler Vergleich*, Campus Verlag, Frankfurt/New York 1988.

Müller-Jentsch, W.: 'Flexibler Kapitalismus und kollektive Interessenvertretung – Gewerkschaften in der dritten industriellen Revolution'. In: W. Müller-Jentsch (Hrsg.): *Zukunft der Gewerkschaften*, Campus Verlag, Frankfurt/New York 1988a.

Nielsen, R.-P.: 'The Case for Cooperative Strategies'. In: *McKinsey Quarterly*, Spring 1987.

Noble, D.F.: 'Social Choice in Machine Design: The Case of Automatically Controlled Machine Tools'. In: A. Zimbalist (ed.): *Case Studies on the Labor Process*, Monthly Review Press, New York/London 1979, pp. 18–50.

Noble, D.F.: *Forces of Production – A Social History of Industrial Automation*, Alfred A. Knopf, New York 1984.

Parker, S.R.; Brown, R.K.; Child, J.; Smith, M.A.: *The Sociology of Industry*, George Allen and Unwin, London 1977.

Parker, M.; Slaughter, J.: *Choosing Sides: Unions and the Team Concept*, South End Press, Boston 1988.

Paulsen, B.: 'Weiterbildung und Organisationsentwicklung in Klein- und Mittelbetrieben'. In: *Berufsbildung in Wissenschaft und Praxis*, Heft 3/4, 1987.

Piore, M.J.; Sabel, C.F.: *The Second Industrial Divide – Possibilities for Prosperity*, Basic Books, New York 1984.

Piore, M.J.; Sabel, C.F.: *Das Ende der Massenproduktion – Studie über die Requalifizierung der Arbeit und die Rückkehr der Ökonomie in die Gesellschaft*, Verlag Klaus Wagenbach, Berlin 1985.

Polanyi, M.: *Implizites Wissen*, Suhrkamp Verlag, Frankfurt 1985.

Popitz, H.; Bahrdt, H.P.; Jüres, E.A.; Kesting, H. (Hrsg.): *Technik und Industriearbeit*, J.C.B. Mohr und Siebeck, Tübingen 1957.

Pries, L.; Schmidt, R.; Trinczek, R. (Hrsg.): *Trends betrieblicher Produktionsmodernisierung – Chancen und Risiken für Industriearbeit*, Westdeutscher Verlag, Opladen 1989.

Pries, L.; Schmidt, R.; Trinczek, R. (Hrsg.): *Entwicklungspfade von Industriearbeit – Chancen und Risiken der Produktionsmodernisierung*, Westdeutscher Verlag, Opladen 1990.

Prinz, W.: *Wahrnehmung und Tätigkeitssteuerung*, Springer Verlag, Berlin/Heidelberg/New York/Tokyo 1984.

Rauner, F.; Ruth, K.: 'Industriekulturelle Determinanten des

CNC-Werkzeugmaschinenbaus im Vergleich Bundesrepublik Deutschland – USA'. In: *Werkstatt und Betrieb*, Heft 12, 121. Jg., 1988, S. 950–2.
Rose, H.: 'Arbeitsbedingungen an Arbeitsplätzen mit Computertechnologie und Anforderungen für einen wirksamen Arbeitsschutz psychomentaler Belastungen'. In: *A & B 82*, München 1982.
Rose, H. (Hrsg.): *Programmieren in der Werkstatt – Perspektiven für Facharbeit mit CNC-Maschinen*, Campus Verlag, Frankfurt/New York 1990.
Rosenbrock, H.H.: 'Social and Engineering Design of an FMS'. In: E.A. Warman (ed.): *CAPE '83*, North Holland 1983.
Rosenfelder, U.: *Das neue Beschäftigungsförderungsgesetz*, Verlag für Verwaltung Franz Rehm, München 1985.
Roszak, T.: *Der Verlust des Denkens – Über die Mythen des Computer-Zeitalters*, Droemer Knaur, München 1986 (original title 'The Cult of Information').
Roth, S.; Kohl, H. (Hrsg.): *Perspektive 'Gruppenarbeit'*, Bund-Verlag, Köln 1988.
Rühle v. Lilienstern, H.: *Zulieferer des Handwerks haben mehr Zukunft – Erfahrungen, Beispiele, Lösungen*, Hans Holzmann Verlag, Bad Wörishofen 1986.
RWI-Handwerksberichte 1987 (Rheinisch-Westfälisches Institut für Wirtschaftsforschung), 34. Jg., Essen 1987.
RWI-Handwerksberichte (W. Lamberts; W. Düring; H. Schrumpf): Die wirtschaftliche Entwicklung im Jahr 1987/8, 35. Jg., 1988.
Ryle, G.: *The Concept of Mind*, Hutchinson, London 1949.
Sauter, E.: *Situation, Probleme und Perspektiven der öffentlich geförderten beruflichen Weiterbildung unter besonderer Berücksichtigung des Arbeitsmarktpolitischen Sonderprogramms der Bundesregierung von 1979*, Bundesinstitut für Berufsbildung, Berlin 1982.
Schank, R.C.: *The Cognitive Computer*, Addison Wesley, Reading, Massachusetts. 1984.
Scharpf, F. W.; Garlichs, D.; Maier, F.; Maier, H.: *Implementationsprobleme offensiver Arbeitsmarktpolitik – Das Sonderprogramm der Bundesregierung für Regionen mit besonderen Beschäftigungsproblemen*, Campus Verlag, Frankfurt/ New York 1982.
Scheer, A.-W.: *CIM – Computer Steered Industry*, Springer Verlag, Berlin/London/ New York 1988.
Schefe, P.: 'Summary of the Second Day'. In: Th. Bernold; U. Hillenkamp (eds): *Expert Systems in Production and Services*, North-Holland/Amsterdam/New York/Oxford/Tokyo 1988.
Schmidt, G.: *Gesellschaftliche Entwicklung und Industriesoziologie in den USA – Eine historische Analyse*, Europäische Verlagsanstalt, Frankfurt/Köln 1974.
Schmidt, G.; Braczyk, H.-J.; Knesebeck, J. v.d.: *Materialien zur Industriesoziologie, Kölner Zeitschrift für Soziologie und Sozialpsychologie, Sonderheft 24*, Westdeutscher Verlag, Opladen 1982.
Schultz-Wild, R.; Asendorf, I.; Behr, M. v.; Köhler, C.; Lutz, B.; Nuber C.: *Flexible Fertigung und Industriearbeit – Die Einführung eines flexiblen Fertigungssystems in einem Maschinenbaubetrieb*, Campus Verlag, Frankfurt/New York 1986.
Schultz-Wild, R.; Nuber, C.; Rehberg, F.; Schmierl, K.: *An der Schwelle zu CIM – Strategien, Verbreitung, Auswirkungen*, RKW-Verlag, Eschborn, Verlag TÜV-Rheinland, Köln 1989.
Schultz-Wild, R.; Weltz, F.: *Technischer Wandel und Industriebetrieb – Die*

Einführung numerisch gesteuerter Werkzeugmaschinen in der Bundesrepublik, Athenäum, Frankfurt 1973.

Schumann, M.; Gerlach, F.; Gschössl, A.; Millhofer, P.: *Am Beispiel der Septemberstreiks – Anfang der Rekonstruktionsperiode der Arbeiterklasse?* Europäische Verlagsanstalt, Frankfurt 1971.

Sellin, B.: 'New Manufacturing Systems: Implications for Initial and Continuing Training'. In: P. Brödner (ed.): *Strategic Options for 'New Production Systems'*, Brussels 1987, pp. 189–93.

Semlinger, K.: 'Kleinbetriebliche Flexibilität zwischen "aktiver Entwicklung" und "passiver Anpassung"'. In: Internationales Gewerbearchiv, Heft 4, 36. Jg., 1988, S. 229–37.

Semlinger, K.: 'Stellung und Probleme kleinbetrieblicher Zulieferer im Verhältnis zu großen Abnehmern'. In: N. Altmann; D. Sauer (Hrsg.): *Systemische Rationalisierung und Zulieferindustrie*, Frankfurt/New York 1989, S. 89–118.

Sengenberger, W.: 'Zur Dynamik der Arbeitsmarktsegmentierung – mit Thesen zur Struktur und Entwicklung des Arbeitsmarktes in der Bundesrepublik Deutschland'. In: C. Brinkmann u.a. (Hrsg.): *Arbeitsmarktsegmentation – Theorie und Therapie im Lichte der empirischen Befunde, Beiträge zur Arbeitsmarkt- und Berufsforschung Nr. 33*, Nürnberg 1979.

Sengenberger, W.: 'Labour Market Segmentation and the Business Cycle'. In: F. Wilkinson (ed.): *The Dynamics of Labour Market Segmentation*, London 1981.

Sengenberger, W.: 'Beschäftigungs- und arbeitsmarktpolitische Anforderungen an betriebliche Weiterbildung'. In: D. Mertens; M. Rick (Hrsg.): *Berufsbildungsforschung*, BeitrAB 66, 1982.

Sengenberger, W.: 'West German Employment Policy – Restoring Worker Competition'. In: *Journal of Economy & Society (Industrial Relations)*, Vol. 23, No. 3, 1984, pp. 323–43.

Sengenberger W.: *Struktur und Funktionsweise von Arbeitsmärkten – Die Bundesrepublik Deutschland im internationalen Vergleich*, Campus Verlag, Frankfurt/New York 1987.

Sengenberger, W.; Köhler, C.: 'Policies of Work-force Reduction and Labour Market Structures in the American and German Automobile Industry'. In: R. Tarling (ed.): *Flexibility in Labour Markets*, Academic Press, London 1987, pp. 245–69.

Sengenberger, W.; Loveman, G.: 'Smaller Units of Employment – A Synthesis Report on Industrial Reorganization in Industrialized Countries – Discussion Paper DP/3/1987 (rev. 88) of the International Institute for Labor Studies', Genf 1988.

Sengenberger, W.; Loveman, G.; Piore, M. (eds): *The Reemergance of Small Enterprise – Industrial Restructuring in Industrialized Countries*, International Institute for Labour Studies, Geneva 1990.

Simon, W.: *Die numerische Steuerung von Werkzeugmaschinen*, 2. Auflage, Carl Hanser Verlag, München 1971.

Sohn-Rethel, A.: *Geistige und körperliche Arbeit*, Suhrkamp Verlag, Frankfurt 1970.

Sohn-Rethel, A.: *Intellectual and Manual Labor – A Critique of Epistemology*, MacMillan and Co., London 1978.

Sonntag, K. (Hrsg.): *Neue Produktionstechniken und qualifizierte Arbeit*, Wirtschaftsverlag Bachem, Köln 1985.

Sorge, A.: 'A European Overview of Work and Vocational Training'. In: Warner, M.; Wobbe, W.; Brödner, P. (eds): New Technology and Manufacturing Management – Strategic Choices for Flexible Production Systems, John Wiley & Sons, Chichester/New York/Brisbane/Toronto/Singapore 1990. pp. 147–57.
Sorge, A.; Hartmann, G.; Warner, M.; Nicholas, I.: *Microelectronics and Manpower in Manufacturing – Applications of Computer Numerical Control in Great Britain and West Germany*, Gower Publishing Company, Aldershot 1983.
Sorge, A.; Warner, M.: 'Manpower Training, Manufacturing Organization and Work Rules in Great Britain and West Germany', Discussion Paper No. 96, IIMV, Berlin 1978.
Spur, G.: 'CIM – Die informationstechnische Herausforderung an die Produktionstechnik'. In: *wt (Werkstattechnik)*, Heft 76, 1986, S. 5–18.
Statistisches Bundesamt Wiesbaden (Hrsg.): *Statistisches Jahrbuch 1989 für die Bundesrepublik Deutschland*, Metzler-Poeschel, Stuttgart 1989.
Statistisches Bundesamt, Industrieberichterstattung 1987, Wiesbaden 1987.
Stone, K.: 'The Origins of Job Structures in the Steel Industry'. In: R.C. Edwards (ed.): *Labour Market Segmentation*, Lexington 1975.
Straus, E.: *Vom Sinn der Sinne, – Ein Beitrag zur Grundlegung der Psychologie*, Springer Verlag, Berlin/Heidelberg/New York/Tokyo 1956.
Streeck, W.: 'Guaranted Employment, Flexible Manpower Use, and Cooperative Manpower Management: A Trend Towards Convergence?' In: S. Tokunaga; J. Bergmann (eds): *Industrial Relations in Transition*, University of Tokyo Press, Tokyo 1984, pp. 81–116.
Streeck, W.: 'On the Social and Political Conditions of Diversified Quality Production'. In: E. Matzner; W. Streeck (eds): *Beyond Keynesianism: The Socio-Economics of Production and Full Employment*, Edward Elgar, 1991 (forthcoming).
Stromberger, C.: 'Numerisch gesteuerte Werkzeugmaschinen – Eine Betrachtung aus europäischer Sicht'. In: *Werkstatt und Betrieb*, Heft 9, 99, Jg., 1966, S. 597–608.
Süddeutsche Zeitung (Beilage Nr. 32): ' . . . bisher immer nur das Scheitern attestiert', 7.2.1991, S. VII.
Szyperski, N.; Kirschbaum, G.: 'Unternehmensgründungen in Nordrhein-Westfalen'. In: *Beiträge zur Mittelstandsforschung*, Göttingen 1981.
Tengler, H.; Hennicke, M.: *Dienstleistungsmärkte in der Bundesrepublik Deutschland*, Poeschel Verlag, Stuttgart 1987.
Teschner, E.: *Lohnpolitik im Betrieb*, Campus Verlag, Frankfurt/New York 1977.
Tokunaga, S.: 'The Structure of the Japanese Labour Market'. In: S. Tokunaga; J. Bergmann (eds): *Industrial Relations in Transition*, Tokyo 1984, pp. 25–55.
Turkle, S.: *Die Wunschmaschine – Vom Entstehen der Computerkultur*, Rowohlt, Reinbek b. Hamburg 1984.
Ulich, D.: *Das Gefühl*, Urban & Schwarzenberg, München/Wien/Baltimore 1982.
Valk, R.: 'Der Computer als Herausforderung an die menschliche Rationalität'. In: *Informatik-Spektrum*, Heft 2, 1987, S. 57–66.
Villa, P.: 'L'evoluzione del mercato del lavoro nell'industria siderurgica', manuscript, Mailand 1983.
Volmerg, B.; Senhaas-Knobloch, E.; Leithäuser, T.: *Betriebliche Lebenswelt – Eine Sozialpsychologie industrieller Lebensverhältnisse*, Westdeutscher Verlag, Opladen 1986.

Volpert, W.: *Zauberlehrlinge – Die gefährliche Liebe zum Computer*, Beltz Verlag, Weinheim 1985.

Volpert, W.: 'Contrastive Analysis of the Relationship of Man and Computer as a Basis of System Design'. In: P. Docherty *et al.* (eds): *System Design for Human Development and Productivity: Participation and Beyond.* Amsterdam: North Holland 1987.

Warnecke, H.J.: 'Von Taylor zur Fertigungstechnik von morgen'. In: *wt (Werkstattechnik)*, Heft 75, 1985, S. 669–74.

Watzlawick, P.: *Die Möglichkeit des Andersseins*, W. Kohlhammer, Bern/Stuttgart 1982.

Weber, W.: 'CNC-Steuerungen für qualifizierte Facharbeit'. In: *Technische Rundschau*, Heft 28, 80. Jg., 1988, S. 15–18.

Weimer, S.: *Arbeitsbedingungen in Klein- und Mittelbetrieben, Rationalisierungskuratorium der Deutschen Wirtschaft (RKW)*, Eschborn 1983.

Weiss, M.: *Labour Law and Industrial Relations in the Federal Republic of Germany*, Kluwer Law and Taxation Publishers, Deventer, Netherlands 1987.

Weizel, G.: 'Bescheidene Beschäftigungswirkungen durch Neugründungen'. In: *IFO-Schnelldienst*, Heft 7, 1986.

Weizenbaum, J.: *Die Macht der Computer und die Ohnmacht der Vernunft*, Suhrkamp Verlag, Frankfurt 1978 (original title 'Computer Power and Human Reason').

Weltz, F.: 'Die doppelte Wirklichkeit der Unternehmen und ihre Konsequenzen für die Industriesoziologie'. In: *Soziale Welt*, Heft 1, 39. Jg., 1988, S. 97–103.

Werkstatt und Betrieb, Nr. 94, 1961.

Wirtschafts- und Sozialwissenschaftliches Institut des Deutschen Gewerkschaftsbundes, Heft 3, Düsseldorf 1986.

Wobbe, W.: 'Introduction'. In: Warner, M.; Wobbe, W.; Brödner, P. (eds): New Technology and Manufacturing Management – Strategic Choices for Flexible Production Systems, John Wiley & Sons, Chichester/New York/Brisbane/Toronto/Singapore 1990. pp. 3–13.

Wood, S.: 'Neue Technologien, Arbeitsorganisation und Qualifikation: Die britische Labour-Process-Debatte'. In: *Prokla* 62, März 1986.

WSI-Mitteilungen (Wirtschafts- und Sozialwissenschaftliches Institut des Deutschen Gewerkschaftsbundes, Düsseldorf): published yearly.

Zimmermann, L. (Hrsg.): *Humane Arbeit – Leitfaden für Arbeitnehmer, 5 Bände*, rororo, Reinbek b. Hamburg 1982.

Index